T0269421

ENERGY POSITIVE NEIGHBORHOODS AND SMART ENERGY DISTRICTS

ENERGY POSITIVE NEIGHBORHOODS AND SMART ENERGY DISTRICTS

METHODS, TOOLS, AND EXPERIENCES FROM THE FIELD

Edited by

A. MONTI
E.ON Energy Research Center, RWTH Aachen University, Aachen, Germany

D. PESCH
NIMBUS Centre for Embedded Systems Research, Cork Institute of Technology, Cork, Ireland

K.A. ELLIS
IoT Systems Research Lab, Intel Labs, Intel Corporation, Ireland

P. MANCARELLA
The University of Manchester, School of Electrical and Electronic Engineering, Manchester, United Kingdom

Amsterdam • Boston • Heidelberg • London • New York • Oxford
Paris • San Diego • San Francisco • Singapore • Sydney • Tokyo
Academic Press is an imprint of Elsevier

Academic Press is an imprint of Elsevier
125 London Wall, London EC2Y 5AS, United Kingdom
525 B Street, Suite 1800, San Diego, CA 92101-4495, United States
50 Hampshire Street, 5th Floor, Cambridge, MA 02139, United States
The Boulevard, Langford Lane, Kidlington, Oxford OX5 1GB, United Kingdom

Library of Congress Cataloging-in-Publication Data
A catalog record for this book is available from the Library of Congress

British Library Cataloguing-in-Publication Data
A catalogue record for this book is available from the British Library

ISBN: 978-0-12-809951-3

For information on all Academic Press publications
visit our website at https://www.elsevier.com/

Publisher: Joe Hayton
Acquisition Editor: Lisa Reading
Editorial Project Manager: Maria Convey
Production Project Manager: Lisa Jones
Designer: Maria Ines Cruz

Typeset by Thomson Digital

CONTENTS

LIST OF CONTRIBUTORS

M. Boudon
EMBIX, Issy-les-Moulineaux, Paris, France

J. Bynum
United Technologies Research Centre Ireland, Cork, Ireland

L. Cupelli
E.ON Energy Research Center, RWTH Aachen University, Aachen, Germany

L. De Tommasi
United Technologies Research Centre Ireland, Cork, Ireland

K.A. Ellis
IoT Systems Research Lab, Intel Labs, Intel Corporation, Ireland

N. Good
The University of Manchester, School of Electrical and Electronic Engineering, Manchester, United Kingdom

T. Greifenberg
Software Engineering, RWTH Aachen University, Aachen, Germany

D. Kelly
IoT Systems Research Lab, Intel Labs, Intel Corporation, Ireland

M. Klepal
NIMBUS Centre for Embedded Systems Research, Cork Institute of Technology, Cork, Ireland

K. Kouramas
United Technologies Research Centre Ireland, Cork, Ireland

E. L'Helguen
EMBIX, Issy-les-Moulineaux, Paris, France

M. Look
Software Engineering, RWTH Aachen University, Aachen, Germany

P. Mancarella
The University of Manchester, School of Electrical and Electronic Engineering, Manchester, United Kingdom

E.A. Martínez Ceseña
The University of Manchester, School of Electrical and Electronic Engineering, Manchester, United Kingdom

A. Monti
E.ON Energy Research Center, RWTH Aachen University, Aachen, Germany

D. Mueller
E.ON Energy Research Center, RWTH Aachen University, Aachen, Germany

D. Pesch
NIMBUS Centre for Embedded Systems Research, Cork Institute of Technology, Cork, Ireland

E.H. Ridouane
United Technologies Research Centre Ireland, Cork, Ireland

M. Schumacher
E.ON Energy Research Center, RWTH Aachen University, Aachen, Germany

C. Upton
IoT Systems Research Lab, Intel Labs, Intel Corporation, Ireland

ABOUT THE EDITORS

Dr. Antonello Monti (Editor)
Professor and Institute Director at RWTH Aachen University, Germany

During his time at the University of South Carolina before joining RWTH, Professor Monti was Associate Director of the Virtual Test Bed (VTB) project, which focused on computational simulation and visualization of modern power distribution systems. His four main areas of research are Simulation of Complex Systems with focus on Real Time and Hardware in the Loop, Distributed Intelligence for Grid Automation, Advanced Monitoring Solution for Distribution Grids, and Development of solution for Smart Home/Smart Cities applications.

Dr. Dirk Pesch (Editor)
Director of the NIMBUS Centre at the Cork Institute of Technology, Ireland

Dr. Dirk Pesch is the Head of the Nimbus Centre for Embedded Systems Research at Cork Institute of Technology, Cork, Ireland. Dr. Pesch received a Dipl.-Ing. degree from RWTH Aachen University, Germany, and a PhD from the University of Strathclyde, Glasgow, Scotland, both in Electrical and Electronic Engineering. His research interests focus on design and performance analysis of heterogeneous wireless networks and cyber-physical systems for energy and building management and smart city applications.

Keith A. Ellis (Editor)
Senior Research Scientist, Energy and Sustainability Lab, INTEL Labs, Ireland

Keith's primary focus is on the context of ICT enablement in the energy and sustainable space—specifically the Built Environment, Districts/Cities. He is a technologist, researcher, and project manager with a wealth of industrial experience. He has led multiple teams focused on technology enabled energy efficient buildings, energy positive neighborhoods, smart grid, smart manufacturing, and water resource management.

Dr. Pierluigi Mancarella (Editor)
Professor of Smart Energy Systems, The University of Manchester, United Kingdom

Dr. Pierluigi Mancarella is Professor of Smart Energy Systems at the University of Manchester, UK, and Chair Professor of Electrical Power Systems at the University of Melbourne, Australia.

He received his MSc and PhD degrees in Electrical Engineering from the Politecnico di Torino, Italy, was a research associate at Imperial College London, UK, and held visiting positions at NTNU, Trondheim, Norway; Ecole Centrale de Lille, France; Universidad de Chile, Santiago, Chile; and National Renewable Energy Laboratories, Colorado, USA.

Pierluigi has led several UK and international projects on technoeconomic modeling of multienergy systems, planning of energy infrastructure under uncertainty, development of business cases for low carbon technologies and district energy systems, and risk and resilience of future networks.

He is author/editor of four books and of over 200 research publications and reports.

Pierluigi is an Editor of the *IEEE Transactions on Smart Grid*, the *IEEE Systems Journal*, and the *International Journal of Electrical Power and Energy Systems*. He is also the Chair of the Energy Working Group of the IEEE European Public Policy Initiative.

PREFACE

There are arguably other research and development areas where interdisciplinarity is as important as in the Smart City domain. We are experiencing an incredible convergence of disciplines that are opening new challenges. The editors of this book had the pleasure to live this experience during to the FP7 project COOPERaTE funded by the European Commission. During 3 years, we had the chance to understand each other's perspective, learn how to work together and how to understand each other. It is not easy to put at the same table construction engineers, electrical engineers, and IT experts but this is what a Smart City is about. We, the editors of this book, had the chance to live this incredibly stimulating experience and at the end of the project, looking back, we understood how much we learned, how much our understanding of the other sectors was grown and we thought it was worth to collect the lesson learned in a book. The world around us is also changing: when COOPERaTE started, the idea of cloud applications for energy was considered at the least exotic, today it is mainstream, just 4 years later. This book has the aspiration to formalize and describe this process. We hope the reader can, after going through this book, quickly grasp the complexity of the process and get to where we are now faster than in 3 years.

ACKNOWLEDGMENT

The editors would like to thank first of all the complete team of the project COOPERaTE. As with any nice collaboration experience, it is mostly created by the great people that work together in an open-minded fashion. Among everybody a special thanks to Lisette Cupelli for the great effort in coordinating everybody in the preparation of this book. Thanks are also due to the European Commission for making all this possible in the framework of the activities called "Positive Energy Neighborhood." In particular, we would like to thank the Project Officers that drove us during the 3 years: Rogelio Segovia, Daniela Rosati, and Hubert Schier. A very sincere thanks to our reviewers that with their feedback made our work more significant: Marta Chinnici, Frank van Obverbeeke, and Fredrik Wallin. Last but not least we would like to thank the team at Elsevier for guidance and suggestions. In particular, thanks to Maria Convey and Lisa Jones for their patience, guidance, and support.

CHAPTER ONE

Introduction

A. Monti*, D. Pesch, K.A. Ellis[†], P. Mancarella[‡]**
*E.ON Energy Research Center, RWTH Aachen University, Aachen, Germany
**NIMBUS Centre for Embedded Systems Research, Cork Institute of Technology, Cork, Ireland
[†]IoT Systems Research Lab, Intel Labs, Intel Corporation, Ireland
[‡]The University of Manchester, School of Electrical and Electronic Engineering, Manchester, United Kingdom

Contents

1 INTRODUCTION

Energy supply relies progressively on renewable sources due to environmental, economic, and political driving force. Further, the same forces are driving increasing electrification of heating and transport. Accordingly, power generation and consumption have to be adapted to this changing situation and the buildings sector and its infrastructure must offer the necessary architecture to address these changes. Since electrical grids cannot be reorganized spontaneously to integrate more Renewable Energy Sources (RES) and electricity loads (such as electric heating and vehicles) without massive network expansion, developing different approaches and architectures are essential steps to achieve the 20-20-20 targets posted by the European Union.

Distributed energy generation is rapidly growing in Europe as well as in other regions of the world. Most of these distributed generation units are based on RES which, together with new electricity loads, create new challenges in balancing energy supply and ensuring stability of the overall system, while operating the network within limits. As a result, an improved communication and information infrastructure at consumer level is necessary. Districts can be seen as complete units with local energy sources management, which can be addressed to sell energy to the overlaid grid if there is excess or, conversely, to buy energy if there is not enough generation at the local level. Therefore, developing a pragmatic and innovative approach of Energy Efficient Neighborhoods and Energy Positive Neighborhoods (EPN), which is a system-level concept where the neighborhood generates more energy than it consumes, with surplus energy being either stored locally or exported, is crucial to address the climate objectives (Howard & Björk, 2008).

According to the EU directive 2010/31/EU, buildings in the European Union account for 40% of the total energy consumption. Thus, the building stock offers a huge energy saving potential once heterogeneity is seen as an opportunity instead of a barrier. Presently, buildings are planned individually, irrespective of the surrounding buildings. An interdisciplinary approach, which would allow an interchange of energy options, is not available. A better understanding of the optimum combination of building-specific and area-specific measures needs to be developed (Eastman, Teicholz, Sacks, & Liston, 2008). There is potential for more efficient buildings and neighborhood energy and performance management, if the intelligence to manage a multigrid system at city quarter level is developed.

Additionally, by visualizing and tracking energy consumption, consumers are sensitized to the high importance of energy topics and develop consumption awareness. Such tendencies are expected to not only impact current consumer behavior significantly, but also to have long-term effect by raising a spirit of consciousness for sustainable environment in general. Therefore, increasing consumer's acceptance and willingness to participate could be essential for dimensions and validity of the results gained.

A significant weakness of most energy saving concepts is the lack of consumer appeal. One possible approach is to add features not directly related to the energy services to raise interest in more customers and countervail withdrawing after the first phase of enthusiasm. The final goal is then a real-time information system at the neighborhood level, helping people to save energy while keeping them connected into a new concept of local community.

Another topic of high actuality is the everyday life improvement potential, which ICT developments enable. Meanwhile cloud computing has reached high popularity due to its abilities to offer flexible dynamic ICT infrastructures. Networking and communications are also increasingly part of life in simplifying various every-day activities. More information about the surrounding environment facilitates the decisions taking process and increases comfort. It is thus evident to combine ICT solutions and cloud computing to have a high potential to provide benefits toward achieving the goals of efficient energy management. Furthermore, a cloud-based management and service platform, integrated in a district, could offer additional services in case of emergency—for example, detailed information about the emergency, together with degree or affected area and exact location. By utilizing a cloud-based platform one supports a more easily transferable blueprint in terms of proliferation.

As stated in a recent report (Anon, 2009), a holistic, interoperable, and well-validated management system supported by innovative ICT technology addressing the monitoring, control, and optimization of the energy generation, consumption, and trading is necessary to facilitate EPN.

Nevertheless, some key technical challenges need to be addressed:

- The adaptation of existing ICT infrastructures, such as communication protocols, monitoring, control, and sensing infrastructures that can provide a sustainable and secure architecture while being cost/effective.

- The existence of an integrated platform that provides a holistic integration of multiple neighborhood supporting systems, enabling their stable and secure interoperability.
- Decentralized monitoring and control systems for power and energy management, including the opportunity to exploit strategies that may include peak shaving or dynamic pricing.
- The business models that would best support the creation of new markets and new incentives that allow new markets to open and new products to be created.
- Public awareness and technology adoption.

To serve the public in this new paradigm of energy efficiency, knowledge of the consumers, their motivations (social proof), and their needs, which is then reflected in the services offered, is fundamental. Another key aspect is that any industry that allows the free flow of information between two parties must be cognizant and protective of privacy and safety. Communication network security is paramount both in terms of privacy issues and to assure its reliability.

While energy efficiency in buildings (individual and campus-level scenarios) is the focus of a range of research programs, and is also now a strategic target for many organizations, energy management at the neighborhood level is still a new topic. The inclusion of local energy generation within the neighborhood is an attractive proposition as it provides opportunities for neighborhoods to be self-sufficient for some time or even be energy positive. The concept of a neighborhood integration and service platform provides industry with an open platform to deliver not just energy management services to neighborhoods but also interoperability with other systems apart from energy, such as security, safety, transport, traffic, etc.

2 BACKGROUND OF THE BOOK

This book is the result of a project experience. All the authors that contribute to this work have been involved in an FP7 Project called COOPERaTE (Control and Optimization for Energy Positive Neighborhoods) (COOPERaTE project; http://www.cooperate-fp7.eu.). This project had a major merit: it forces three different business sectors to sit together at the table and discuss the issue of EPN. These sectors are: building, energy, and ICT.

The experience earned in the interaction is what is captured in this book. This book is not about the project but it is about the knowledge built, thanks to this interaction, among business sectors. This knowledge allowed the authors of this book to earn a complete vision of the challenges ahead as well as a process to address them.

COOPERaTE had originally the goal to deliver an open platform as mentioned in the introduction: at the end, it turned out that the most important result was however the definition of a process of implementation and of the tools that support the process.

The idea of this book is exactly to capture this process and to make it public so that the implementation of an EPN can happen in reality.

3 STRUCTURE OF THE BOOK

The different chapters of this book address a precise aspect of the EPN process. Readers may use this book in a variety of ways. It is definitely advisable to read all the chapters one after the other but, on the hand, each chapter has enough self-consistency that it can be also approached alone.

Chapter 2, "Energy Positivity and Flexibility in Districts," sets the scene. While as an intuition it may seem very simple to define what an EPN is, the reality is quite different. This chapter deals with the definition and the consequences of such a definition. In principle, it is not so difficult to create a neighborhood that can generate more energy than what it consumes, but the reality is that what we have in mind when we refer to this concept is way more complex. A simple example can illustrate the challenge: does positivity mean involvement of renewable? In principle we could say not, but in reality we know that this is a prerequisite people have in mind when they refer to EPN.

Chapter 3 is dedicated to the definition of the process. The main question is how to build an EPN Roadmap? Such a roadmap is quite an interdisciplinary challenge, involving all the three main areas mentioned previously: building, ICT, and energy. The process depends also on the point of view: the owners versus the technology providers, for example. This chapter describes a formalized step-by-step solution capable of solving the EPN problem from a technoeconomical perspective.

Key elements of the process described in Chapter 3 are Simulation and Optimization methods: this is the content of Chapter 4. The idea is that the design of a future EPN cannot be performed without adequate modeling. The modeling will be used to quantify the energy impact of different design decisions. The modeling then is combined with Optimization methods. Optimization is under vast element of the process, highly depending on the characteristics of the neighborhood. For example, a single-owner neighborhood will have a completely different approach to the optimization with respect to a multiowner situation.

At the end though the key element of the transition is ICT. Chapter 5 presents the option available for the ICT infrastructure in a neighborhood. A key point is behind the proposed philosophy: it is quite unlikely that only one ICT player will be active in one neighborhood. As a result and in the main interest of the customers, interoperability in the ICT solution is the key ingredient.

As it was said in relation to Chapter 3, the process makes sense only if it is approached with a technoeconomic approach. The EPN transformation can be guided by environmental targets but must be economically viable. Chapter 6 presents an analytical approach to the evaluation of the business cases that can be adopted in an EPN. Not only are examples of revenue opportunities discussed but also the way the business case can be assessed.

Chapter 7 presents some real life experience. EPN is not only theory: there are concrete experiences that can be reported to understand how it may work in real life. The results of these experiences, even being related to early adopters, are anyhow extremely promising.

In conclusion Chapter 8 reviews the challenges to the process and specifies barriers in front of the path of the EPN, and smart, demand side interventions generally. While the EPN is today possible, the climate for its adoption can be improved, and this chapter makes multiple recommendations to facilitate this process.

REFERENCES

Directive 2010/31/EU of the European Parliament and of the Council of 19 May 2010 on the energy performance of buildings, 2010. Available from http://eur-lex.europa.eu/legal-content/EN/ALL/?uri=CELEX%3A32010L0031

Eastman, C. M., Teicholz, P., Sacks, R., & Liston, K. (2008). *BIM handbook*. Hoboken, New Jersey: John Wiley & Sons, Inc.

Howard, R., & Björk, B. -C. (2008). Building information modelling—experts' views on standardisation and industry deployment. *Advanced Engineering Informatics*, *22*(2), 271–280 [Network methods in engineering].

Towards an ICT infrastructure for energy positive neighbourhoods. *Report from ELSA Thematic Working Group on ICT for Energy Efficiency*, October 2009.

CHAPTER TWO

Energy Positivity and Flexibility in Districts

N. Good, E.A. Martínez Ceseña, P. Mancarella
The University of Manchester, School of Electrical and Electronic Engineering, Manchester, United Kingdom

Contents

1 INTRODUCTION

Energy positivity is an ill-defined term and potentially subject to controversial interpretations. In the context of a neighborhood (or, equivalently, district), it can be understood to mean "the generation of more electricity, by a neighborhood, than it consumes." This definition has parallels with the concept of Zero-Energy Buildings (ZEB), or Net-Zero-Energy Buildings (NZEB) (Marszal et al., 2011; Sartori, Napolitano, & Voss, 2012). However, such a simplistic definition needs clarification on several significant points. In existing literature (Sartori et al., 2012; Torcellini, Pless, & Deru, 2006; Reichl & Kollmann, 2011; Marszal et al., 2011), the appropriate *metrics*, the *temporal frame*, and the *types of energy use* to be considered have been identified as key factors in assessment of buildings (and by extension, neighborhoods or districts). These are all certainly relevant for the assessment of an Energy Positive Neighborhood (EPN) also. However, they do not systematically address what may be considered the key defining features of future energy systems. These can be summarized as:

1. increasing penetration of low carbon electricity generation [both variable, such as many types of Renewable Energy Sources (RES), and relatively inflexible, such as nuclear];
2. decreasing availability of traditional, flexible, thermal generators;
3. increasing penetration of electric heating and transport.

These features are significant as they all result in increasing scarcity of a resource which is crucial for the operation of electricity systems, given the need to balance the electricity system second-by-second (i.e., *flexibility*[1]). Increasing penetration of low carbon electricity generators will reduce flexibility as they are either nonschedulable as well as partly nondispatchable [e.g., in the case of solar Photovoltaic (PV) and wind generation], or are (traditionally) not designed for frequent and fast modulation (e.g., nuclear). Meanwhile, increasing penetration of large-scale (variable) RES, will increase demand for flexibility at the system level, just as the availability of traditional providers of flexibility (thermal generation) is decreasing. At the same time, increasing penetration of small-scale RES as well as of electric heating and transport is increasing demand for flexible resources at the distribution network level, to prevent voltage and congestion issues (Navarro-Espinosa & Ochoa, 2015; Navarro-Espinosa & Mancarella, 2014; Mancarella, Gan, & Strbac, 2011), which can motivate expensive network reinforcement (Martínez-Ceseña, Good, & Mancarella, 2015). While there are various types of flexible resources available, flexibility from the demand-side, from entities, such as neighborhoods, have been identified as particularly attractive sources (Strbac et al., 2012). Hence, there is clearly motivation for neighborhoods to become, as a primary objective, *flexible*, rather than *energy positive* (besides the ambiguity of the terminology itself).

Naturally, given the variable and stochastic nature of demand for energy services (e.g., heating or air conditioning), RES, and plant and networks outages, the demand for operation of flexible resources is variable and dynamic. Hence, to achieve flexibility (rather than energy positivity, in the classic sense) focus should be on strategic adoption of flexible resources and enabling ICT. This motivates methods of neighborhood/district assessment which fundamentally reflect this focus on flexibility. This requirement is, however, not straightforward. The value of flexibility varies substantially, both temporally and spatially, but also according to the preferences of the evaluator. For example, at times of low system demand, flexibility may be important especially in cases when there is little inertia in the system (as it may be in the presence of large volume of power electronics-connected RES). On the other hand, for periods of high system demand, flexibility may have substantial value as the system is more sensitive to shocks due to the increased penetration of inflexible technologies. Similarly, on congested distribution networks with little spare capacity the value of flexibility may be significant to avoid network reinforcement. Finally, whatever the physical status of the electricity system and networks, the value of flexibility may be different to different parties. For example, a consumer with a requirement for high reliability in their electricity supply (e.g., a hospital), may value reliability highly. In contrast, a relatively poor (hence price sensitive) domestic customer may be willing to accept a lower level of electricity supply reliability if this provides a

[1]By *flexibility* we intend here the ability of a system to provide supply and demand balance over different time scales in an economic and reliable way, including response to unforeseen events.

discount on their bill. Similarly, users with an alternative energy-service supply (e.g., back-up electricity generation or an alternative supply of fuel/alternative energy service providing device) may be willing to accept lower electricity supply reliability. In this context, reliability [and to some extent even resilience (Panteli & Mancarella, 2015)] could be provided by flexible neighborhood-level demand-side resources that might respond to economic incentives to decrease their electricity consumption (or increase their production in the case of embedded generation).

Accounting for these varying, dynamic conditions and preferences is not an easy task for one central assessor, making assessment of flexibility difficult. However, in a liberalized system, where (generally variable and dynamic) price signals indicate the value of agent behavior, economic value can be considered the most suitable proxy metric to measure flexibility (Chapter 6). Although it is not a perfect measure, as it relies on well-functioning (i.e., cost-reflective) markets, which may not always be the case, it is the best available. Further its efficacy as a measure of flexibility should only increase as technology increases the penetration of variable, dynamic pricing (Faruqui, Harris, & Hledik, 2010), and reforms (such as including the price of disconnections and voltage reduction in imbalance penalties, and making the penalties reflect the marginal cost of balancing actions) increase the cost reflectiveness of prices (Flamm & Scott, 2014).

Given the identification of economic value as the appropriate proxy measure of the flexibility of a neighborhood, Section 2 examines how the economic value metric of neighborhoods varies given objectives of "classic" energy positivity and flexibility. Quantitative examples demonstrate the material difference in economic value which results from each objective. Section 3 then considers, through further quantitative examples, how optimization of a neighborhood with respect to various objectives which are commonly conflated (i.e., economic value, CO_2 emission, and self-sufficiency) are, in fact, not equivalent. Recognizing that decision-makers may want to evaluate neighborhoods according to many criteria, rather than solely according to flexibility, Section 4 introduces the concept of multicriteria analysis, outlining a framework for assessment of interventions according to multiple criteria. Section 5 offers some concluding remarks.

2 ECONOMIC VALUE OF AN EPN

An EPN will be able to achieve flexibility through exploiting various flexible resources. However, the amount of value will depend on the objective they follow, and the degree to which this objective aligns with the requirements of the system. As explored in more detail in Chapter 6, flexible resources can be classified as storage, substitutable, or curtailable resources (Althaher, Mancarella, & Mutale, 2015). Storage within the EPN can enable the *shifting* of grid electricity consumption in time. This storage may be of electricity (in the form of a battery or other electrical storage medium), of some other derived energy vector (such as heat in thermal storage), or be of some product within a process (such as clothes within a washing machine) (Zhou, Mancarella, & Mutale, 2016). Substitutable appliances (using an alternative energy vector for fuel) for the provision of the energy service can enable the *substitution* of one energy vector for another (Capuder & Mancarella, 2014; Mancarella & Chicco, 2013). For example, heat provision may be switched from an Electric Heat Pump (EHP) to a Combined Heat and Power (CHP) unit, resulting in a substitution of electricity with gas (Capuder & Mancarella, 2014; Mancarella & Chicco, 2013), while more general, "multigeneration" cases could also involve cooling (Mancarella & Chicco, 2009). Curtailable resources can, given the willingness of the consumer to put a price on the utility of the energy service that they are consuming, enable the *trading* of consumer *utility* (Good, Karangelos et al., 2015; Althaher et al., 2015). For example, pricing thermal comfort would enable optimized trading between thermal comfort and overall utility (Good et al., 2015a). How a neighborhood exploits these resources, and how much economic value they produce, depends on the objective they follow, as shown in subsequent sections.

2.1 Economic Value of "Classic" Definition of *Energy Positivity*

The "classic" definition of *energy positivity* is taken here as "the generation of more electricity, by a neighborhood, than it consumes," also implicitly intending that this electricity should be coming from local resources that bring also environmental benefits (e.g., RES)—even though it may not always be the case. Assuming that a neighborhood will aim to be as (electrically) energy positive as possible, and that measurement is over a year, the objective of an EPN under this definition will be to maximize its net electricity generation, or, equivalently, minimize its net grid electricity consumption, over the year.

With the aim of illustrating the potential effects on economic value associated with pursuing the classic definition of energy positivity, it is useful to consider the effects of employing the three identified sources of flexibility: *shifting*, *substitution*, and *utility trading*. Accordingly, the effects of employing the three types of flexibility will be explored with reference to *two notional neighborhoods*, situated in the north of England:

- *Neighborhood 1* is made of 50 well-insulated flats each heated by an Air Source Heat Pump (ASHP) and each with 1.9 kW of solar PV panels, and a 2 kWh/0.2 kW electrical battery.
- *Neighborhood 2* is made up of 50 well-insulated flats each heated by a gas CHP, and each with a 145 L hot water Thermal Energy Storage (TES).

ASHP and CHP are sized according to peak loads (as in Good et al., 2016), resulting in average size of 1.9 $kW_{electrical}$ for ASHPs and 4 kW_{gas} for CHPs. The ASHPs Coefficient-Of-Performance (COP) varies linearly with the air temperature, taking value of 1.97 at 0°C and 3.51 at 20°C. The CHP is a Stirling Engine type, with thermal efficiency of 0.78, and electrical efficiency of 0.13. Building set temperatures are sampled from a normal distribution with mean 21.6°C and standard deviation of 2.4°C (Shipworth et al., 2010), with set temperatures valid only during times of active occupancy (i.e., the occupants are present and active) (Good, Zhang et al., 2015). The after-diversity peak demands for the neighborhoods are 140 $kW_{thermal}$ and 17 $kW_{electrical}$,[2] while the annual demands are 231 $MWh_{thermal}$ and 48 $MWh_{electrical}$.[2]

With respect to *shifting*, as the classic definition of energy positivity is over the period of 1 year, there is actually no benefit from shifting electricity demand/generation, using electrical/thermal storage within daily time scales, as in typical applications. In the case of neighborhood 1 there is no motivation to use any battery storage to shift electricity, as energy positivity is measured over a long period. So, whether the battery storage is employed or not makes no difference considering an objective of energy positivity only. In fact, if the battery is utilized at all the neighborhood will become *less* energy positive, as battery losses will result in net increase in electricity consumption compared with the no storage case. This inherently disregards the potential economic and environmental benefits that could be accrued by shifting energy consumptions to periods of lower energy prices or carbon intensity.

In the case of neighborhood 2, the objective of maximizing energy positivity provides an incentive to keep the flats at maximum temperature (respecting thermal comfort constraints), to stimulate more CHP operation, and hence more electricity generation. As a result, there is no incentive to match electricity consumption and generation on a settlement period[3] by settlement period basis, as energy positivity is defined over the year. To demonstrate how the operation of the energy neighborhood varies when optimized under a classic energy positivity objective, compared with an economic objective, results in this section illustrate the relevant operation of the neighborhood's resources in terms of CHP electricity production, electricity imports, and TES temperature under both objectives. Here optimization is undertaken for neighborhood 2 on a typical winter day in the United Kingdom. Outdoor temperature (a major determinant of energy demand) and energy prices (not including retailer costs and profit margin, which are socialized over all settlement periods and hence do not produce a time varying signal to drive economic operation of the local resources[4]) are as in Fig. 2.1 and reflect typical values for the winter period. Settlement periods are half-hourly, as in the UK market.

The CHP operation and electricity imports under the classic energy positivity objective are shown in Fig. 2.2, where it can be seen that the storage is not used (indicated by the lack of variation in the TES temperature), with the CHP operation skewed toward the early morning and late evening, when it is colder and there is less active occupancy (and hence no limits on indoor temperature), to try to produce as much electricity as possible, to get as close as possible to "energy positivity" over the day (for the

[2]Electrical demand here refers to "base," non-ASHP related electricity demand.

[3]A settlement period is the unit of time (usually 5 min to 1 h), in which electricity is traded in markets.

[4]We are assuming here that the neighborhood is subject to time-varying price signals to be able to deploy economic-driven flexibility.

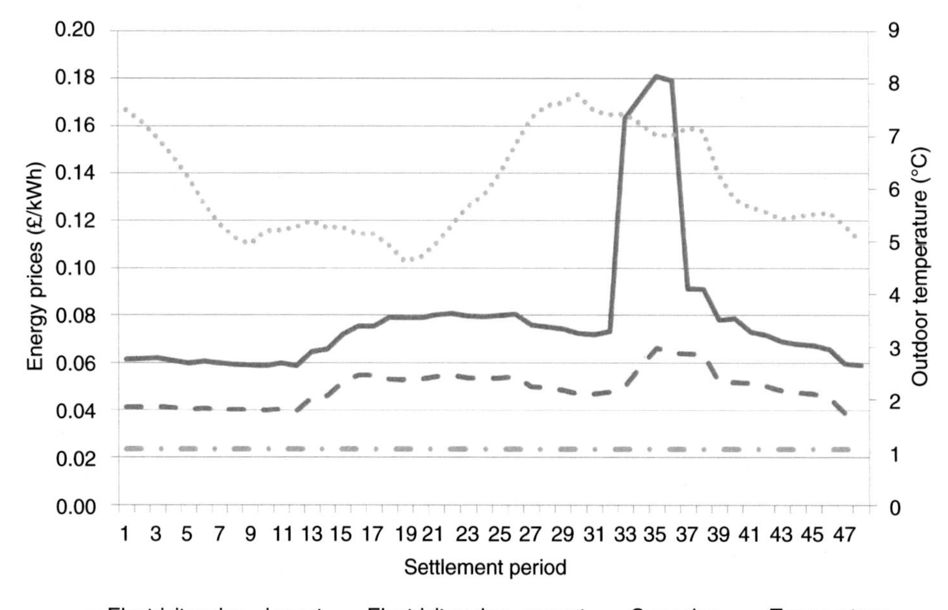

Figure 2.1 ***Typical winter energy prices and temperatures.***

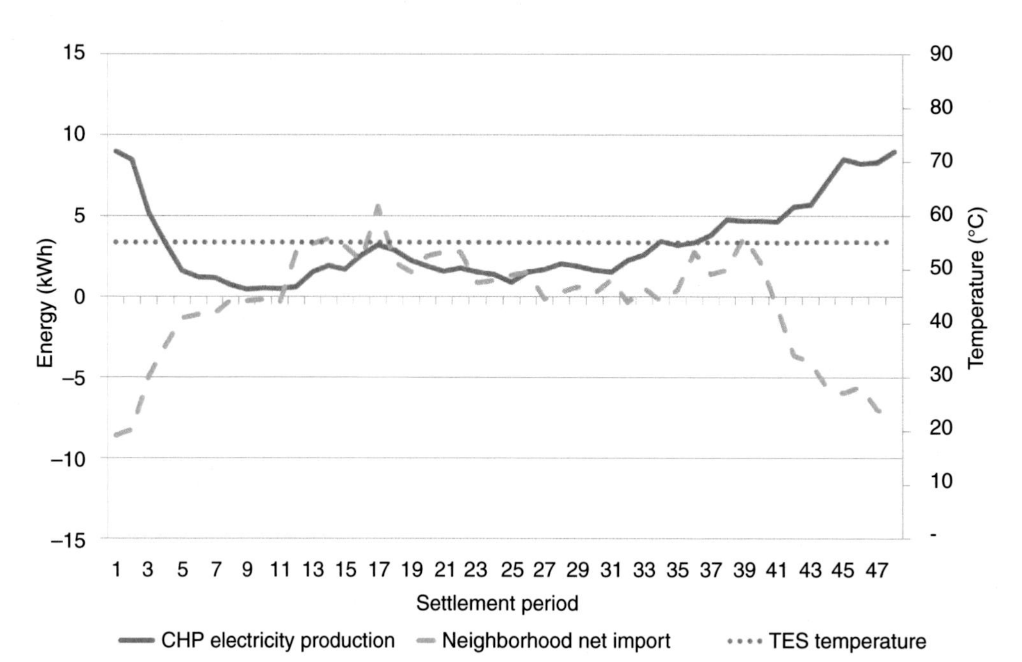

Figure 2.2 ***Neighborhood 2 behavior under a classic energy positivity objective, typical winter day, United Kingdom.***

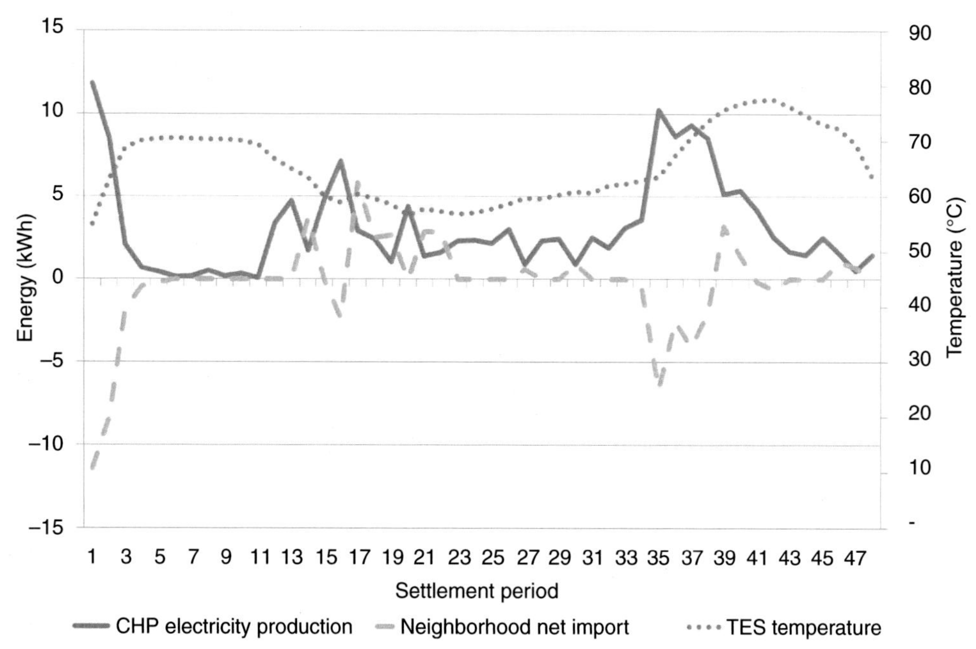

Figure 2.3 ***Neighborhood 2 behavior under an economic objective, typical winter day, United Kingdom.***

purposes of illustration, energy positivity is enforced in this case over a day rather than a year as in the definition given previously). Alternatively, under an economic objective, the thermal storage would be used more selectively, particularly to better match CHP electricity generation to neighborhood electricity demand (maximizing self-sufficiency in electricity, which could potentially increase economic value by reducing electricity imports during the most expensive periods). The CHP operation and electricity imports optimized based on the objective of minimizing economic costs is shown in Fig. 2.3.

An analysis of the key metrics of net energy imports over the day and the relevant costs reveals the misalignment between the classic energy positivity objective and the economic objective, as the energy positive objective produces net imports of −18.7 kWh and costs of £23.75, while the economic objective produces net imports of −12.7 kWh and costs of £22.32.

With respect to *substitution*, there can be a benefit, in terms of the classic definition of energy positivity, from substituting electricity generating devices for electricity consuming devices (e.g., substituting a CHP for an EHP), or from substituting a nonelectrical device for an electricity consuming device (e.g., a gas boiler for an EHP). Considering neighborhood 1, again on a typical UK winter day, as defined in Fig. 2.1, the addition of a CHP unit may improve energy positivity (according to the classic definition) by substituting EHP heat generation with CHP heat generation, which would reduce electricity

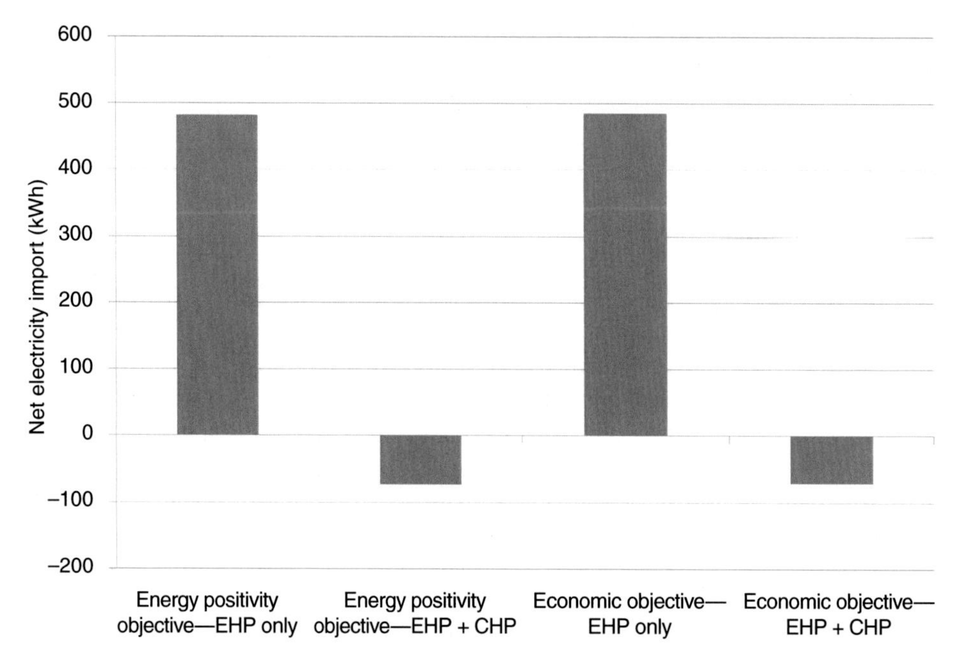

Figure 2.4 ***Net electricity import under various objectives, with different heating technologies, neighborhood 1, typical winter day.***

consumption. This is demonstrated in Fig. 2.4 which shows the net electricity consumption of neighborhood 1 when heated with EHP heaters or EHP and CHP heaters, under both energy positivity and economic objectives. However, it is important to note that the alternative, to substitute electricity consumption with gas imports, may not increase the economic value of the neighborhood if the electricity price is relatively low compared with the gas price. Under certain conditions (such as low electricity prices due to, for instance, high penetration of PV and wind generation) the electricity price may be such that an increase in energy positivity does not result in energy cost reductions. This is the case in Fig. 2.5 where the addition of the CHP actually increases energy cost under both the energy positivity and the economic objectives. What is worse, of course in this analysis there is no consideration for either environmental assessment (it is possible that fuel substitution increases emissions, which again exposes the criticality of having an ill-defined, electricity oriented definition of energy positivity) or capital investment (attempts to increase electrical energy positivity may result in substantial investment cost that might again make the whole EPN economics shaky).

With respect to *utility trading*, energy positivity may be increased by trading utility related to electricity consumption (such as illumination from electric lighting or thermal comfort from electric heating). In effect, this means reducing lighting or heating, to increase energy positivity. In this case the energy positivity and economic objectives

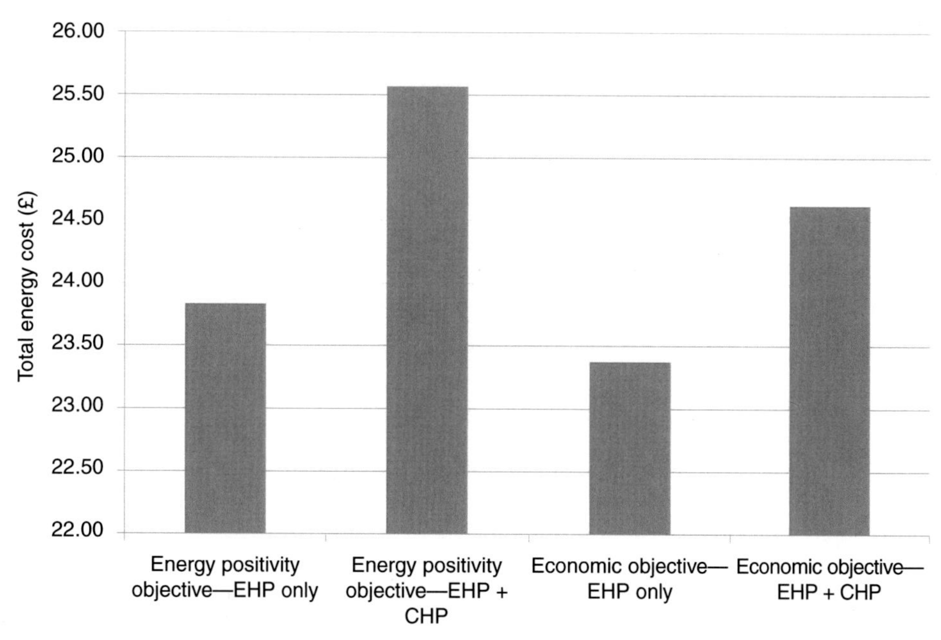

Figure 2.5 ***Net total energy cost under various objectives, with different heating technologies, neighborhood 1, typical winter day.***

coincide, as a reduction in user utility results in both less energy consumption and expense. However, energy positivity could also be increased while increasing supply of some service/commodity, for example, if heating is provided by CHP. In this case, increasing CHP operation will generate additional heat, and hence increasing CHP electricity generation, increasing energy positivity. This is demonstrated by implementing an energy positivity objective on neighborhood 2, again for a typical UK winter day, with a +1°C and a +2°C window around the building set temperature, allowing the temperature to vary freely with the ranges (T_{set}, T_{set} + 1°C), and (T_{set}, T_{set} + 2°C), where T_{set} is the building set temperature. As shown in Fig. 2.6, the higher flexibility window allows the buildings' temperatures to rise as CHP generation is increased to maximize net electricity export (from −17.35 kWh in the +1°C case to −29.40 kWh in the +2°C case). As, compared with gas prices, electricity prices are not high enough for the neighborhood to realize a profit from the increased CHP operation, the cost of energy for the neighborhood rises from £23.65 in the +1°C case to £24.93 in the +2°C case.

In order to assess the optimal trade-offs between customer utility and improved energy positivity, it can be attractive to express both utility and energy positivity in widely used and well understood economic terms (Good et al., 2015a). However, using such an approach can be arguable as the value of energy positivity is yet to be understood. In fact, this is a fundamental issue with the classic definition of energy positivity, as discussed

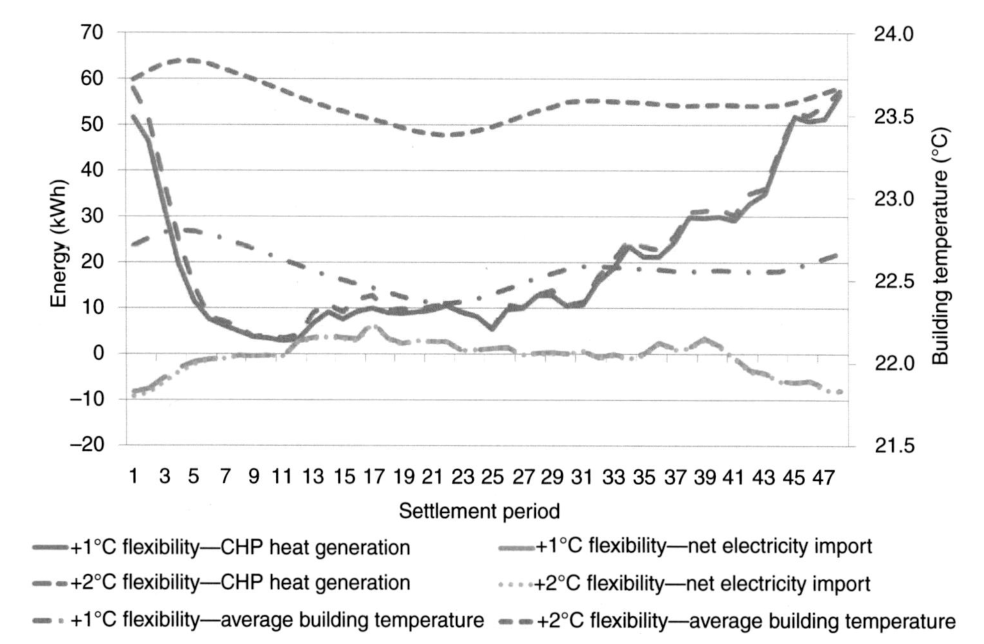

Figure 2.6 ***Neighborhood behavior with a +1°C and a +2°C flexibility window, typical winter day, United Kingdom.***

insofar: *energy positivity, under the classic definition, may have relatively little or no economic value in itself, and therefore is not compatible with a framework which is fundamentally concerned with economic value, as the modern, liberalized energy system is.*

In light of this, while, as demonstrated, energy positivity can produce economic value, the extent to which this is true is case specific. Indeed, further to the situations detailed previously, the economic value produced by following an energy positivity objective may be substantially affected by several factors. First, the nature of the imports and exports tariffs to which the neighborhoods are subjected can be central to the economic value of an EPN. In particular, the difference between the imports and exports prices (through a retailer or directly from the market) can be crucial. A wide difference between these prices may increase the economic value of energy positivity if the electricity generation within the EPN is well-matched to its demand (e.g., given CHP utilization). This may be especially true if the neighborhood operates following microgrid principles from an economic standpoint (Good et al., 2016),[5] with the neighborhood owning the local electricity distribution network so that flexible resource operation can

[5]In the sense that the EPN may not be an *actual* "*physical*" microgrid, meaning having the capability to operate in islanded mode under certain conditions, but an "*economic*" microgrid; in that it is allowed to internally trade energy and may therefore have incentives to maximize local generation and consumption balance.

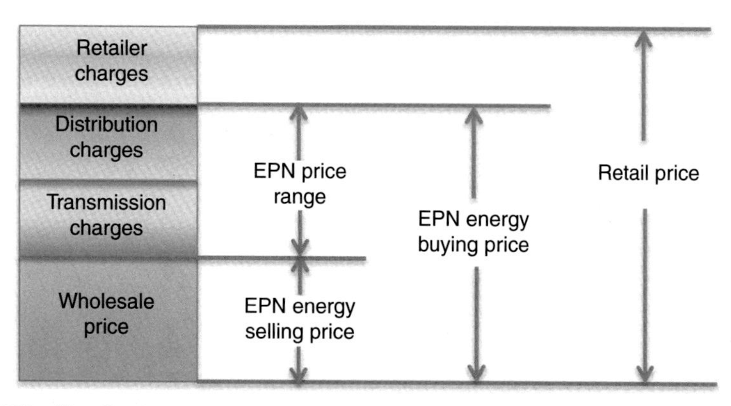

Figure 2.7 ***EPN selling/buying price.***

be coordinated with the objective that most electricity generation is used within the EPN rather than exported,[6] besides other potential technical and environmental objectives (Schwaegerl et al., 2011). This is because, commercially, such an arrangement would result in avoidance of substantial imports at high prices that include use-of-system (UoS) charges, as well as avoidance of substantial exports at low prices (although favorable prices may encourage the EPN to export electricity) (Fig. 2.7).

However, if electricity generation is not well matched to demand (e.g., solar PV without cooling demand, and therefore with higher electricity demand that may occur in mornings or evenings as in the United Kingdom), then a wide difference in imports and exports prices may not significantly increase economic value. This is because electricity would be generated at times of relatively low electricity demand (i.e., the middle of the day, in the United Kingdom; resulting in electricity export at times of low electricity prices), and grid electricity would be consumed at times of low electricity generation (i.e., early mornings and evenings, in the United Kingdom; resulting in electricity imports at times of high electricity prices).

Another factor which may influence the economic value of energy positivity may be the availability of subsidies for low carbon generation. These subsidies tend to encourage renewable but, in most cases, also inflexible technologies, such as PV and wind generation. More pertinently, such subsidies can be applicable to CHP generation, providing incentives to maximize operation, which on the one hand is in line with the electrical energy positivity objective, while, on the other hand, may encourage heat spilling in the case of absence of sufficient local heat demand. Subsidies may be also applicable to (low

[6]Besides possessing enough electricity generation capacity to become self-sufficient (as EPNs under the classical definition), microgrids also possess islanding capabilities, which are optional for EPNs.

carbon) EHP heat generation, which may provide a disincentive to energy positivity, by rewarding EHP electricity consumption.

2.2 Economic Value of *Flexibility*

Following a *flexibility* objective can be considered to be particularly suitable for the realization of economic value as it recognizes the desirability of increasing use of both local and renewable energy sources, as well as of contributing to the efficiency and security of the wider grid, on which an EPN will rely for electricity imports (when required) and transport of exported electricity. The explicit recognition of these factors is more suited for the realization of economic value as, in a rational, well-designed and liberalized energy system, underlying price signals and incentives are likely to encourage EPN operation in line with increasing local and renewable energy usage and supporting the efficient and secure operation of the grid.

To illustrate how energy price signals can motivate behavior that contributes toward economic value (and therefore toward system efficiency and security), the rationale behind the variable electricity wholesale price signals will be discussed later. Particular emphasis will be placed on the contribution of electricity wholesale price signals to the various components of the energy positivity definition (i.e., *local* generation, *renewable* generation, grid *efficiency, and* grid *security*).

Generally speaking, electricity prices rise as electricity demand increases, as more expensive plant are brought on-line. High grid prices may motivate the use of cheaper options, such as *local* flexible generating plant (e.g., CHP). In addition, if prices are expected to remain generally high, this will also motivate investments in *local* plant, including *renewable* plant. High grid prices also contribute to grid *efficiency* and *security* by motivating operation of demand-side resources [e.g., local generation, storage, and Demand Response (DR)], which can help to maintain system reserves and capacity, thus aiding *security* and more generally *reliability* (Zhou, Mancarella, & Mutale, 2015), and avoiding use of inefficient peaking plant, thus aiding *efficiency*. Low prices may also encourage carbon emissions reductions, as these low grid prices can be due to a substantial amount of renewable generation available in the market. Therefore, increasing grid consumption at these times will boost *renewable* generation [that may also include electrified heating (Capuder & Mancarella, 2014)], and also aid grid *security* by avoiding the instability that the combination of low demand and high renewable generation can bring (as more synchronized generators are pushed off the grid).

However, for these price signals to contribute toward the flexibility objective, the price signals must be partially or wholly visible to the EPN (i.e., the retail prices to which the EPN are subject should be dynamic). Under current arrangements, retailers manage all energy trades for EPNs using flat price signals that isolate the neighborhood from the dynamic market price signals. On the one hand, such price arrangements protect EPNs from risks related to price volatility and balance responsibilities. On the other hand, the flat signals obstruct the attainment of the flexibility objective by cutting the EPN off from valuable price signals that would direct the EPN toward desirable behavior. Given the general demonstrated coincidence of flexibility and economic value, such fixed prices also deprive the EPN of the opportunity of employing its flexibility, effectively providing DR (Losi, Mancarella, & Vicino, 2015), to increase its economic value.

Clearly, prices faced by the EPN must be more dynamic for the EPN to achieve flexibility. As also discussed in Chapter 6, this can be achieved by lowering or abolishing regulatory and cost barriers to market participation by small parties, to allow the EPN to partake directly in the relevant markets. In practice, this means deploying a net cost minimization objective, explicitly taking into account price signals from energy markets, ancillary services (such as system reserve) markets, capacity markets/mechanisms, grid fee regimes, low carbon incentive regimes, and taxes. Other solutions, which encourage flexibility but fall short of full market exposure, are also possible. Such solutions may involve a retailer offering more refined retail prices, such as time-of-use, or critical-peak pricing (Six et al., 2015). Such pricing schemes can pass on some of the price signals presented by the various relevant markets, while ensuring some protection and predictability for the EPN. However, such schemes will also tend to reduce the degree of flexibility, and hence economic value available to the EPN (compared to full exposure to markets). This is because, while risk is correlated with price dynamism (more dynamic prices mean more risk), so is potential value (more dynamic prices offer more potential value) for those parties with the necessary flexibility and optimization and communication capabilities (as demonstrated in the COOPERaTE project). This relationship between risk and potential reward is illustrated in Fig. 2.8.

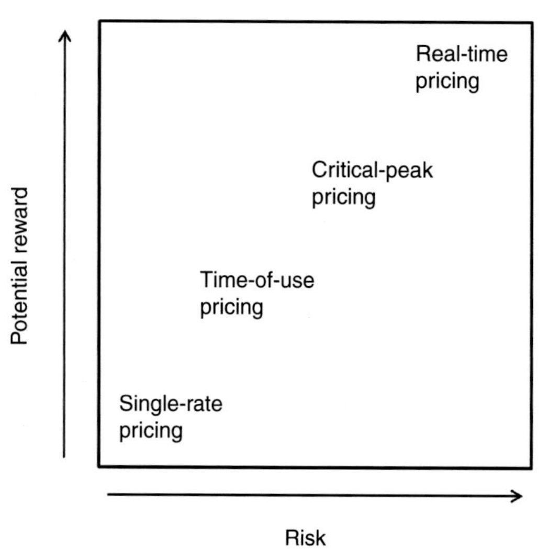

Figure 2.8 ***Risk and reward (for the EPN) of various pricing schemes.***

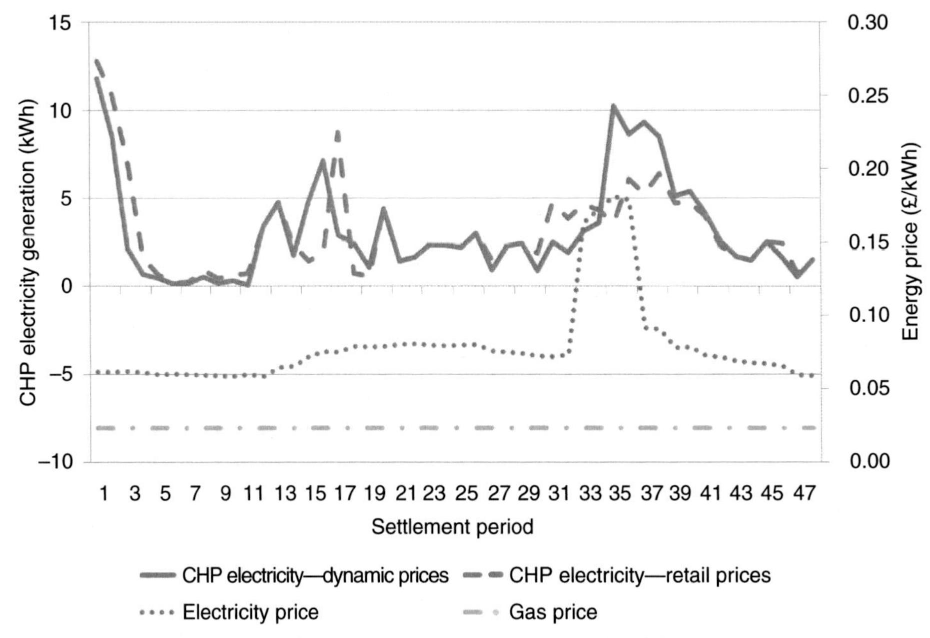

Figure 2.9 ***Neighborhood CHP electricity generation under retail and dynamic prices.***

To demonstrate how dynamic prices can motivate behavior more in-line with the flexibility objective, the optimal CHP electricity generation regime subject to retail and dynamic prices[7] is further analyzed for neighborhood 2 (Section 2.1) on a typical UK winter day (Fig. 2.9). Fig. 2.9 shows how visibility of dynamic prices, which directly reflect the prices in the various applicable markets, encourages shifting CHP operation as much as possible (given the limited flexibility afforded by the associated hot water thermal energy stores) toward times of high price. This reaction to price signals results in a reduction of total energy cost from £13.80 when the operation of CHP is optimized based on retail signals to £12.68 when CHP is optimally operated based on dynamic prices.

2.2.1 Further Aligning of Flexibility and Economic Value Objectives

Besides becoming exposed, to a greater or lesser extent, to the various market price signals, for maximum efficacy in pursuing the flexibility objective it is necessary that market designs and the designs of regulated charging regimes (such as for grid fees) are in line. That is, that they encourage local and renewable generation, grid efficiency, and system security.

[7]Retail and dynamic prices have been created here so as to be averagely consistent.

Based on the aforementioned, for example, recognition of the benefits of local generation (e.g., reduced losses, increased reliability, and lower carbon emissions), in the form of payments or grid fee rebates, would further encourage the local generation component. An increased carbon price would similarly encourage grid-level renewable generation (given the low-carbon status of renewable generation). Extending the carbon market to the local level, or increasing subsidies for local small-scale renewable (electricity and heat) generation would also encourage the renewable component at the local level. With respect to grid efficiency and system security, several measures could be taken to further align the proxy economic value objective with the true flexibility objective. Capacity markets or mechanisms could be introduced (if not already in place) to motivate cost–effective supply adequacy (Zhou et al., 2015), and the design of system frequency response and reserve products could be tailored to remove barriers to demand-side resources, such as may be found in an EPN. At the local level, efficient and secure operation may be encouraged and aligned with the economic objective through incorporation of a Distribution Network Constraint Management (DNCM) service (Chapter 6), and through monetization of the benefits of the EPN concept in terms of increased reliability (brought, e.g., through implementation of microgrid capability) (Syrri, Martínez-Ceseña, & Mancarella, 2015; Schwaegerl et al., 2011). At both transmission and distribution levels, grid fees could improve grid efficiency and security by becoming more cost reflective and, thus, recouping more network costs at times of highest network stress (i.e., from customers whose operation results in network stress) (Martínez-Ceseña et al., 2015).

3 ALTERNATIVE OBJECTIVES/CRITERIA

Although the focus in this chapter has been on discussing the key advantages associated with maximizing the flexibility inherent in energy neighborhoods, it is of interest to consider that this objective is only one of the many objectives that may be pursued (Schwaegerl et al., 2011). For example, a neighborhood may wish to follow a specific CO_2 minimization objective in order to burnish their green credentials. This would result in increased operation of flexible low-carbon plants such as CHP, employment of storage to better match demand to low-carbon generation (to shift grid consumption to low-carbon periods) and, possibly, trading of utility at high-carbon times. Alternatively, a neighborhood may wish to follow a local energy utilization maximization objective. This will involve matching local generation to local demand as much as possible, without regard to prices or CO_2 emission rates. This would result in increased operation of local flexible plants, such as CHP or diesel generators, and operation of any storage to assist in matching local generation and demand.

However, it is important to note that while the aforementioned objectives may motivate economically valuable behavior, given the general coincidence of economic and low-carbon drivers (i.e., high CO_2 generation is often expensive), and of economic and

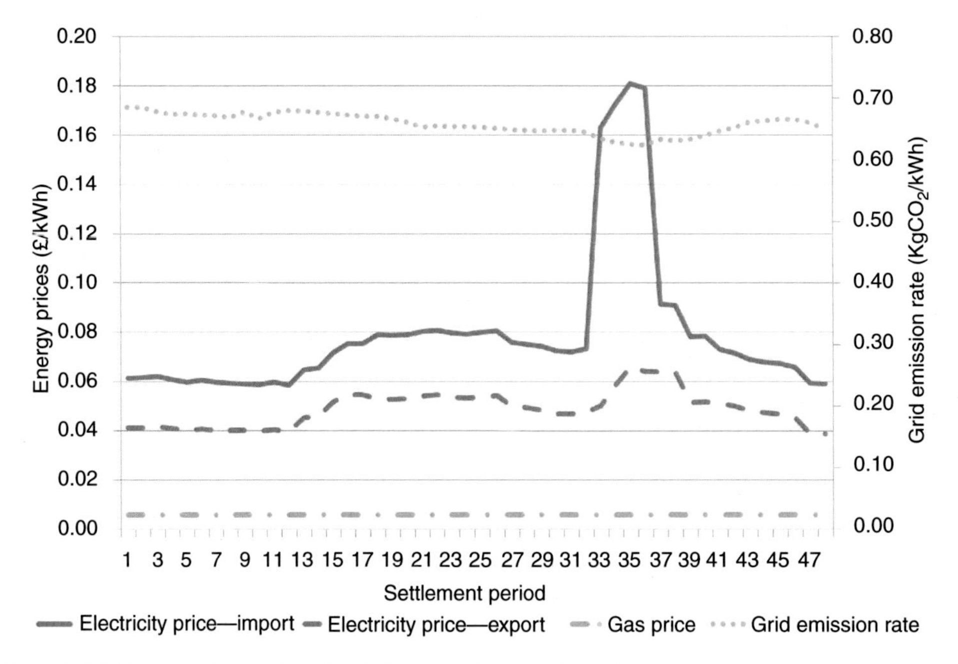

Figure 2.10 ***Energy prices and grid emission rate, typical winter day.***

local energy drivers (as on-site generation avoids UoS costs and can avoid some elements of taxation), pursuing a noneconomic objective may result in a suboptimal economic value for the EPN.

To illustrate this, consider as an example that the operation of neighborhood 2 is optimized for a typical winter day with respect to three different criteria, which could be the criteria of an employed energy optimization engine (such as that explored in Chapter 4):

1. minimize CO_2 emission (both local and grid);
2. maximize local energy utilization (self-sufficiency); and
3. maximize economic value.

The neighborhoods are optimized for a typical winter day, with energy price and grid CO_2 emission rates (based on the UK context) as in Fig. 2.10.

For each case, the relevant CO_2 emission (including both local and grid emissions), the amount of energy traded with the grid (as a measure of neighborhood self-sufficiency) and the total energy cost are quantified. The results for the three studies are shown in Table 2.1.

Results in Table 2.1 clearly highlight that while some optimization criteria may produce results which are desirable with respect to other metrics, a result will be generally suboptimal with respect to a given criteria unless it is optimized specifically with respect to that criteria.

Table 2.1 Optimization results, neighborhood 2

	CO_2 emission (kg)	Cost (£)	Grid usage (kWh)
Economic value maximization	241.34	22.24	72.21
CO_2 emission minimization	225.06	22.56	22.08
Self-sufficiency maximization	227.06	22.78	20.00

More detail is shown in Figs. 2.11–2.13, which show CHP electricity generation, local and grid CO_2 emission, and energy cost in the cases of optimization with respect to economic value, CO_2 emission, and self-sufficiency, for neighborhood 2, respectively. As shown through comparison of Figs. 2.11 and 2.12 the profiles for CO_2 minimization and self-sufficiency maximization are quite similar, as both the CO_2 minimization and self-sufficiency maximization objectives are served by minimizing imports from the grid. The slight difference arises as CO_2 emissions are minimized by directing grid consumption to a later period, where grid emission rates are lower. In contrast, the grid is used more often when maximizing economic value, as although there are penalties associated with grid usage (as UoS fees and taxes are charged on import, but not accrued as profits during export), these do not produce as high an incentive as in the previous cases to avoid electricity imports (and hence export, during other periods). The relationship

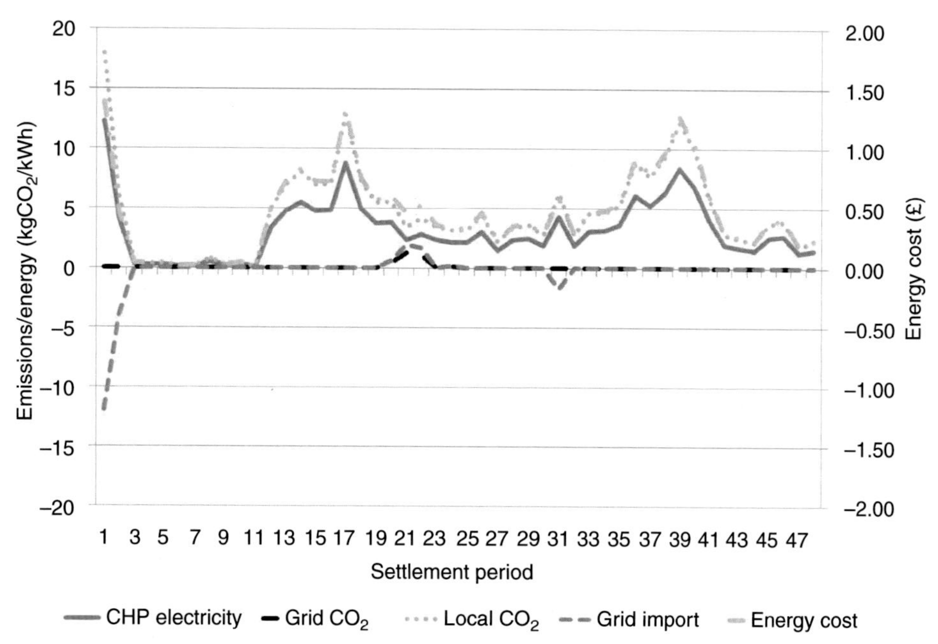

Figure 2.11 ***Neighborhood 2 behavior, CO_2 minimization case.***

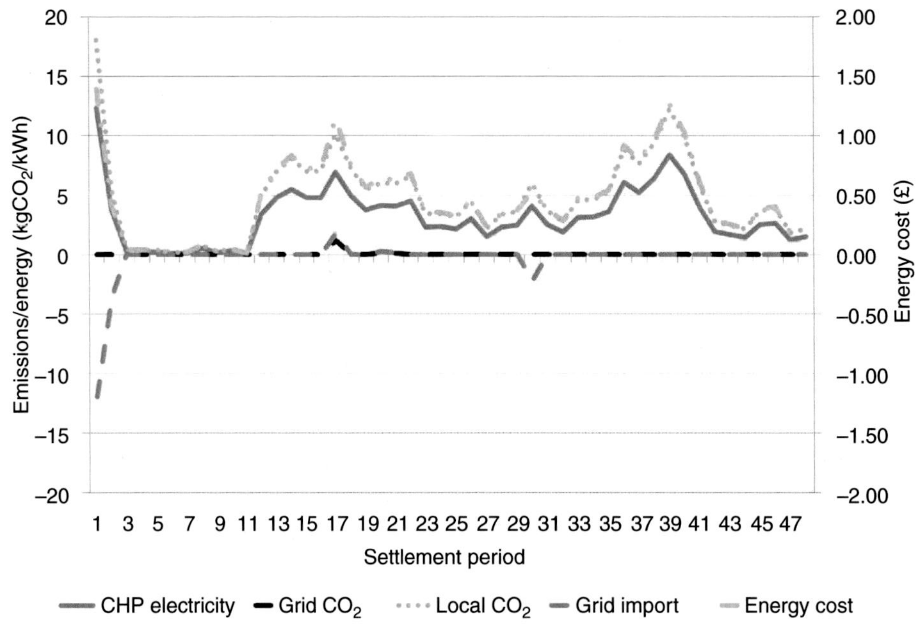

Figure 2.12 ***Neighborhood 2 behavior, self-sufficiency maximization case.***

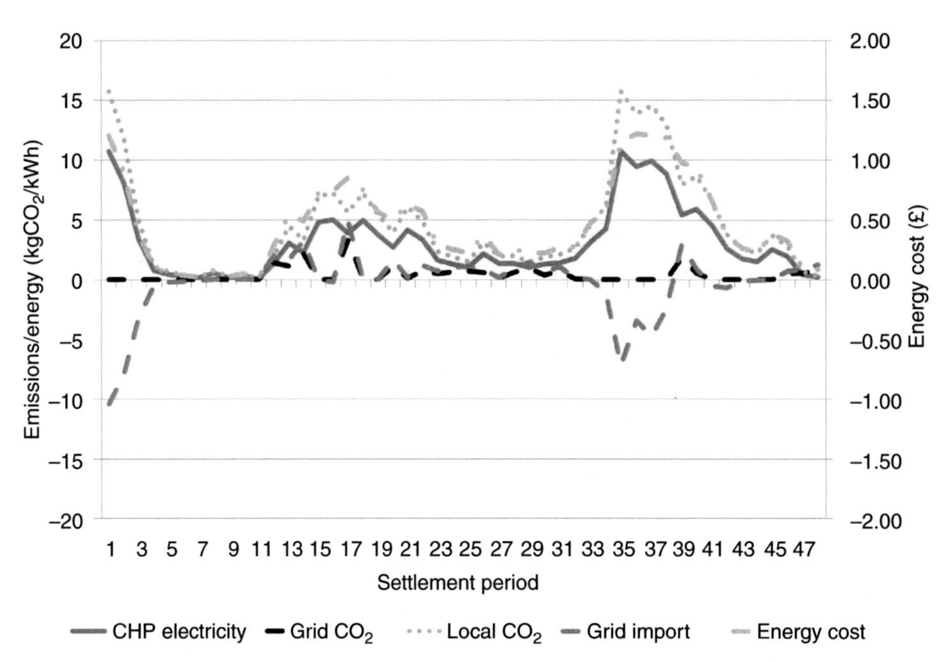

Figure 2.13 ***Neighborhood 2 behavior, economic value maximization case.***

between electricity prices and grid emission rates means that operating the neighborhood based on optimal economic gain also produces reasonably low emissions (as the CHP is motivated to operate at high price times, reducing the neighborhood grid consumption and hence grid emissions). It is worth noting that a neighborhood primarily heated by EHP would have a different result, as correlation of high grid emission rates and low prices would motivate consumption at times of high grid emissions.

4 MULTICRITERIA ANALYSIS

In practice, parties are likely to be interested in several criteria (such as economic value, CO_2 emissions, or self-sufficiency, as in Section 3) when assessing an intervention such as a new operational regime for the neighborhood, installation of CHP or TES units, and so forth. In the planning phase of an EPN, a multicriteria analysis can be undertaken to assess the suitability of an intervention, considering various criteria. This may be useful when an investor wants to assess and compare interventions according to several metrics at once, such as the CO_2, energy positivity, and cost metrics in Section 3. For this purpose several techniques are available, such as the direct analysis, linear additive model, and hierarchy method, among others (Zopounidis, Pardalos, & Fallis, 2010). The outcome of a multicriteria analysis is the identification of an intervention that is deemed to perform better than the rest with regard to all criteria (a ranking of the projects may also be possible). The development of a multicriteria Cost Benefit Analysis (CBA) is particularly relevant for this work, as EPNs may want to follow various objectives and assess interventions according to various economic, environmental, and energy criteria. This section briefly overviews classical multicriteria analysis techniques that may be used for the assessment of the EPN concept. For such purpose, a small illustrative example is presented, on assessment of intervention options.

A multicriteria CBA is meant to compare different intervention options based on relevant criteria. In the context of this book, the interventions may represent a vision of the EPN (e.g., investment in particular infrastructure and provision of a specific portfolio of services) from the perspective of an actor (e.g., consumers, retailers, and so forth) under a given market framework. For this purpose, it is convenient to present the performance of the interventions according to different criteria in the form of a matrix (performance matrix) as shown in Table 2.2. In the examples later, performance criteria

Table 2.2 Example of a multicriteria performance matrix

Intervention	NPV	Payback time	IRR	CO_2 emissions
Option A	Attractive	Attractive	Attractive	Attractive
Option B	Unattractive	Attractive	Attractive	Attractive
Option C	Attractive	Attractive	Unattractive	Unattractive

include Net Present Value (NPV), payback time, Internal Rate of Return (IRR), and CO_2 emissions.[8] The performance of each intervention option has, for illustration purposes, been set arbitrarily and, for the sake of simplicity, each criterion is expressed as either attractive or unattractive. In practice, each particular criterion would typically be assigned a numerical value.

The basic technique (and thus first step) to perform the multicriteria analysis is *direct analysis*. *Direct analysis* consists of inspecting if any of the options performs as well or better than all the other options according to each criterion. In this example, this would be option A, which is attractive according to all criteria under consideration. That is, this option offers high benefits (high NPV) on the short term (low payback times), high premium for the capital invested (high IRR) and low environmental impact (low CO_2 emissions), all of which are attractive for the corresponding criteria. In practice, there might not be an option that meets the direct analysis requirements, thus other multicriteria techniques have to be used.

If it can be reasonably assumed that all criteria are independent of each other (e.g., the perception of the NPV will not change regardless of the payback time) and can be assigned a weight, the *linear additive model* may be applicable for the analysis. This technique consists of assigning a weight to each criterion and multiplying the score of each project by the weight. This would produce a single criterion for each intervention option based on which the best alternative can be determined. Such an approach can also be used to form a multicriteria objective, in the operational domain. Such an objective could be used to weight the economic value, CO_2 and self-sufficiency objectives discussed in Section 3. In order to illustrate this, assume that each option is credited 1 point per criterion if its performance is attractive. Accordingly, option B is worth 3 points (for its performance according to the payback time, IRR, and CO_2 criteria) and is deemed a better alternative than option C, which would only be credited 2 points (for its performance according to the NPV and payback time criteria), while option A would be deemed the best option given its 4 points (for its performance according to all criteria). Clearly the weight assigned to each criterion is a key parameter, as it has significant impact on the outcome of this technique.

Now, if it is not reasonable to assume that all criteria are independent of each other (e.g., a high NPV is desired but short payback times are preferred) an *analytical hierarchy process* could be used. In this case a pairwise comparison approach (based on a particular analysis or judgment) is used to determine the proper weights for the different criteria and intervention options. In order to illustrate this process, consider the pairwise weights shown in Table 2.3. Table 2.3, which illustrates how a given criterion (row) is valued with respect to other criteria (column) (e.g., the NPV criterion is deemed twice as valuable as the IRR criterion and thrice more valuable than the CO_2 emissions criterion).

[8]More information on the economic metrics is given in Chapter 6.

Table 2.3 Example pairwise weight for the different criteria

	NPV	Payback time	IRR	CO_2 emissions
NPV	1	0.5	2	3
Payback time	2	1	1	2
IRR	0.5	1	1	2
CO_2 emissions	0.333	0.5	0.5	1

Table 2.4 Criteria weights according to an analytical hierarchy process

Criterion	Weight
NPV	0.31
Payback time	0.34
IRR	0.23
CO_2 emissions	0.12

The weights for each criterion associated with the example of pairwise comparison are presented in Table 2.4. As it can be seen, the payback time is deemed the most valuable criterion, followed by the NPV, IRR, and CO_2 emissions. An explanation of the mathematics required to process this matrix to obtain the weight for each criterion is beyond the scope of this chapter. It will only be mentioned that the weights can be determined as the eigenvector associated with the maximum eigenvalue of the matrix (for more information, see Weisstein, 2016).

Following the weighting procedure undertaken previously, a similar procedure is performed for the assessment of the intervention options. For the sake of simplicity, option A is not considered in this weighting procedure (as it would be deemed the best in this analysis due to its attractive performance according to all criteria) and an investment option is deemed twice as valuable as another if its performance under a given criterion is better according to Table 2.2. The results of such a pairwise comparison and the associated weights are shown in Tables 2.5 and 2.6, respectively.

Finally, the weights for both criteria and intervention options are combined (i.e., the sum of the weight of the option multiplied by the weight of the corresponding

Table 2.5 Pairwise comparison of intervention options

	NPV		Payback time		IRR		CO_2 emissions	
Options	B	C	B	C	B	C	B	C
B	1	0.5	1	1	1	2	1	2
C	2	1	1	1	0.5	1	0.5	1

Table 2.6 Example intervention option weights according to an analytical hierarchy process

	NPV	Payback time	IRR	CO_2 emissions
Option B	0.33	0.5	0.67	0.67
Option C	0.67	0.5	0.33	0.33

Table 2.7 Results for the example according to an analytical hierarchy process

	NPV: 0.31	Payback: 0.34	IRR: 0.23	CO_2: 0.12	Ranking
Option B	0.33	0.5	0.67	0.67	0.51
Option C	0.67	0.5	0.33	0.33	0.49

criterion). The results show that option B is marginally a better option than option C under the selected criteria and weights (Table 2.7).

In cases where district assessors are interested in multiple assessment criteria, such as energy positivity and flexibility, but extensible to any other criteria (such as CO_2 emissions and economic value, as discussed in Section 3), the methods presented here are clearly useful, in the EPN planning phase. Further, the linear additive model (in which criteria are assigned weights equaling unity) may be employed for operational optimization. This may enable simultaneous consideration of multiple criteria (such as economic value, CO_2 emissions, and self-sufficiency, as in Section 3). In all cases, assessors must determine the weights that should be assigned to various criteria. Assuming districts are primarily concerned with minimizing costs, most weight is likely to be given to the economic objective. Thus, if price signals can be considered an effective indicator of the value of flexibility, economic value is demonstrated to be a suitable proxy for flexibility (Section 1).

5 CONCLUSIONS

This chapter has introduced the concept of flexibility (of energy resources) as an alternative to the inflexible, static, classical concept of energy positivity for a neighborhood. The key defining features of future energy systems (which both decrease the availability of flexibility in the electricity system, and increase the demand for it) have been reviewed, and the demand for flexibility, and hence its value, was shown to be varying and dynamic. Furthermore, the preferences of consumers (specifically in relation to electricity supply reliability) was also shown to be varying and dynamic, increasing complexity, as such changing preferences mean that the value of flexibility varies not only spatially and temporally, but also by individual. The economic value of both the classical energy positivity definition and of flexibility was demonstrated through quantitative case studies. In these case studies the result of using flexible resources (i.e., storage, substitutable and curtailable resources) to pursue an energy positivity objective and an economic (flexibility) objective was presented. In particular the nonequivalence of the two objectives was demonstrated. Subsequently, measures that would further align the flexibility and economic objectives, to ensure the economic objective better motivates behavior beneficial to the system, are reviewed. Possible alternative objectives and criteria were then examined, and the general principle that optimization with respect to one objective will result in behavior which is suboptimal with respect to other criteria was asserted, again through quantitative case studies. Methods for multicriteria analysis of interventions, for use when assessors need to balance the multiple, possibly competing, objectives in the EPN planning phase were then presented.

Through development and explanation of the concept of flexibility, comparison to the inflexible, static, classical definition of energy positivity, and through quantitative studies, this chapter demonstrates the suitability of flexibility, and, indeed, the inadequacy of the concept of energy positivity, especially when ill-defined and focused on the electrical side only, as measures of desirable behavior in current and future electricity systems. However, in a power system context, the key message is that the value of energy services and electricity supply reliability varies by individual, and that the value of demand and generation, and hence system and network redundancy, vary in time and space. Hence, neighborhoods should be optimized and assessed using objectives and metrics which reflect this. Naturally, this requires suitable models and optimization algorithms (as detailed in Chapter 4) and suitably advanced ICT for monitoring, communication and to undertake complex optimization (Chapter 5). Another key message is that, in the absence of any other practical objective/metric, economic value should, in a liberalized electricity system, be used as a proxy for flexibility. Further, this objective/metric can be made a better proxy through improving the cost reflectivity of prices. In this context, economic value should be assessed through quantification of the district *business case*, which is discussed in detail in Chapter 6.

REFERENCES

Althaher, S., Mancarella, P., & Mutale, J. (2015). Automated demand response from home energy management system under dynamic pricing and power and comfort constraints. *IEEE Transactions on Smart Grid*, *6*(4), 1874–1883.

Capuder, T., & Mancarella, P. (2014). Techno-economic and environmental modelling and optimization of flexible distributed multi-generation options. *Energy*, *71*, 516–533, Available from http://linkinghub.elsevier.com/retrieve/pii/S0360544214005283.

Faruqui, A., Harris, D., & Hledik, R. (2010). Unlocking the €53 billion savings from smart meters in the EU: how increasing the adoption of dynamic tariffs could make or break the EU's smart grid investment. *Energy Policy*, *38*(10), 6222–6231.

Flamm, A., & Scott, D. (2014). *Electricity Balancing Significant Code Review—Final Policy Decision*, Ofgem, Available from https://www.ofgem.gov.uk/sites/default/files/docs/2014/05/electricity_balancing_significant_code_review_-_final_policy_decision.pdf.

Good, N., et al. (2016). Techno-economic and business case assessment of low carbon technologies in distributed multi-energy systems. *Applied Energy*, *167*, 158–172, Available from http://www.sciencedirect.com/science/article/pii/S0306261915012155.

Good, N., Karangelos, E., et al. (2015a). Optimization under uncertainty of thermal storage based flexible demand response with quantification of residential users' discomfort. *IEEE Transactions on Smart Grid*, *6*(5), 2333–2342, Available from http://www.scopus.com/inward/record.url?eid=2-s2.0-84923658441&partnerID=tZOtx3y1.

Good, N., Zhang, L., et al. (2015b). High resolution modelling of multi-energy domestic demand profiles. *Applied Energy*, *137*, 193–210.

Losi, A., Mancarella, P., & Vicino, A. (2015). *Integration of demand response into the electricity chain: challenges, opportunities, and Smart Grid solutions*. London, Hoboken: Wiley-ISTE.

Mancarella, P., & Chicco, G. (2009). *Distributed multi-generation systems: Energy models and analyses*. Hauppauge, NY: Nova Science Publishers, Available from https://www.novapublishers.com/catalog/product_info.php?products_id=7335.

Mancarella, P., & Chicco, G. (2013). Real-time demand response from energy shifting in distributed multi-generation. *IEEE Transactions on Smart Grid*, *4*(4), 1928–1938.

Mancarella, P., Gan, C.K., & Strbac, G. (2011). Evaluation of the impact of electric heat pumps and distributed CHP on LV networks. In *2011 IEEE PowerTech*, Trondheim, Norway, pp. 1–7.

Marszal, A. J., et al. (2011). Zero Energy Building—a review of definitions and calculation methodologies. *Energy and Buildings*, *43*(4), 971–979Available from http://linkinghub.elsevier.com/retrieve/pii/S0378778810004639.

Martínez-Ceseña, E. A., Good, N., & Mancarella, P. (2015). Electrical network capacity support from demand side response: techno-economic assessment of potential business cases for small commercial and residential end-users. *Energy Policy*, *82*, 222–232, Available from http://www.sciencedirect.com/science/article/pii/S0301421515001184.

Navarro-Espinosa, A., & Mancarella, P. (2014). Probabilistic modeling and assessment of the impact of electric heat pumps on low voltage distribution networks. *Applied Energy*, *127*(1), 249–266.

Navarro-Espinosa, A., & Ochoa, L. F. (2016). Probabilistic impact assessment of low carbon technologies in LV distribution systems. *IEEE Transactions on Power Systems*, *31*(3), 2192–2203.

Panteli, M., & Mancarella, P. (2015). The Grid: stronger, bigger, smarter? *IEEE Power and Energy Magazine, April*, 58–66.

Reichl, J., & Kollmann, A. (2011). The baseline in bottom-up energy efficiency and saving calculations—a concept for its formalisation and a discussion of relevant options. *Applied Energy*, *88*(2), 422–431Available from http://linkinghub.elsevier.com/retrieve/pii/S0306261910000723.

Sartori, I., Napolitano, A., & Voss, K. (2012). Net zero energy buildings: a consistent definition framework. *Energy and Buildings*, *48*, 220–232.

Schwaegerl, C., et al. (2011). A multi-objective optimization approach for assessment of technical, commercial and environmental performance of microgrids. *European Transactions on Electrical Power*, *21*, 1271–1290.

Shipworth, M., et al. (2010). Central heating thermostat settings and timing: building demographics. *Building Research & Information*, *38*(1), 50–69. Available from http://www.informaworld.com/openurl?genre=article&doi=10.1080/09613210903263007&magic=crossref||D404A21C5BB053405B1A640AFFD44AE3.

Six, D., et al. (2015). Techno-economic analysis of Demand Response. In A. Losi, P. Mancarella, & A. Vicino (Eds.), *Integration of demand response into the electricity chain: Challenges, opportunities, and Smart Grid solutions* (pp. 296). Wiley-ISTE.

Strbac, G. et al. (2012). *Understanding the Balancing Challenge*. Available from http://www.nera.com/publications/archive/2012/understanding-the-balancing-challenge.html

Syrri, A.L.A., Martínez-Ceseña, E.A., & Mancarella, P. (2015). Contribution of microgrids to distribution network reliability. In *Proceedings of PowerTech*. Eindhoven.

Torcellini, P., Pless, S., & Deru, M. (2006). Zero Energy Buildings: a critical look at the definition. In *ACEE Summer study*. Available from http://www.nrel.gov/docs/fy06osti/39833.pdf

Weisstein, E.W. (2016). Eigenvector. *MathWorld a Wolfram Web Resource*. Available from http://mathworld.wolfram.com/Eigenvector.html

Zhou, Y., Mancarella, P., & Mutale, J. (2015). Modelling and assessment of the contribution of demand response and electrical energy storage to adequacy of supply. *Sustainable Energy Grids and Networks*, *3*, 12–23, Available from http://www.sciencedirect.com/science/article/pii/S2352467715000387.

Zhou, Y., Mancarella, P., & Mutale, J. (2016). A framework for capacity credit assessment of Electrical Energy Storage and Demand Response. *IET Generation Transmission & Distribution*, 1–17.

Zopounidis, C., Pardalos, P. M., & Fallis, A. (2010). *Handbook of multicriteria analysis*. Berlin, Heidelberg: Springer-Verlag.

CHAPTER THREE

Description of the Process Needed To Achieve EPN Roadmap

E.H. Ridouane*, K.A. Ellis, K. Kouramas***
*United Technologies Research Centre Ireland, Cork, Ireland
**IoT Systems Research Lab, Intel Labs, Intel Corporation, Ireland

Contents

1 SUDA IMPROVEMENT METHODOLOGY OVERVIEW

Undertaking a neighborhood energy improvement can seem overwhelming. Even individuals familiar with the construction process may feel a little lost at first when considering new elements, such as measurement and verification (M&V), conservation measures, and ICT. The purpose of this chapter is to describe a framework and logical process for achieving what is a complex undertaking that is an Energy Positive Neighborhood (EPN). This framework applied to the EPN context leverages the expertise and heuristics the COOPERaTE consortium brought to and developed over the course of the project. The proposed framework is based on SUDA and consists of four important phases in the development of an EPN. Fig. 3.1 later illustrates the four phases of the

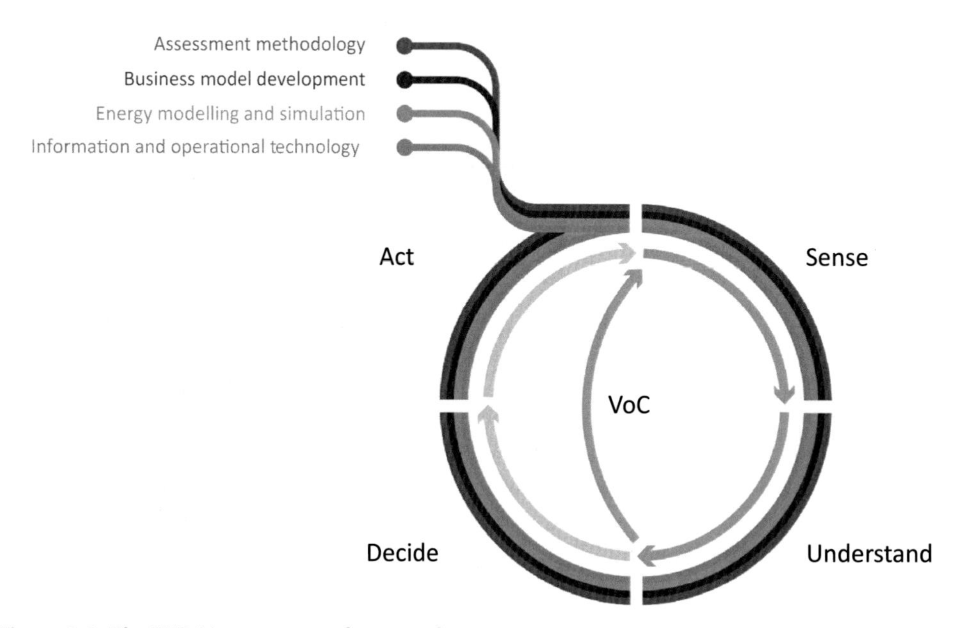

Figure 3.1 ***The SUDA improvement framework.***

SUDA framework. SUDA is posited as a generic framework for defining a problem/opportunity, for selecting a proposed improvement and for testing the efficacy of same. The framework is cyclical in nature and moves from scoping, to improvement selection, to improvement refinement. SUDA is based on Kolb's learning cycle (Kolb, 1984) and is used throughout to assist in orientating practitioners as they navigate the more complex collective of methodologies and tools required to achieve the defined improvement goal.

The intent is not to suggest that the end goal or process is simple but rather to offer a simple means of orientation as one progresses from start to finish. Effectively the SUDA phases of "sensing, understanding, deciding, and improving" are used to bucket methods, techniques, and tools that can be used to navigate toward the end goal. In all phases the methodologies and tools applied will depend on the point at which a practitioner is in the overall cycle, for example, in the sensing phase the tools used to scope a problem/opportunity on first pass will not be the same as those used to assess any implemented improvement.

Where a problem and/or proposed solution is multifaceted, the process can be considered as a grouping of associated loops. In other words each aspect of a multifaceted solution will most likely involve its own progression through the SUDA phases. This is the case with EPNs. Fig. 3.1 shows five themes/pillars required for delivering on the concept, namely—assessment methodology, energy modeling and simulation, business models, OT, and IT. With a sixth voice of the customer (VoC) central to all. For example, when choosing an assessment methodology for reviewing a posited improvement one will navigate the cycle in understanding options and in selecting the most apt approach. When deciding on business models one will sense what models, markets exist, try to understand what barriers/enablers apply, decide on recommendations, implement a suggested solution (perhaps in simulation) and then test the outcomes, adjusting ones understanding and approach as required.

Additionally, an "Improvement" can itself be thought of in relation to the SUDA loop. This is particularly true in information based systems where the improvement is itself sense and control based, that is, where the system is dynamically sensing stimuli, developing some understanding, supporting a decision, and then acting accordingly. In short, the SUDA-loop is posited as a useful construct for guiding a practitioner, with its true value resting in the identification of a menu of useful tools and methods at each phase. When coupled with heuristic capture, this can be very powerful in assisting fellow practitioners. The overall process as applied to the EPN context is summarized in Section 2. While what follows gives a generic description of each phase.

1.1 Sensing

This phase is the entry point of the process. It is a check point on subsequent iterations of the framework that is sensing if an improvement has had the desired effect. However, is best described as the point of data acquisition with respect to informing the "Understanding" phase. In practice these two phases are very closely linked with many of the methods and tools being common to both, this is represented by the inner loop between the two in Fig. 3.1. On first pass this phase is focused on identifying the problem/opportunity statement and gathering as much information as possible so as to build an accurate picture. A good starting point is to use one or several of the tools of this phase as part of a scoping exercise. Table 3.1 below illustrates the "5W1H" method, which outlines some of the exemplar questions one could initially ask. One could also use rich pictures, direct observation, interviews, storyboarding, PEST analysis, SWOT

Table 3.1 An exemplar scoping method 5W1H

Who	What	Where	When	Why	How
Are the stakeholders	Are our goals	Else has this been done	Do we start/finish	Is this important	Is this currently done
Are the experts	Points of reference do we have	Is the system of interest	Do we communicate	Can't it be done	Could this be done
Needs to know	Resources do we have	Do we communicate	Will we know we are done	Now	Should we communicate
Should we communicate with	Equipment is required	To find relevant data		Us	Will we assess/measure

analysis, CATWOE analysis, system maps, stakeholder analysis matrix, 5 Why's, after action review template, and so on.

On subsequent iterations of the loop, that is, where a possible improvement has been identified, these initial scoping methods and tools will give way to specific assessment methodologies and tools for assessing the efficacy of the posited solution. However, some generic methods/tools may still prove useful, for example, an after action review (AAR).

1.2 Understanding

This phase is about shaping the information acquired in the sensing phase and in identifying gaps in information. There is typically a lot of iteration between the sensing and understanding phases and the line between each is nuanced with respect to tool/method selection. Primarily the focus is to establish who and what is important and to understand what could be important. Again in a multifaceted scenario this means that one is trying to understand what is important in respect of several themes. *It involves understanding the on-the-ground reality and a PEST or STEEPLE analysis is often a useful tool to apply at this stage to understanding potential barriers/enablers to ones given context.* Additionally, interviews and workshops will often be employed to gain greater understanding in identifying the actors and aspects of most interest. Any of the tools methods from, but not limited to, Table 3.2 can be used. Additionally, as the process matures statistical analysis and data analytics will increasingly be utilized. While data analytics will typically form a component of any proposed IT improvement/solution.

In relation to the EPN options in terms of OT, IT, energy services, and business models are formed through iteration between the "sensing" and "understanding" phase. This involves understanding missing elements, interactions between the neighborhood components, subsystems, and ICT infrastructure. It involves understanding the actors involved, the potential for local energy generation, energy exchange, data metering, and actuation. Again a STEEPLE analysis can be helpful, for example, in identifying any national and/or regional/local regulations that must be accounted for. Functional and technical requirements are gathered for the neighborhood energy management services and supporting ICT infrastructure. As stated, on subsequent iterations ones assessment methodology and/or statistical modeling will be increasingly used to assess the efficacy of the identified improvements whereby ones understanding is tested and adjusted as required.

Table 3.2 Exemplar tools and methods at the "understanding phase" of the SUDA loop

PEST	Benchmarking	Concept maps
STEEPLE	9 Windows analysis	Triz
Cause and effect analysis	5 Forces	Simulation
Affinity diagrams	Value Stream Mapping	Statistical Modeling
4Ps or 7Ps analysis	SIPOC analysis	Flow Charts

Table 3.3 Exemplar tool and methods at the "deciding phase" of the SUDA loop

ROI analysis	Ansoff Matrix	Modeling
CBA	Critical Path analysis	Six thinking hats
RACI analysis	Ideal Final Result	Prototyping
Risk assessment matrix	Solution selection matrix	Decision Tree analysis
RAPID analysis	TRL assessment	Simulation

1.3 Deciding

This phase builds on the understanding developed to date and is focused on deciding on possible actions to take. The phase involves both ideation and solution selection, although ideation sits somewhat between understanding and deciding. Any or several of the tools from, but not limited to, Table 3.3 can be used to decide what improvement actions can be taken. The phase will also establish if the actions are within the remit of the practitioners or if additional stakeholder interaction is required, or indeed if the existing market or regulatory conditions support the defined action. Exiting this phase the practitioner will have a well-shaped understanding of what they are to implement in terms of improvement actions and what the expected outcome is envisaged to be.

Again in the case of the EPN, while improved energy efficiency and flexibility is obviously the overarching goal, there are several ways to achieve this and the kind of intervention also depends on regulations and the actors involved. A review of potential benefits of improvements will help define apt intervention. The stakeholders need to select from the potential business models for the neighborhood. That will impact site-adaptation that is OT and IT adoption, adaptation, and or/ development selection. The process will in addition have defined a number of core energy and data services (optimization, real-time monitoring, forecasting) and these services may now need refinement based on the business models context and/or OT availability.

1.4 Acting

At the point of action one has decided upon preferred improvement interventions. This phase from a framework perceptive is about initiating, planning, executing, controlling, and closing with respect to the proposed solution, that is, one is into a classic project management life cycle. There is also strong emphasis now on linking back into the SUDA phase of sensing in order to sanity check the value of the deployed improvement and this is the home of ones selected assessment methodology. There may also be significant emphasis now on change management to mitigate resistance to the proposed solution(s). Table 3.4 lists exemplar tools that may be useful at this phase. In summary, the overall process when applied to a concrete use-case and augmented with heuristic capture can be very powerful in assisting fellow-practitioners by acting as a blueprint for improvement. Section 2 that follows is a specific application of the framework to the EPN context.

Table 3.4 Exemplar tools and methods at the "act phase" of the SUDA loop

PMBOK	DICE Framework	Kotter's 8 step change model
Project management tools	McKinsey 7s model	Direct Observation
Proof-of-Concept	Prototyping	Simulation
Wireframe tools	Survey tools	Modelling

2 SUDA PROCESS FOR ENERGY POSITIVE NEIGHBORHOODS

This section describes in detail each of the core pillars (Fig. 3.1) that constitute the multifaceted concept of an EPN.

2.1 Assessment Methodology

Much of the description that follows discusses the assessment methodology chosen within the COOPERaTE project, which itself aligns to the SUDA cycle. This is naturally the case because the very act of assessment involves sensing some observation, understanding how that observation relates to an established condition and then deciding on whether some action is required. However, one must first choose an assessment methodology and that process involves sensing and understanding what options are available then deciding if one will adopt, adapt, or develop a means of assessing ones envisaged improvement. The action phase in this case being the application of the chosen methodology.

While a solution selection matrix can be helpful, one needs a certain level of knowledge in defining the scoring criteria. For example, a prime considerations for choosing a methodology in the EPN case was acceptance/credibility within the "built environment" community as they were the main owners of the assets to be controlled by the proposed "smart grid" energy services. That type of insight essentially comes from domain expertise, desk based research, and qualitative interviews with relevant stakeholders. Table 3.5 details the short list of methods and tools deemed useful.

Given the multifaceted definition of an EPN, as adopted in this project (Section 1), there is no appropriate assessment metric. Instead those wishing to assess a particular neighborhood should refer to a portfolio of KPIs, which can indicate any neighborhood's degree of "positivity," on a number of scales. As discussed in Chapter 7, there are many appropriate KPIs that can be of interest. These include more traditional measures,

Table 3.5 Choosing an assessment methodology

Task description	Tools/methods	Output/outcome
Decide upon a suitable assessment methodology	Desk based research, Qualitative interviews, Surveys, Solution Selection Matrix	An agreed fit for purpose assessment methodology

Table 3.6 List of KPIs for a neighborhood

Energy/power
Total energy consumption (kWh/day)
Local energy production (kWh/day)
Max. power demand (kW)
Max. power demand from grid (kW)
Environment
Energy savings (%, kWh)
CO_2 emission savings (kg CO_2)
Local resource utilization (%)
Economics
Grid energy consumption and cost (kWh, €)
Energy sold to the grid (%, €)
Energy cost savings (%, €)

such as energy saved or local energy produced versus total consumption. In addition, economic indicators can be relevant as a useful proxy for optimization and security aspects of the energy positivity definition, while CO_2 emission reduction and/or carbon intensity can be valuable outcomes of the enhanced definition. A suggested list of KPIs is given in Table 3.6.

Determination of energy savings requires both accurate measurement and a replicable methodology, known as a M&V protocol. The long-term success of energy management projects is often obstructed by the inability of project partners to agree on an accurate M&V plan. The International Performance Measurement and Verification Protocol (IPMVP) (The ICT PSP, 2011), adopted by COOPERaTE, discusses procedures that, when implemented, help stakeholders to agree on a M&V plan required to quantify savings from Energy Conservation Measures (ECM). The purpose of the IPMVP is to increase investment in energy efficiency and renewable energy by:

- Increasing energy savings through accurate determination of savings, which gives facility owners and managers valuable feedback on the operation of their facility, allowing them to adjust facility management to deliver higher levels of energy savings.
- Encouraging better project engineering through good M&V practices related to good design of retrofit projects. Good energy management methods help reduce maintenance problems in facilities allowing them to run efficiently.
- Helping to demonstrate and capture the value of reduced emissions from energy efficiency investments. The IPMVP provides a framework for calculating energy reductions before (baseline) and after the implementation of projects. The IPMVP can help achieve and document emissions reductions from projects that reduce energy consumption and help energy efficiency investments be recognized as an emission management strategy.

- Increasing public understanding of energy management as a public policy tool through improving credibility of energy management projects. M&V increases public acceptance of energy efficiency improvements.

An M&V plan mainly focuses on meter installation, calibration, and maintenance; data gathering and screening; computation of measured data and reporting. This Protocol defines broad techniques for determining savings from both a "whole facility" and an individual component. It is applicable to a variety of facilities including residential, commercial, and industrial buildings. In IPMVP, four options A–D are given for M&V. Stakeholders need to review these options and choose the appropriate option for their neighborhood. More details on the methodology are available in (The ICT PSP, 2011).

2.1.1 Baseline Assessment

As touched upon aforementioned, establishing the energy baseline of a neighborhood is very important to understand the neighborhood energy performance before and after the implementation of any EPN improvements. This is a key aspect to evaluate whether or not a neighborhood has achieved its energy performance objectives. In addition, the energy baseline specification is fundamental in informing more detailed understanding of the neighborhood consumption and in helping to identify opportunities for improvement. Although a number of methodologies have been established for defining the energy baseline for single buildings, no such method exists for establishing the energy baseline of neighborhoods.

A range of energy baseline and benchmarking methodologies exist for building energy rating assessment, such as CIBSE TM46 (CIBSE, 2008), Energy Consumption Guide (ECG) 19 (ECG, 2000), and the Energy Savings Measurement methodology adapted from IPMVP. Here, the latter (IPMVP) is leveraged and adopted for EPN baseline assessment, following the key steps as outlined in Fig. 3.2.

The process as outlined in Fig. 3.2 and later is essentially about sensing, understanding, and deciding upon scope, appropriate duration, and variables to adjust for. This is followed by action whereby one senses/collects data used to establish a baseline over the specified duration.

2.1.2 Step 1: Define the Scope

The first step defined in the IPMVP protocol is to define the scope of the energy study. This scope is defined for the entire project. It is within this scope that the baseline will be defined and the energy impact of the project will be analyzed. The tools described in Section 1 can be useful in this regard.

2.1.3 Step 2: Baseline and Reporting Period

The second important step is to define the various stages of study. The study is divided into two major periods:

- Study period for the baseline definition
- Study period after the implementation of the EPN program

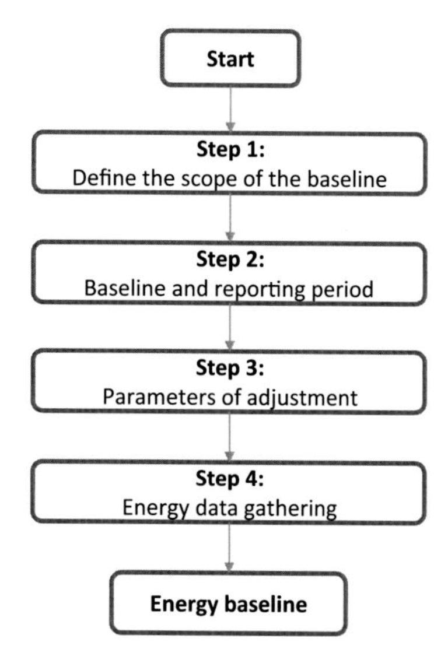

Figure 3.2 ***Energy baseline approach for Energy Positive Neighborhoods*** **(EPN).**

To represent all operating modes of the facilities of a site, as energy use is affected by weather conditions, the baseline period should be a full year. This period will show a full operating cycle from maximum to minimum energy use. The reporting period for the EPN program will start after the site adaptation and implementation of EPN identified equipment (e.g., battery storage, electrical vehicle charging stations, wind turbine, energy submetering). This period will continue till the end of the project.

2.1.4 Step 3: Parameters of Adjustment

The term "adjustments" relates to identifiable physical facts about energy governing characteristics of equipment within the measurement boundary. Two types of adjustments are possible and mentioned in IPMVP protocol:

- Routine adjustments: expected to change regularly during the reporting period, such as weather, occupancy of buildings. A method that was put in place by COOPERaTE to define and measure these parameters during the reporting period. They are defined for both sites in Chapter 7.
- Nonroutine adjustments: factors which are not expected to change, such as the facility size, design of installed equipment, and so on. These static factors must be monitored for change during the reporting period.

2.1.5 Step 4: Energy Data

Data from submetering or utility meters can be used to define the energy baseline depending on IPMVP option selected. Monitoring and energy optimization systems will be installed on each site as part of the system upgrade (Deliverable 4.2, 2014). Data from the reporting period will come from these monitoring and energy optimization systems. It is at this stage that improvements can be selected and introduced, while benchmarking activities can now be undertaken given a credible baseline on which assessment can be based. Benchmarking informs stakeholders about how they use energy, where they use it, and what drives their energy use. It is a key step in identifying opportunities for improvements to lower energy costs.

2.2 Business Model Development

The first step in selecting business models is to inspect the regulation of the system in which the prospective EPN is situated (sensing), to enable an understanding of the system to be built up. This is an undertaken that includes desk-based research and qualitative interviews. System maps, 5W1H, and STEEPLE analysis are useful in understanding the system of interest and in asking initial questions, like who can participate? Is participation of small-scale parties allowed, for example?

This is a relevant question at both the commercial and technical level. For example, electricity system operators may have technical minimum size requirements which may preclude participation of EPNs in the necessary markets. Similarly, relevant market operators may require that bids and offers into a market are above a certain threshold. Such limits on party size in markets may overcome by aggregation of EPNs with other resources. However, regulation should be inspected to be sure such aggregation is allowed. Even if EPNs satisfy the technical and commercial rules of system/market operators, potential EPNs should generally assess whether markets recognize the value they can deliver (Chapters 6 and 8). Without full recognition of the value of EPNs, which may require markets that don't currently exist (e.g., distribution network constraint management, or reliability services, see Chapter 6 on business cases, achievable revenue of a business case may not be satisfactory). Even if the necessary markets exist, regulation may be such that the prices in such markets are not cost reflective. This is the case, for example, for electricity distribution use of service costs in both French and Irish contexts (Chapter 6), which are heavily regulated.

If regulation is satisfactory, potential EPNs should seek to understand the markets in which they hope to partake in more detail. As discussed in Chapter 6, the value of the EPN concept is heavily dependent on exploitation of EPN flexibility. If the market in question, despite being cost reflective, does not offer sufficient reward for flexibility deployment (i.e., the variability of price through the day is small), then the business case should not be pursued.

With regard to the prospective EPN itself, the amount of flexibility that can be exploited should be understood. As described, value is derived from deployment of flexibility, so if there is insufficient flexibility within the EPN, sufficient revenue cannot be achieved. Flexibility may be in energy storage (electrical or thermal), or may be derived from the ability to shift provision of an energy service (e.g., heating) from one energy vector to another. Further, flexibility may be achieved through trading of end-user utility (e.g., thermal comfort), for reduced costs (i.e., through reduction in heat, hence thermal comfort, provided).

Lastly, the attitudes, preferences, and capabilities of possible/requisite partners should be assessed. For example, if a suitably accredited partner is needed to access the necessary markets, is such a partner available? If third party control of EPN flexible devices by an aggregator is deemed necessary, does a party with the combination of necessary technical and commercial capabilities exist to fulfil the aggregator role? As it will be demonstrated in Chapter 6, revenue is likely to increase given some collaboration between parties within the EPN. Are all relevant parties open to such collaboration?

Even with a good understanding of the regulation, markets and EPN flexibility, the complexity of the physical and economic systems, and their interactions, mean that it is unlikely to be clear whether any proposed business case is profitable. To reliably assess this, and decide on appropriate business cases, a suitable techno-economic model is required. This model should be able to capture the demand for the relevant energy vectors (electricity and gas). This information can be collected from suitably granular meter data, or can be the product of sophisticated modeling, as described in work package 2. Further the model should be capable of modeling all energy-related devices (e.g., electricity generators, battery storage, thermal storage). This is necessary to be able to justify acting on any business case, to exploit the flexibility of flexible electricity generators and energy storage, which are the resources which create the value in an EPN.

It should be noted that business case selection is not a one-off, but, instead, should be an ongoing process (following the SUDA loop), which the prospective EPN can, where appropriate, feed back into. For example, if regulation (e.g., minimum size requirement) is a barrier to a business case, the EPN may seek to have the relevant regulation reviewed. If prices in a market are insufficiently variable to justify pursuing a business case, the market (e.g., wholesale electricity market) should be monitored, in case developments (e.g., increased renewable penetration) increase price variability. If the EPN is insufficiently flexible to take advantage of otherwise promising business cases, flexibility could be increased (through investment in storage or on-site generation).

While the specific techno–economic model aforementioned is fundamental to business model selection, many of the more generic tools of Section 1.1, Table 3.2 and Table 3.3 are particularly useful in the process of business model development. None more so then desk based research, qualitative interviews, stakeholder analysis, PEST, value flow mapping, CBA, and so on.

2.3 Energy Modeling and Simulation: Core Services

Sensing ones system of interest: Before starting any modeling and simulation task it is important to be clear about why you are modeling, what the model is for, who will use it, what type of information is required, and in what format the output will be needed.

Table 3.7 Generation and storage technologies to consider

	Electrical	**Thermal**
Generation	Adjustable (variable): CHPs Fuel cells Not adjustable (parameter): Wind turbines PV cells	Adjustable (variable): HP CHPs Gas boilers Not adjustable (parameter): Solar thermal heating
Storage	Battery	Water tank storage PCM storage
Network	Electrical connection	Thermal loop

The sensing phase of the third pillar of the EPN process involves desk-based research in the form of a literature review focused on similar energy services so as to learn from previous experiences in informing ones understanding.

System understanding: For the definition of suitable modeling and simulation environments for neighborhood services (optimization and control concepts), it is essential to understand your neighborhood infrastructure and capabilities. Again one can use affinity diagrams, system diagrams, and so on to understand the various components/assets. This helps capture all deployed thermal and electrical generation and storage technologies in the neighborhood. Table 3.7 shows some exemplar assets to consider mapped against the generic categories "generation" (adjustable and nonadjustable), "storage," and "network."

Additionally, engineering specifications and CAD drawings are good source of information to help understand the systems. Mathematical formulation of component behavior and their neighborhood-specific parameterization provide the basis for custom energy management solutions, as well as providing the possibility to create physical models for the sake of testing and validation.

Deciding on requirements and modeling environment: Following the business model selection, service requirements, and use-cases can now be refined. This refinement of requirements coupled with choosing an appropriate modeling environment effectively constitute the deciding phase. Depending on the business model selected, one or several use-cases can be selected from the following list to specify one or a set of core functionalities and scenarios for the neighborhood [(Deliverable D1.1, 2013) and Section 5.3]:

- UC1: Real-time monitoring of the consumption of a neighborhood
- UC2: Energy demand and power generation forecasting
- UC3: Optimization of power purchases versus on-site generation
- UC4: Demand response

For example, if the business model to optimize the purchase on the wholesale market is selected, then the first three core use-cases will need to be specified for the neighborhood scenarios as they describe the key functionalities needed to realize this business model. The key functional requirements can be derived then from the definition of these use-cases [(Deliverable D1.1, 2013) and Chapter 5] which will then be mapped into detailed technical requirements for the NEM system.

Neighborhoods are multiphysics systems comprising thermal and electrical elements which are interconnected through devices, such as Combined Heat and Power (CHP) generation units or heat pumps (HP). When studying such systems, simulation proves to be a powerful tool to analyze the interactions between thermal and electrical components before carrying out field tests.

Various modeling tools are available to model both thermal and electrical parts of a neigborhood. These include building energy simulation tools (EnergyPlus, eQUEST), Modelica, TRNSYS, and others. In the scope of COOPERaTE, we have identified and modeled the components of both test sites, each featuring specific technologies. Table 3.8 presents an example list of energy components that can be available in an EPN—the list is not exhaustive and more energy components can be added depending

Table 3.8 Example of energy components in an EPN

Feature	Data requirement, parameters
Building physics • Walls • Floors • Windows	• Floor and piping plan • Wall structure • Insulation standards
Heating system • Boiler • CHP • Solar thermal panel • HP • Storage • Radiator	• Schematic • Technical specification of heat generators, storages and radiators
Electrical grid	• Topology • Cable type • Load profiles
Battery storage system	• Total capacity • Power of interface converter
Photovoltaic (PV) installations	• Total installed power and area • Location • Temperature data • Solar irradiation data
Electric Vehicle (EV) charging stations	• Charging type and power
Wind Turbine	• Rated power • Technology • Wind velocity data

on the EPN. The list here is taken from the COOPERaTE project and describes all available energy components of each test-site of the project with the definition of relevant modeling data. The proposed modeling approach for this work and a list of detailed component models is presented later in Chapter 4.

Modeling and Simulation actions: The acting phase is about the development of the multiphysics modeling and implementation of optimization scenarios of the neighbor hood energy systems. Due to the interdisciplinary character of the systems, some considerations need to be made for the setup of a realistic analysis. Instead of conducting separate simulation scenarios for the thermal and electrical domains, they should be integrated in one simulation in order to study joint effects. At the same time, their very different dynamic characteristics must be taken into account in the modeling and simulation approach, to avoid neglecting fast effects or to carry out a disproportional amount of calculations for slower phenomena. Chapter 7, will describe the experience of modeling and simulation for two real-life neighborhoods.

Having identified business cases for the neighborhood and having defined and performed a suitable benchmarking process, a neighborhood level objective can be formulated in terms of optimization. Depending on the neighborhood, there may be different levels of willingness for different users to provide information on their consumption to the energy management service. The mutually agreed level of transparency can roughly be divided into the following for each party:

- All generation, storage, and consumption data available to the energy management service (high level of transparency)
- Only anonymous or encrypted data, such as generation schedules and aggregated or residual consumption data available (lower level of transparency)

Once the availability of equipment and consumption data has been specified, a two-level optimization framework can be defined featuring a centralized and decentralized optimization as seen in Fig. 3.3.

The proposed optimization framework consists of two main optimization approaches: the centralized and decentralized. *The decentralized optimization* reflects the possibility for single actors to perform an optimization according to their own goals, as long as they can offer some flexibility to the centralized optimizer. *The centralized optimization* then uses this flexibility as well as all other known parameters to target the neighborhood objective. The selected equipment schedules within the flexibility offered by the decentralized optimizations are then communicated back to the decentralized level.

In Chapters 4 and 7, the two optimization approaches will be described for two real-life neighborhoods: the Challenger Campus, France and the Bishopstown, Ireland. These two neighborhoods have the following different characteristics:

- The Challenger campus is a singleowner neighborhood type with office buildings;
- The Bishopstown campus is a multiowner type of neighborhood with research/office, residential, and recreational buildings

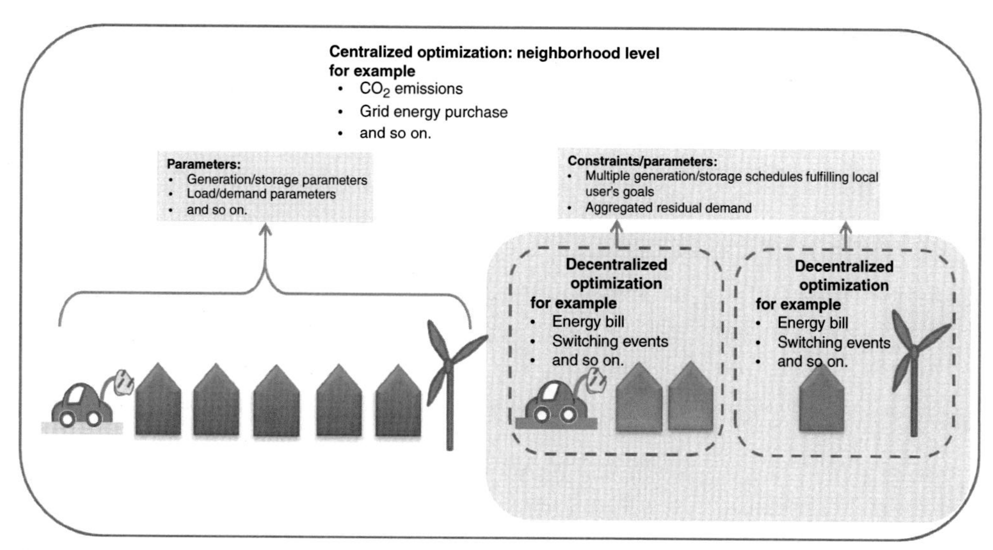

Figure 3.3 *A two-level optimization framework for neighborhoods.*

The optimization algorithms for these two types of neighborhood are respectively, as described previously, a centralized optimization algorithm, and a two-level hierarchical optimization algorithm.

2.4 Information Technology: Systems and Services

IT/Information Communication technology (ICT) underpins the delivery of an EPN as defined in Section 1. As outlined in the COOPERaTE Description of Work the ICT goal of the project was to "Utilize the power and flexibility of cloud computing to integrate diverse local monitoring and control systems to achieve energy services at a neighborhood scale." What follows describes the highly iterative process of the ICT System-of-Systems (SoS) approach.

Sensing and Understanding: As with the other core pillars IT navigated its own SUDA-loop in sensing, understanding, and deciding how the concept of EPN was to be best enabled and supported. Any proposed IT improvement would need to be researched given an understanding of existing solutions and SUDA naturally lends itself to that research process. The sensing phase in essence began in the project proposal stage with the desk-based research element of the SOTA. Therefore, from the project outset there was a sense that ICT considerations would be focused around the following themes:

- *Scalability*: EPNs are concerned with the district scale not buildings. Hence the proposed focus on "cloud" computing technologies. In relation to WP3 there was also a focus on the role of embedded compute termed "edge compute."

- *Service development*: Services needed to be developed based on WP2 optimization and prediction algorithms and supported at scale.
- *Data acquisition, aggregation, and actuation*: Primarily relating to service provision, it was understood that access to sensors and actuators would be required as the algorithms would assume certain datasets, while optimization recommendations would assume control of certain assets.
- *Data interoperability*: Again primarily relating to service provision it was understood that for data exchange to occur common understanding needed to be shared and there was a planned focus on adopting, adapting, or developing a suitable data model.
- *Security and Privacy*: While an ever-present consideration security and privacy innovations were not envisaged within the project, the assumption being current SOTA would apply. This changed in relation to privacy over the project lifetime.
- *Voice of the Customer/User*: Another central theme is that of user experience and user centerd solutions. Often lip-service is played to user experience but adoption is a concern when dealing at the district level, so there was an inherent sense that users would be central to our approach.

There is much iteration between sensing and understanding and Table 3.9 aims to give an overview of the progression path and the tools/methods used. Some elements were worked in parallel, while some elements were dependant on the maturity of other themes, such as the optimization and prediction algorithm development.

The project also had two validation sites with different partner platforms in situ, namely—the UrbanPower platform in the Challenger campus, Paris, the NICORE, and the E2E platforms in Bishopstown, Cork. This meant different cloud architectures/implementations were utilized and also meant parallel service development relevant to the four main use-cases.

Deciding and acting: The sense and understand phases represent the wider end of the research funnel and understandably involve a wider array of tools and methods. These initial phases greatly inform the decision and action phases honing the focus area considerably as reflected by Table 3.10. In terms of defining, a specific architectural direction, such as STEEPLE analysis, has to be utilized. The main considerations in solution selection related to system complexity due to heterogeneity in respect of the myriad of systems and the multiownership of same.

Addressing data access and data interoperability would be of prime importance in understanding how district scale impact could be achieved and solution selection matrix weightings reflected this. In Fig. 3.4 "a loosely coupled" SoS approach is presented utilizing a NIM (Chapter 5) as a basis for data interoperability. This approach is recommender based and largely relies on existing OT within the EPN, such as existing cloud platforms, Building Management Systems (BMS), Facility Management Systems (FMS),

Table 3.9 Information technology sensing to understanding: themes, tools and outcomes

Task description	**Tools/methods**	**Output/outcome**
Scalability: As an EPN is about district level services we set about identifying existing best practice rescalable cloud based architectures	Desk-based research, qualitative interviews. View model diagrams, Domain model diagrams	identified key generic architectural considerations possible approaches and technologies
Service development: As an EPN was a new concept we need to consider possible/likely services and associated requirements. This would involve working with internal consortium partners and external stakeholders	Qualitative interviews, workshops, system diagrams, data flow mapping, PEST, Stakeholder analysis, Use-case model, UML diagrams coupled with use-case templates, sequence diagrams Six-hats used for service ideation. Affinity diagrams Wire-framing techniques used for application prototypes, ethnographic direct observation techniques	4 core use-cases identified, initial functional and nonfunctional requirements defined, Initial architectural concepts formulated. possible services identified Realization that existing OT and IT would play a more fundamental role than initially expected, partly due to reluctance to use systems not owned by stakeholders in question, specifically cloud
Data acquisition, aggregation and actuation: Iteration in trying to understand the various OT elements, their current operation and ownership from an information systems perspective. What data formats, transfer mechanisms, semantics apply etc., where the gaps where and what IT could be deployed. Also assessing edge versus cloud means of processing, data storage, and aggregation	Desk-based research, system diagrams, data flow mapping, qualitative interviews. Refinement of logical component diagrams, sequence diagrams Brainstorming regarding approaches to heterogeneous data exchange, existing standards Performance testing on partner own platforms completed, regarding data acquisition, persistence, and so on.	Stakeholders identified, systems map complete with subsystems identified. Realization that control, plus sensing to a degree, was highly likely to remain within the remit of existing systems, any improvement would most likely need be recommender based. Data privacy became a specific issue/opportunity in relation to determining edge versus cloud based data processing
Data interoperability: Focused on data model requirements given a district level heterogeneous data context	Desk-based research encompassing standards review, Partnership—eeBDM community. Solution selection matrix employed to score potential data models to build upon Requirement tables—Current data model mapping, compare to service requirements, sanity check against proposed NIM for completeness Basic NIM PoC deployed on one test-site data model	Defined model requirements, namely: • Flexible, extensibility at runtime • Handle static data and dynamic data, for example, near-real-time sensor data • Privacy by design, based on agreed Usage, physical Location, and expiry Data • Allow for domain specific language translations, thus targeting ease of adoption/use

Table 3.10 Information Technology deciding to acting: tasks, tools and outcomes

Task description	Tools/methods	Output/outcome
Decide upon a proposed ICT approach for enabling *district level* energy services delivery	STEEPLE, Solution Selection Matrix, Qualitative expert interviews.	Loosely coupled SoS approach posited
Decide upon specific energy service implementations, sense/understand use-case output had narrowed to 4 use-cases, but how to implement for different components of the test-bed required refinement	Qualitative interviews, surveys, Selection Matrix, wireframes, and prototyping Agile techniques effectively rapidly moving through action, ⋙ sense/test ⋙ understand ⋙ action	RT energy & environmental data monitoring & presentation Optimization/prediction service Demand response/flexibility service

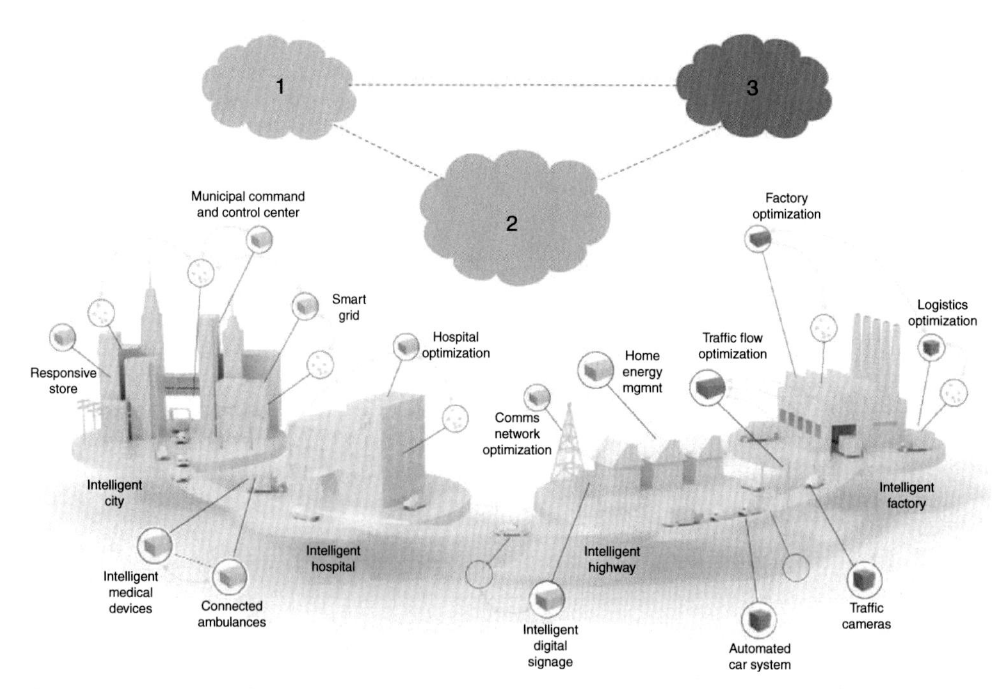

Figure 3.4 ***A System-Of-Systems approach to EPN delivery.***

electrical meters, gas meters, programmable logic controllers (PLCs), and so on to sense and control the physical assets of the neighborhood.

3 EXAMPLES OF NEIGHBORHOOD ADAPTATIONS IN COOPERaTE

This section describes two real-life examples of EPN process applied to the COOPERaTE demosites.

3.1 Challenger Adaptation Augmentation

The Challenger campus, in the Paris region, is one of the two validation sites of the project. The Challenger campus is about 68,000 m^2, mainly offices. About 3,000 people work there every day for Bouygues Construction. The buildings were renovated between 2012 and 2015 and a solar farm was installed. It produces about 2,200 MWh/year. This renovation clearly reduced the energy bill of Challenger Campus but a lot of work was yet to be done in order to transform Challenger into an EPN. What follows describes the process of accessing, deciding, and acting on a means of delivering on the EPN concept as defined in Section 1.

3.1.1 Sensing and Understanding

The energy baseline methodology as outline in Section 2.1 was primarily utilized in understanding the energy profile of Challenger. The scope of the study included the entire Campus, composing of:

- A main entrance, the *Atrium* with a restaurant, several facilities dedicated to the employees (shop, hairdresser, bank, travel agency...) and a fitness club.
- A *Technical Building*
- Office buildings with open offices, meeting rooms, and shared cubicles : *North wing, South wing, North triangle, South triangle*
- A solar farm and distributed solar panels
- A phyto-purification treatment area
- Electrical Vehicle substations
- Diesel backup generators

While related to Section 2.1, this pillar focuses on the process of assessing the on the ground situation with respect to the following questions:

- What OT, that is, BMS, FMS, RES, metering, and so on currently exist?
- What OT assets are missing or cannot be accessed? Why?
- What OT assets might be deployed? Where? When?
- How do we integrate the OT and IT?
- Who are our stakeholders? Decision makers?

For this we used 5W1H and diagramming techniques (rich pictures and concept maps)

Post IT implementation (see deciding/action), we set about utilizing our IT platform to more effectively understand:

- Total site consumption, reported to calculate the baseline.
- The site utility bills and the contract with the French energy supplier.
- The operation of the thermal and electrical production of the site, mainly with the facility manager of the site.

This increased insight led to further OT investment (e.g., battery storage units) and further action, see later.

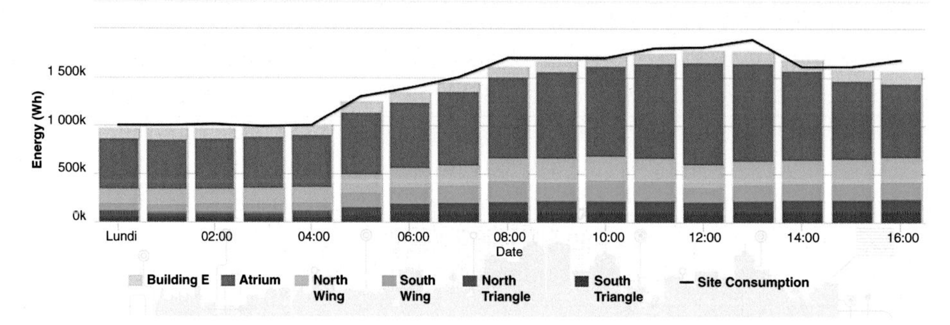

Figure 3.5 ***Example of data collected in Urban Power: energy consumption per building.***

3.1.2 Deciding and Acting

An obvious first decision was to augment and integrate existing OT with IT in this case the Urban Power platform (UPP), designed by EMBIX. This connected hitherto unmonitored assets and offered greater data granularity in relation to lighting, heating, and grid electricity imports and so on (Fig. 3.5).

Moving back into the sense/understand cycle the visualization and statistical analysis tools offered by UPP were utilized to further understand the operation of Challenger as described previously.

At that point the stakeholders/actors involved needed to be consulted in deciding which COOPERaTE EPN actions/services were going to be implemented.

First, we looked for anomalies (Chapter 7): for instance, a cooling system that was constantly on or the boilers that were triggering at the same time every evening. If the anomaly was not taken into account by the renovation of Challenger, we addressed it.

As per Fig. 3.6 anomalies were identified during the understand phase which resulted in a controlling module implementation.

Moreover, it was also decided to develop different modules in order to manage the batteries storage available on the site (66 kWh). UPP enables us to communicate with

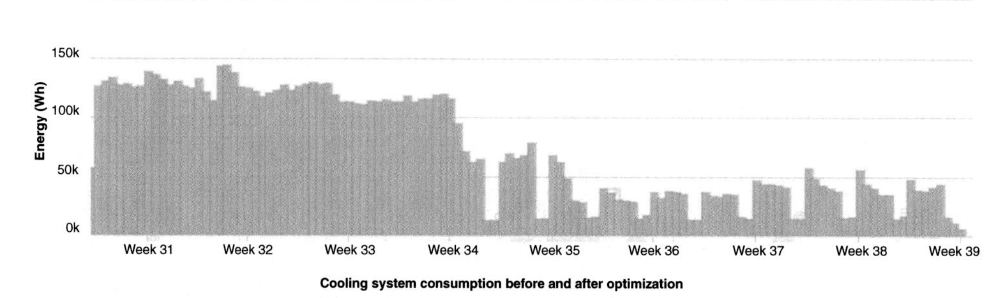

Figure 3.6 ***Improvement done concerning the cooling system.***

the batteries, both sense and control, to manage demand response and to showcase the different consumptions of the site.

Apart from implementing the COOPERaTE actions decided in the previous phase (controlling the cooling system for instance), we implemented platforms and modules able to create value from the collected data:

- IT-cloud based infrastructure for energy management, optimization
- Interface to the Demand Response market
- More sophisticated visualizations showing energy consumption and production in real-time.
- The evaluation of KPIs, described in Chapter 7, was also a good way to have a better understanding of the outcomes of COOPERaTE in terms of savings in order to improve the program in a next cycle of SUDA.

3.2 Bishopstown Adaptation Augmentation

Bishopstown campus is one of the two validation sites for the project. The campus neighborhood includes the National Sustainable Building Energy Test-bed (NSBET) located at the NIMBUS building, the Leisure World indoor sports facility as well as the Parchment Square student accommodation.

In this section we discuss the steps of the proposed process for EPN in relation to this demonstration site. As with the Challenger campus case, certain steps of the EPN process have already been applied for the adaptation of the Bishopstown demo-site, which are summarized here.

In addition, this section describes how the COOPERaTE consortium envisages the implementation of the steps of the EPN process that have not been yet applied in this demo-site.

3.2.1 Sensing and Understanding

The demonstration site at Cork Institute of Technology (CIT) Bishopstown has 3 distinct and is different from a building adaptation point of view, elements. (1) The Nimbus Building with its own Building Management System (BMS) monitoring and controlling the building environment, and a Supervisory Control and Data Acquisition (SCADA) system monitoring and controlling the Microgrid. (2) The Leisure World complex with its own BMS. (3) Parchment square, no BMS system and no onsite monitoring and control of energy at the apartment level.

The IPMVP method was followed to develop the energy baseline and benchmarking of the Bishopstown demonstration site. Monthly gas and electricity consumption data for a period of a year were gathered from building owners/facility managers and the existing BMS/SCADA metering systems of the demonstration site. The results of the energy baseline activity are described in detail in (Deliverable D5.1, 2013) "Report on measurement methodology and energy baseline for the validation sites." The results revealed a number of improvements for optimization as well as retrofit opportunities for the demo site.

As Leisure World and Parchment Square was owned and operated independently of CIT a number of workshops initially took place with the facility managers (FM) of the operations using in particular the "5W1H" methodology and SWOT analysis.

These series of workshops were seen as especially important due to the nontechnical expertise of the FMs, the need to explain the concepts of COOPERaTE, what an EPN was, and to get a handle on how the facilities are managed and operated, particularly from an Energy Management point of view. As these were distinct site it was particularly important to understand:

- Who are the main stakeholders?
- What OT assets exist, and their suitability for monitoring and control?
- Where are the gaps? How do they get filled?
- What IT infrastructure exists? How suitable is it for integration into the NEM?
- How do we ensure compatibility between systems?
- How do we do the systems integration?
- How do we ensure the minimal disruption to operations?

Subsequently flow mapping techniques were adapted in mapping the data flows required to support the identified use-cases. This was also used to identify gaps in data and potential need for additional metering solutions and so on.

It was subsequently decided to take different appropriate actions for each of the three elements within the Bishopstown site:

1. *Nimbus*: Augment the current infrastructure by improving and increasing the data coming from the preexisting energy meters (Fig. 3.7).
2. *Leisure World*: Employ a system's integrator to augment the infrastructure by adding additional control and monitoring hardware and software, and enable access to the CIT cloud management platform software (Fig. 3.8).
3. *Parchment Square*: Design and install a systems monitoring platform at the apartment level, accessible via the CIT cloud management platform software. The Intel E2E solution incorporating a home gateway for each apartment was deployed in this case.

Post implementation the data was collated and used to monitor consumption at the various sites, and used to help calculate the baseline. It was used to monitor the real-time operation of the sites and used to help in decision making of the FMs.

3.2.2 Deciding and Acting

Through the use of NiCORE the CIT cloud management platform, the gathering, collation, and use of data, between diverse buildings with distinct IT infrastructures was facilitated.

The NiCORE BMS connector services were installed and configured to interface Nimbus, Leisure, and Parchment Square buildings. While the Nimbus and Leisure World

EM01 CHP Real Values

Energy Meter Comms				Energy Monitoring							
Comms Healthy	True			Active Energy +	LVEM01_JT_450780	68301	kWh	Harmonic I2 9th Harmonic	LVEM01_XT_451560	0.020	%
Response Time	LVEM01_XT_2001	208.00	ms	Reactive Energy +	LVEM01_JT_450782	19020	kvarh	Harmonic I3 9th Harmonic	LVEM01_XT_451561	0.018	%
Real Time Values				Apparent Energy	LVEM01_JT_450784	75244	kVAh	Harmonic In 9th Harmonic	LVEM01_XT_451562	0.477	%
Voltage 1-2	LVEM01_ET_450514	415.99	V	Active Energy -	LVEM01_JT_450786	2542.00	kWh	Harmonic I1 11th Harmonic	LVEM01_XT_451563	0.015	%
Voltage 2-3	LVEM01_ET_450516	419.71	V	Reactive Energy -	LVEM01_JT_450788	3472.00	kvarh	Harmonic I2 11th Harmonic	LVEM01_XT_451564	0.006	%
Voltage 3-1	LVEM01_ET_450518	415.06	V	**Total Harmonic Distortion**				Harmonic I3 11th Harmonic	LVEM01_XT_451565	0.016	%
Voltage 1-N	LVEM01_ET_450520	239.17	V	THD Voltage 1-2	LVEM01_ET_451536	0.021	%	Harmonic In 11th Harmonic	LVEM01_XT_451566	0.094	%
Voltage 2-N	LVEM01_ET_450522	241.67	V	THD Voltage 2-3	LVEM01_ET_451537	0.021	%	Harmonic I1 13th Harmonic	LVEM01_XT_451567	0.009	%
Voltage 3-N	LVEM01_ET_450524	241.28	V	THD Voltage 3-1	LVEM01_ET_451538	0.020	%	Harmonic I2 13th Harmonic	LVEM01_XT_451568	0.004	%
Frequency	LVEM01_ST_450526	50.04	Hz	THD Voltage Phase 1	LVEM01_ET_451539	0.021	%	Harmonic I3 13th Harmonic	LVEM01_XT_451569	0.011	%
Current Phase 1	LVEM01_IT_450528	43.73	I	THD Voltage Phase 2	LVEM01_ET_451540	0.021	%	Harmonic In 13th Harmonic	LVEM01_XT_451570	0.056	%
Current Phase 2	LVEM01_IT_450530	37.95	I	THD Voltage Phase 3	LVEM01_ET_451541	0.020	%	**Harmonic - Voltage**			
Current Phase 3	LVEM01_IT_450532	52.17	I	THD Current Phase 1	LVEM01_IT_451542	0.273	%	Harmonic V1 3rd Hamonic	LVEM01_XT_451766	0.004	%
Current Neutral	LVEM01_IT_450534	11.53	I	THD Current Phase 2	LVEM01_IT_451543	0.32[illegible]	%	Harmonic V2 3rd Hamonic	LVEM01_XT_451767	0.000	%
Active Power	LVEM01_JT_450536	27.13	kW	THD Current Phase 3	LVEM01_IT_451544	0.217	%	Harmonic V3 3rd Hamonic	LVEM01_XT_451768	0.001	%
Reactive Power	LVEM01_JT_450538	-17.14	kVAr	THD Current Phase N	LVEM01_IT_451545	2.689	%	Harmonic V1 5th Hamonic	LVEM01_XT_451769	0.020	%
Apparent Power	LVEM01_JT_450540	32.08	kVA	**Harmonics - Current**				Harmonic V2 5th Hamonic	LVEM01_XT_451770	0.021	%
Power Factor	LVEM01_JT_450542	-0.85		Harmonic I1 3rd Harmonic	LVEM01_XT_451547	0.097	%	Harmonic V3 5th Hamonic	LVEM01_XT_451771	0.018	%
Active Power Phase 1	LVEM01_JT_450544	9.49	kW	Harmonic I2 3rd Harmonic	LVEM01_XT_451548	0.098	%	Harmonic V1 7th Hamonic	LVEM01_XT_451772	0.006	%
Active Power Phase 2	LVEM01_JT_450546	7.35	kW	Harmonic I3 3rd Harmonic	LVEM01_XT_451549	0.075	%	Harmonic V2 7th Hamonic	LVEM01_XT_451773	0.005	%
Active Power Phase 3	LVEM01_JT_450548	10.29	kW	Harmonic In 3rd Harmonic	LVEM01_XT_451550	2.626	%	Harmonic V3 7th Hamonic	LVEM01_XT_451774	0.006	%
Reactive Power Phase 1	LVEM01_JT_450550	-4.40	kVAr	Harmonic I1 5th Harmonic	LVEM01_XT_451551	0.252	%	Harmonic V1 9th Hamonic	LVEM01_XT_451775	0.000	%
Reactive Power Phase 2	LVEM01_JT_450552	-5.50	kVAr	Harmonic I2 5th Harmonic	LVEM01_XT_451552	0.304	%	Harmonic V2 9th Hamonic	LVEM01_XT_451776	0.000	%

Figure 3.7 ***Increased resolution of Combined Heat and Power* (CHP) *energy meter.***

connectors interface BMS systems directly, the parchment Square connector communicates with Intel system installed at Parchment Square through the neighborhood information model (NIM).

In order to gain better insight of building operation behavior a NiCORE MATLab BMS extension was developed allowing real time data analyzing and BMS control directly from the MATLab environment taking full advantage of existing MATLab analytical toolboxes.

That combined with the installation and augmentation of sensing, monitoring, and control assets, and their subsequent aggregation has allowed for greater understanding of the dynamics of their interaction. This has provided the FMs with more tools to inform their decision making processes.

4 SUMMARY AND RECOMMENDATIONS

4.1 Summary

This Chapter presents an overview of the process for enabling EPNs and is proposed as a best practice guide for enabling a neighborhood to transform to an EPN. It's

Leisureworld Bishopstown energy monitoring project

Cable schedule and points list

Sensor or Service Description	Model Number/reference	Quantity	Cable Schedule	Comment
Room Air Temperature Sensor	TT-1000	10		Wireless to be fitted by the electrical contractor
RH sensor	RH-1000	4	2 pair screened belden controls cable	
Cylon OPC Server	1MASW250	1		
Cylon WebLink	1MASW230	0		
HEAT METERS				
Heat Meter	UKM50P	5		Solar Panel return pipe 25 mm diameter. AHU 1 (main pool) 100 mm diameter, AHU 2 (wet changing) 65 mm diameter, AHU 3 (gym) 32 mm diameter, AHU 4 (gym) 32 mm diameter, AHU 5 (dry changing) 65 mm diameter, AHU 6 (18 m pool) 80 mm diameter. All meters to be placed on return line from coils prior to three way bypass branch to ensure coil flow only is monitored.
Prelims		5		
Power Supplies		5	230V power supply	
Meter comms	Kamstrup 601 CAL	5	2 pair screened belden controls cable	
Testing and commissioning of meters		5		
BMS interface for split units in gym and office areas		21	2 pair screened belden controls cable	electric meters need to be installed.
BMS verification of CHP electrical meter output kWh and kW		1		electric meters need to be installed.
BMS datalogs to be set up for kW meter readings		1		
Meter re-calibration and BMS mapping verification		1		All new and existing meters to be re-calibrated and BMS mapping verified
ELECTRICITY METERS				
Electrical meter total grid import		1		
electric meter	Diris A20 Energy Meter (48250A20)	28		7 Plus 21 for AC units
A20 communication module	Diris A20 (48250082)	28	2 pair screened belden controls cable	To be fitted to each distribution board refer to specification
MF Meter	Diris A40 Energy Meter(48250A40)	2		main incoming supply. CHP power generation
A40 communication module	Diris A40 (48250092)	2	2 pair screened belden controls cable	
Split Core	400:5 CT (192T4640)	6		Transformer
Split Core	150:5 CT (192T4640)	42		
A20 communication module	Diris A20 (48250082)	1	2 pair screened belden controls cable	
Misc contracting service		6		
Commissioning per day on site		2		
Power & commu cabling	Cabling (198174)	10	2 pair screened belden controls cable, 230V power supply	
Drop link terminals	Drop links terminals (CTDL)	27		
Electric meter with comm module	Diris A10 (48250011)	14		To be fitted to each distribution board refer to specification
Moulded Case CT	Moulded Case CT (192T2022)	42		To be fitted to each distribution board refer to specification
Power & commu cabling	Cabling (198174)	14	2 pair screened belden controls cable, 230V power supply	
Drop link terminals	Drop links terminals (CTDL)	42		

Figure 3.8 ***Cable Schedule for Leisure World.***

acknowledged that undertaking a neighborhood energy improvement can seem overwhelming. Even individuals familiar with the construction process may feel a little lost at first when considering new elements, such as M&V, conservation measures, and ICT. The report provides a framework and logical process for achieving this complex undertaking that is an EPN. The process is based on the SUDA framework posited as a generic framework for defining a problem/opportunity, for selecting a proposed improvement and for testing the outcomes. The framework is cyclical in nature and moves from scoping, to improvement selection, to improvement refinement. It's used throughout to assist in orientating practitioners as they navigate the more complex collective of methodologies and tools required to achieve the defined improvement goal.

A SUDA overview outlining the objectives, proposed methods, and tools, and expected outcomes of each phase of the process was first generically explained, then applied to the EPN context. The framework leverages the expertise and learning the COOPERaTE consortium brought to and developed over the course of the project. Five core pillars of EPN were identified and discussed providing stakeholders all the necessary details to establish each of these pillars. SUDA was followed in completing each of these pillars offering tools and methodologies to be used including alternatives to provide some flexibility to practitioners as they make their choices.

4.2 Recommendations

Summarizing the overall EPN process the following recommendations can be made:

- *Pick an improvement methodology*: An initial step to implementation is to agree on an improvement methodology which will guide with the overall process. Multifaceted projects are complex and a team will be glad they took time to agree a process when in the middle of that complexity. SUDA was used here but others could be utilized, such as the DMAIC framework of Lean Six Sigma (https://www.tx.ncsu.edu/sixsigma/faq.cfm; https://www.tx.ncsu.edu/sixsigma/faq.cfm), Hard Systems Method (Checkland, 1978), Soft System Method (Checkland, 2001), Triz (http://www.triz-journal.com/triz-what-is-triz/), Capability Maturity Model (http://www.sei.cmu.edu/smartgrid/tools/) and so on. What is important is that the methodology is agreed and adhered to, but not to the point of paralyses, if something does not work, try another way, move on and iterate. Employee a multidisciplinary team and trust in its collective expertise.
- *Sense, scope, and define*: The first step of an implementation project should always be to sense, scope, and agree on the overall goal of the initiative. While improving energy positivity is the specific overarching goal here, there are typically several ways to achieve it or any goal and the kind of intervention depends on regulations and actors involved. Try using tools and methods that help with scoping the problem and leverage the diversity of the team. See the problem from different sector viewpoints and persona. Get talking with stakeholders.
- *Obtain buy-in*: active support from stakeholders is essential for exploiting the potential for improvement. Knowing the main drivers and potential constraints for an EPN

project will make it easier for practitioners to work with stakeholders on in the project. Stakeholders need to be aware that benefits from energy efficiency may not be immediate. This is important to ensure that projects are not cut before they can get off the ground.

- *Map, measure, and understand*: Understanding of neighborhood capabilities and potential is essential. Establishing the energy baseline of a neighborhood is very important to understand the neighborhood energy performance before and after the implementation of any EPN improvements. This is a key aspect to evaluating whether or not a neighborhood has achieved its energy performance objectives.
- *Transparent decisions*: Involve stakeholders in deciding which core ICT technologies, services, and business models to model or implement. This comes after the understanding, which is usually done through benchmarking your neighborhood against reference neighborhood. Clear requirements for the selected services are necessary. Make a decision based on the evidence, act on it, and evaluate it.
- *Act:* ensure action is taken whether its simulation, modeling, physical prototyping, or proof-of-concepts. "Perfection is the enemy of good," so act, evaluate, and iterate.
- *Double loop learning*: ensure that one is questioning the beliefs, values, and assumption carried into the project and do not be afraid to change them based on evidence and data. A less generic but somewhat related point pertains to the classical definition of energy positivity. The recommendation within COOPERaTE is that this needs to be revisited in the context of district level energy management. The focus should not be solely energy consumption or efficiency but also account for value creation and carbon intensity of the energy while accounting for stability and security of the energy network.
- *Value focused*: Start with business models where possible, then conceive the services and then adoption, adaption, or development of technology to meet defined requirements. Be value focused and think beyond immediate value streams.

REFERENCES

Checkland, P. B. (1978). The origins and nature of "Hard" systems thinking. *Journal of Applied Systems Analysis*, *5*(2), 99–110.

Checkland, P. B. (2001). Soft Systems Methodology. In J. Rosenhead, & J. Mingers (Eds.), *Rational Analysis for a Problematic World Revisited*. Chichester: Wiley.

CIBSE TM46 Energy Benchmarks, 2008. http://www.cibse.org/knowledge/cibse-tm/tm46-energy-benchmarks

Deliverable 4.2, Second version infrastructure adaptation and deployment, COOPERATE, 2014.

Deliverable D1.1, Report on requirements and use-cases specification, COOPERATE, 2013.

Deliverable D5.1, Measurement methodology and energy baseline for the demonstration sites, COOPERATE, 2013.

Energy Consumption Guide (ECG) 19, 2000. http://products.ihs.com/Ohsis-SEO/572574.html

Kolb, D. A. (1984). *Experiential Learning experience as a source of learning and development*. New Jersey: Prentice Hall.

The ICT PSP methodology for energy saving measurement—a common deliverable from projects of ICT for sustainable growth in the residential sector, ESEH, 2011. http://esesh.eu/outputs/

CHAPTER FOUR

Simulation Tools and Optimization Algorithms for Efficient Energy Management in Neighborhoods

L. Cupelli*, M. Schumacher*, A. Monti*, D. Mueller*, L. De Tommasi, K. Kouramas****

*E.ON Energy Research Center, RWTH Aachen University, Aachen, Germany
**United Technologies Research Centre Ireland, Cork, Ireland

Contents

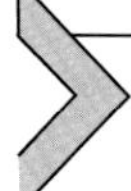

1 INTRODUCTION

1.1 Motivation

As detailed in the EPN process of Chapter 3, the Neighborhood Energy Manager (NEM) and owner/owners select the potential business models for the Neighborhood and have available a number of energy services, such as optimization, real-time monitoring, and forecasting to select for deployment on their Neighborhood energy management platform.

Having identified business cases for the Neighborhood and having defined and performed a suitable benchmarking process, a Neighborhood-level objective can be formulated in terms of optimization accounting for both the economic criterion and environmental impact (energy cost and CO_2 emissions). In this context, modeling and simulation tools can be used as decision support tools for planning and operation of energy systems in buildings as well as proof of concept of energy optimization algorithms before carrying out field tests.

1.2 Our Contributions

In this chapter different schemas for energy optimization in Neighborhoods are proposed and a Modelica-based library is used to efficiently simulate and test the optimization actions in city districts.

The Neighborhood energy optimization algorithms perform an energy price driven scheduling of the generation and storage equipment to reduce the Neighborhood netload seen from the grid side while keeping Neighborhood pollution emissions below a given threshold.

The Modelica-based library supports the field tests with more extensive assessments and offers numerous components for a multiphysics simulation of Neighborhood energy system. While a first version of the library developed by RWTH is available in GitHub at https://github.com/RWTH-EBC/AixLib, a second version enriched with the electrical elements from RWTH and Dynamic Phasors will be offered in the same server as a separate release.

2 PRELIMINARIES

Prior to formulating the energy optimization algorithm, the availability of equipment and consumption data must be specified. It is essential to understand your Neighborhood infrastructure and capabilities. Table 4.1 shows some exemplary assets to consider. They are mapped against the generic categories "generation" (adjustable and nonadjustable), "storage," "consumption," and "network." Readers interested specifically in modeling and simulation of such assets can jump directly to Section 3.

Table 4.1 Thermal and electrical generation, storage, and consumption considerations

	Electrical	Thermal
Generation	Adjustable (variable): • CHPs • Fuel cells Nonadjustable (parameter): • Wind turbines • PV cells	Adjustable (variable): • Heat pumps • CHPs • Gas boilers Nonadjustable (parameter): • Solar thermal heating
Storage	Battery	Water tank storage PCM storage
Consumption	Electrical Load profiles	Thermal Load profiles
Network	Electrical connection	Thermal loop

Application of optimal control concepts to achieve energy efficiency in buildings and optimal exploitation of local resources is not widely extended to Neighborhoods or districts.

We consider a Neighborhood including multiple buildings, which agree to operate their energy resources and loads in a coordinate manner, such that some global goals are achieved while individual optimization goals can still be pursued. In such case, a building's local resources should be coordinated with a top-level optimization engine to balance achievement of local optimization goals and Neighborhood-level goals.

To address the aforementioned scenario in the most general case of multiple-ownership Neighborhoods, a hierarchical optimization algorithm can be designed to split the optimization into the building level and the Neighborhood level in such a way that the bottom level manages the individual building objectives whereas the top level addresses the Neighborhood-level objective. In particular, the building-level energy management will offer flexibility to the top-level optimizer and will receive recommendation on the control actions to implement to drive the Neighborhood toward the fulfillment of its goals. Typical objectives applicable to energy optimization are: energy cost and pollution emission (e.g., CO_2).

Business models determine the reward which each building owner receives from the Neighborhood when he contributes within his flexibility band to the achievement of Neighborhood goals. A relevant example of that is a scenario where: (1) each building owner minimizes the operating cost of energy by scheduling the local resources and storage while guaranteeing the electrical and thermal load supply; however, minimization of such cost can be performed by using different load profiles corresponding to different levels of comfort and CO_2 emissions; (2) a Neighborhood optimization agent selects for each building the load profiles such that the total level of Neighborhood CO_2 emissions is below a Neighborhood limit and offers a reward to those building owners who are willing to reduce their thermal loads (within the occupants comfort band) to contribute to the reduction of environmental pollution. Such a reward can be, for example, in the form of economic incentive or energy price discount.

The optimization algorithm combines economic dispatch and load shifting; furthermore, it performs an energy price driven scheduling of energy storage (if available) and enables to support the grid reducing the net-load (load seen from the grid side) when the energy price is high. To achieve these goals, it exploits historical energy consumption data to forecast or estimate 24-h ahead of the electrical/thermal Neighborhood consumption. Furthermore, depending on the contractual agreement with the energy supplier, both energy price forecasts (based on real-time recorded data) and fixed two stages (day/night) or three stages (day/night/peak) tariffs can be used to optimize local generation and storage in line with the business models. A schematic including the main functional blocks of the energy optimization algorithm is shown in Fig. 4.1.

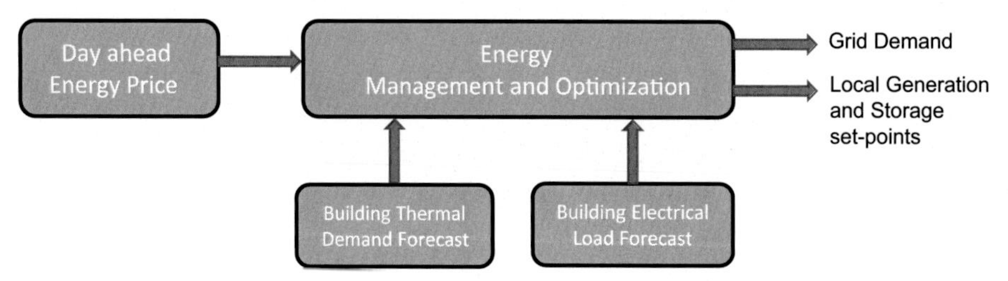

Figure 4.1 ***Energy Optimization Algorithm Schematic.***

In addition to the possible presence of multiple objectives in the energy optimization formulation, there is also the issue of different levels of willingness on the part of different users to provide information on their consumption to the energy management service. The mutually agreed level of transparency can be roughly divided into the following for each party:

- All generation, storage, and consumption data available to the energy management service (high level of transparency).
- Only anonymous or encrypted data, such as generation schedules and aggregated or residual consumption data available (lower level of transparency).

According to the business models applicable (see Chapter 6) and the level of transparency, a centralized optimization algorithm can be defined for a Neighborhood with high level of transparency and where aggregated objectives are applicable, such as a single-ownership Neighborhood providing full access to energy data; moreover, a hierarchical two-level optimization algorithm can be formulated for a Neighborhood with lower level of transparency where both individual and aggregated objectives are applicable, such as a multiple-ownership Neighborhood with access to aggregated consumption. This classification is illustrated in Fig. 4.2.

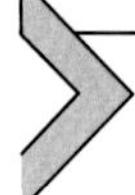

3 SIMULATION MODELS FOR ENERGY POSITIVE NEIGHBORHOODS

When studying Neighborhoods, simulation proves to be a powerful tool to analyze the interactions between thermal and electrical entities before carrying out field tests. In this section, models for different assets at the Neighborhood level will be presented that capture deployed thermal and electrical generation and storage technologies.

Furthermore, a simulation-based assessment of the Neighborhood optimization algorithms described in the following sections of this chapter will be performed to support the decision-making process toward energy positivity.

Due to the interdisciplinary character of the systems, some considerations need to be made for the setup of a realistic analysis. Instead of conducting separate simulation

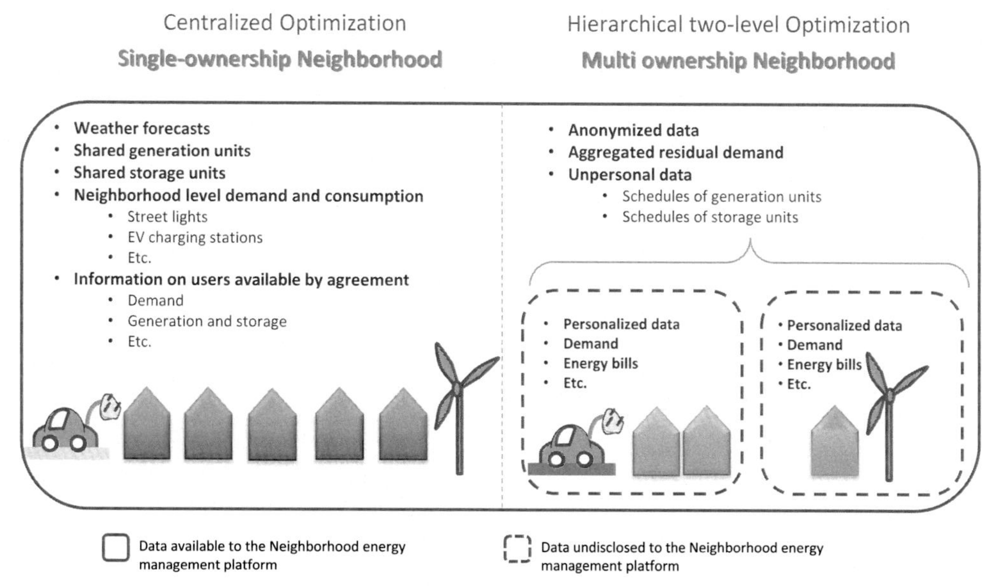

Figure 4.2 ***Optimization algorithms classification.***

scenarios for the thermal and electrical domains, they should be integrated in one simulation in order to study joint effects. At the same time, their very different dynamic characteristics must be taken into account in the modeling and simulation approach, to avoid neglecting fast effects or to carry out a disproportional amount of calculations for slower phenomena.

3.1 Time Constants

In Neighborhoods containing buildings and energy generators, different types of dynamics must be considered:

- Electrical transient effects
- Electrical steady-state effects
- Thermal/thermohydraulic effects

While thermal processes such as heat exchange are rather slow and changes take place in the range of minutes, electrical effects are much faster and range from 10^{-5} s for electromagnetic transients to a few seconds for electromechanical or longer term transients. The temperature changes of the building mass which are evoked by air temperature changes (external) or core activation (internal) take place in the magnitude of several hours. Faster sampling frequencies can occur in heat generators, for example, condensing boilers, though even here the fluctuations happen in a time scale of several minutes.

3.2 Modeling Language

Various modeling tools are available to model both thermal and electrical parts of a Neighborhood. These include building energy simulation tools EnergyPlus, eQUEST, Modelica, TRNSYS, and others. The Modelica language was chosen for the modeling of components. Modelica is a freely available, object-oriented language for modeling of physical systems. It is suited for multidomain modeling, for example, mechatronic models in robotics, automotive and aerospace applications involving mechanical, electrical, hydraulic, and control subsystems, process-oriented applications, and generation and distribution of electric power. Models in Modelica are mathematically described by differential, algebraic, and discrete equations. No particular variable needs to be solved manually. Within a Modelica development tool (e.g., Dymola, OpenModelica, JModelica) physical models are graphically put together, mathematically reformulated and translated to executable C-Code. Afterward the solving process is conducted; rather than interpreting the causality of variables depending on the type of equation, the tool purely considers the relation between them. Modelica is designed for the efficient handling of large models with more than one hundred thousand equations (https://www.modelica.org/).

The Modelica language is a textual description to define all parts of a model and to structure model components in libraries, called packages. In the scope of EU FP7 project COOPERaTE, the authors identified and modeled the components of two demonstration sites in Modelica; however, the component library modeled for COOPERaTE is not limited to the demonstration sites, but can be applied to Neighborhoods with similar features or extended to accommodate further equipment. Table 4.2 shows all available energy components of each test site with the definition of relevant modeling data.

3.3 Thermal System

The thermal modeling of energy systems can be performed either in a very detailed or a simplified way depending on the purpose of the model. Regarding the level of detail in modeling there is a trade-off between model accuracy and acceptable computation and modeling effort. This trade-off has to be resolved in a reasonable manner. A large computational effort comes with fluid dynamic calculations for the heating medium. Using the standard Modelica fluid library the state of the current (laminar or turbulent) is calculated in every time step, burdening the computational time. For detailed considerations of single components which exist in energy systems, for example, a heat-exchanger validation, those calculations may be beneficial to create sophisticated simulation models.

For more holistic investigations of energy systems, as performed in the scope of COOPERaTE, the calculation of in-pipe turbulence and pressure losses is negligible and would not lead to reasonable benefit. Instead of a thermohydraulic fluid connector,

Table 4.2 Available energy components of COOPERaTE demonstration sites

Feature	Data requirements and parameters	Challenger	Bishopstown
Building physics			
Walls	Wall structure	✓	✓
Floors	Floor and piping plan	✓	✓
Windows	Insulation standards	✓	✓
Heating system			
Boiler	Technical specifications	✓	✓
Combined Heat and Power (CHP)	Technical specifications	✓	✓
Solar thermal panel	Technical specifications	✓	✓
Heat Pump (HP)	Technical specifications	✓	✓
Storage	Technical specifications	✓	✓
Radiator	Technical specifications	✓	✓
Electrical system			
Grid	Topology, Cable type and Load profiles	✓	✓
Battery storage system	Total capacity and Power of interface converter	✓	✓
Photovoltaic (PV) installations	Total installed power and area, Location, Temperature data, Solar irradiation data	✓	
Electric Vehicle (EV) charging stations	Charging type and power	✓	
Wind Turbine	Rated power, Technology and Wind velocity data		✓

that connects different components within a simulation model, an enthalpy connector approach is considered in this project (Stinner, Schumacher, Finkbeiner, Streblow, & Müller, 2015). This connector does not calculate pressure losses but contains all relevant information of the heating medium, namely temperature T, specific enthalpy h, and mass flow, where the enthalpy is calculated in accordance to the following equation using the specific heat capacity c_p of the fluid:

$$h = T \cdot c_p.$$

The values for each variable at the beginning of the simulation are set during the model initialization and can be adjusted as a simulation parameter. With this approach the energy balances for the heating medium passing through a component, for example, a gas boiler, can be easily calculated according to the following equation:

$$\dot{Q} = \dot{m} \cdot c_p \cdot (T_{out} - T_{in}) = \dot{m} \cdot (h_{out} - h_{in})$$

where $\dot{m}$ is the mass flow rate of the fluid and T_{in} and T_{out} are the entering and leaving fluid temperatures, respectively.

3.4 Electrical System

For the electrical part of the system, a quasidynamic approach is chosen as opposed to the more typical steady-state load flow analysis. In this way, electrical transient behavior can also be studied, thus enabling the analysis of the effects which occur when interfacing the faster electrical system with the slower thermal system. At the same time, a quasidynamic approach does not capture the faster electromagnetic effects, which would be out of scope for control algorithms and would represent an additional computational burden due to significantly smaller simulation time steps.

For this purpose, a custom library was developed with the *dynamic phasor approach* (Mattavelli, Verghese, & Stankovic, 1997).

This principle is based on the fact that in a time interval $I = (t-\Theta, t]$ a time-domain waveform $x(\tau)$ with $\tau \in I$ can be represented by a Fourier series:

$$x(\tau) = \sum_{k=-\infty}^{\infty} X_k(t) e^{jk\omega\tau},$$

where $\omega = 2\pi/\Theta$ and the time varying k-th coefficient $X_k(t)$, also called dynamic phasor, is defined as an average over the interval I sliding in time:

$$X_k(t) = \frac{1}{\Theta} \int_{t-\Theta}^{t} x(\tau) e^{-jk\omega\tau} d\tau =: \langle x \rangle_k (t).$$

As $\langle x \rangle_k$ corresponds to the k-th harmonic of the waveform, the representation is truncated at the index of the maximal harmonic of interest, thus resulting in a finite number of equations.

Due to the fact that complex numbers are introduced in the representation, equations for the real part and the imaginary part, are obtained, respectively. One may argue that the dynamic phasor approach therefore increases the computational effort; however, for a small number of harmonics this is not the case compared to the drastically smaller time steps of purely dynamic simulations, and the benefit of covering a large span of phenomena, including large parts of both transient and steady-state behavior (Demiray, 2008).

In order to study the transient phase as well as the steady state of the first harmonic, the focus is placed on the first Fourier coefficient,. This corresponds to the intended type of application, which is to identify control and optimization strategies.

The models can be extended to represent further harmonics by adding the higher Fourier coefficients if desired. However, in this case smaller simulation time steps would be necessary than for the first harmonic and more equations would arise.

Now some aspects which are significant for the modeling process of electrical systems are highlighted:

1. When substituting the original waveforms of voltage and current, the Kirchhoff laws are maintained. This is explained by the additive nature of the Fourier coefficients, resulting from the linearity of integrals:

$$\langle x+y\rangle_k = \langle x\rangle_k + \langle y\rangle_k .$$

2. The dynamic behavior in terms of Fourier coefficients can be derived directly from the original mathematical model of each component thanks to the relation below which can be seen by substituting each expression in the integral definition of the coefficients.

$$\left\langle \frac{d}{dt}x \right\rangle_k = \frac{d}{dt}\langle x\rangle_k - jk\omega\langle x\rangle_k ,$$

Following the previously mentioned logic, a library of basic electrical elements for single-phase and multiphase was created, starting from the modeling of a new type of single-phase connector (pin) which supports a complex flow variable (current) and a complex across variable (voltage), Fig. 4.3 displays the schematic of a single-phase connector.

The multiphase connections (plugs) are obtained from multiple single-phase connectors. It is possible to introduce phase differences and to connect elements to a single phase or to all phases. Fig. 4.4 shows the principle of a three-phase component.

The basic three-phase element with two plugs does not specify whether a delta or wye connection should be used. Both of these connection logics are available as a basic component which is connected to the three-phase element to obtain the desired connection type.

1. Wye connector: All three connectors of one plug are connected to one single pin (which can act as a voltage reference). The three free pins of the other plug are the branches.
2. Delta connector: The single-phase elements are connected in a cyclic way.

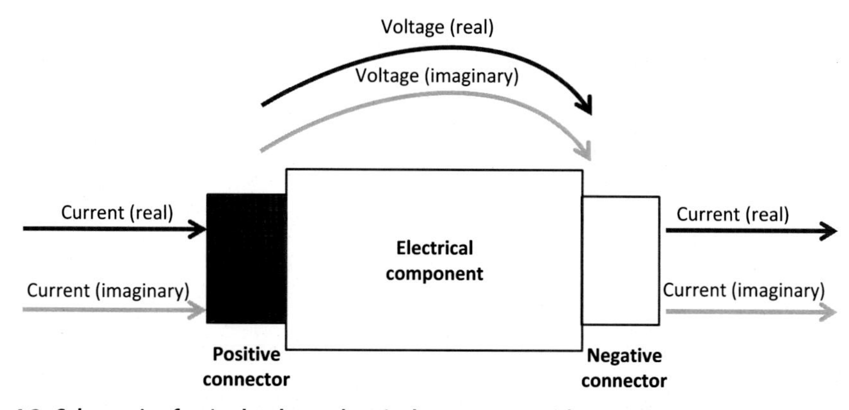

Figure 4.3 ***Schematic of a single-phase electrical component with two pins.***

Apart from a library of elementary single-phase and multiphase components, such as resistors and capacitors or sensors, several models of the site-specific technologies which were listed at the beginning of Chapter 4 in Table 4.1 were implemented. These will be described in more detail in the following section.

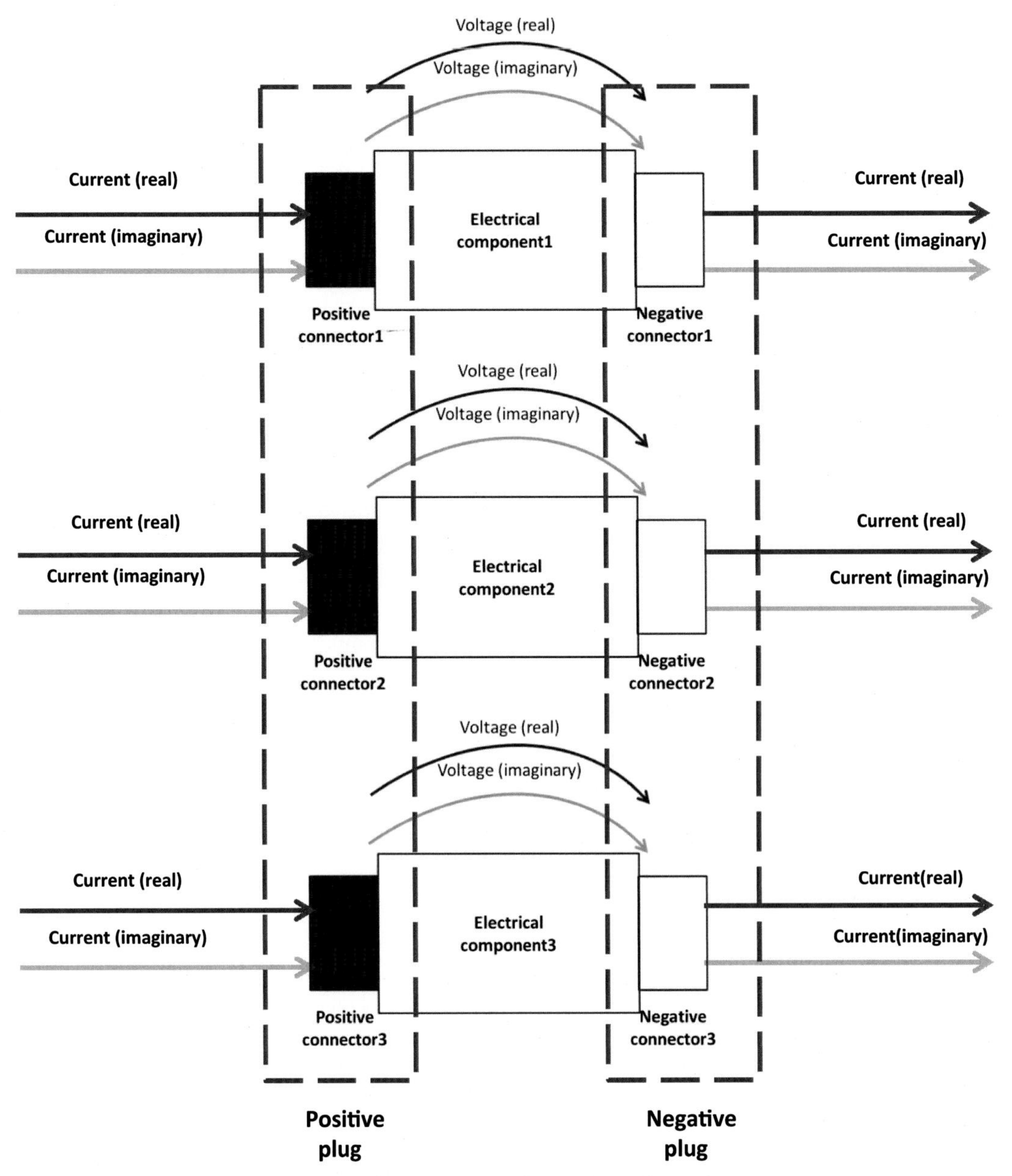

Figure 4.4 ***Schematic of an electrical component with two three-phase plugs.***

3.5 Model Library

The subsequent sections describe the thermal and electrical libraries developed for the COOPERaTE platform. For each component, a standard template is used, defining the following properties:

- A *description* of the component.
- The *parameters* of the component.
- The *interfaces* of the component.
- The technical *formulation* of the component.

3.5.1 *Thermal Components*

The following list summarizes all thermal components:

- thermal zone model and building physic,
- boiler,
- heat pump,
- CHP generator,
- solar thermal panel,
- thermal storage,
- heat exchanger,
- radiator,
- pump, and
- three-Way valves.

3.5.1.1 Thermal Zone Model and Building Physics

Description	The building physics library comprises all relevant components to represent a building through a simulation model. Basic components are outer and inner walls, air exchange, equivalent air temperature, solar irradiation gains, and radiation exchange. Based on these components a reduced multizone model of a random building can be built which also contains heat exchange interfaces, that is, radiators that represent the connection between the building mass and enthalpy flow within the building. The thermal connection to external heat generation and storage facilities is achieved via enthalpy connectors.
Parameters	Inner and outer walls • Initial temperatures (°C) • Resistance (K/W) • Heat capacity (J/K) • Area of walls/floors (m^2) • Heat transfer coefficient [W/(m^2 K)] • Emissivity (−)

Windows
- Factor for convective part of radiation through windows
- Window area (m^2)
- Emissivity (−)
- Energy transmittance (−)

Room air load
- Volume of air in zone (m^3)
- Density of air (kg/m^3)
- Heat capacity of air [J/(kg K)]

Interfaces
- Total solar irradiation (beam and diffusive)
- Weather (Real Input)
- Infiltration temperature (Real Input)
- Ventilation infiltration (Real Input)
- Internal gains (Real Input)

Formulation

The thermal zone model is based on the simplified control engineering model which is described in the German VDI 6007 standard. The basic idea behind the previously mentioned model is the equivalence between the differential equation of the thermal conduction in a wall and the processes in an idealized electrical cable, as shown in the following equation:

$$\frac{\partial T(t,x)}{\partial t} = \frac{1}{R \cdot C} \cdot \frac{\partial T^2(t,x)}{\partial x^2},$$

where T is the temperature, R is the thermal resistance of the wall layer, and C is the heat capacity of the wall layer.

According to the norms 6007 and 6020 individual building components can be divided into three basic groups:
- Outside surfaces, such as external walls, roof, outside, windows, etc.
- Internal walls for which the temperature conditions and radiation conditions prevailing in the adjoining rooms are practically equal (adiabatic loading).
- Internal walls which adjoin rooms with different temperature and radiation conditions, such as a cellar.

The different wall types can be modeled using the equivalence explained earlier, leading to a 2-transfer functions model of a thermal zone, depicted in Fig. 4.5. The resistances and capacities of the walls are calculated depending on the wall's structure and its geometry. The model representation of a thermal zone in Modelica according to VDI 6007 is shown in Fig. 4.6 (Lauster, Streblow, & Müller, 2012).

Outer wall
Indoor
Inner wall
$\dot{Q}_{sunbeam}$
$\dot{Q}_{IL,rad}$
$R_{rad,o}$
$R_{rad,i}$
R_{Rest}
R_{1o}
R_{1i}
$\vartheta_{A,eq,av}$
$R_{conv,o}$
$R_{conv,i}$
$\dot{Q}_{IL,conv}$
$\Delta\vartheta_{A,eq,av}$
C_{1o}
$R_{AirExchange}$
C_{Air}
C_{1i}
$\vartheta_{AirExchange}$
$\Delta\vartheta_{AirExchange}$
ϑ_R

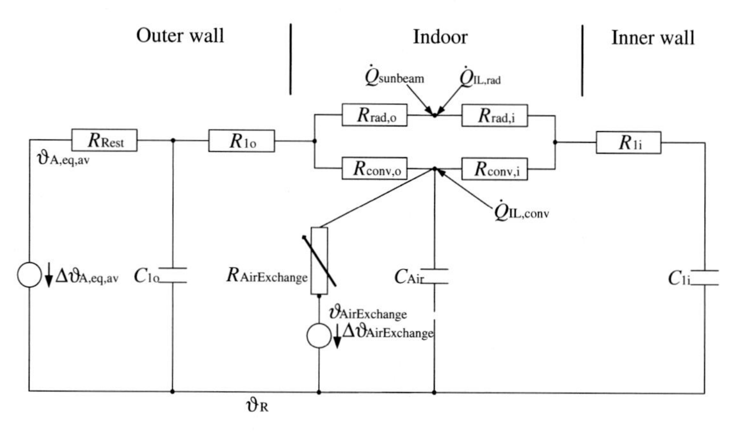

Figure 4.5 ***Schematic of the 2-transfer function model.***

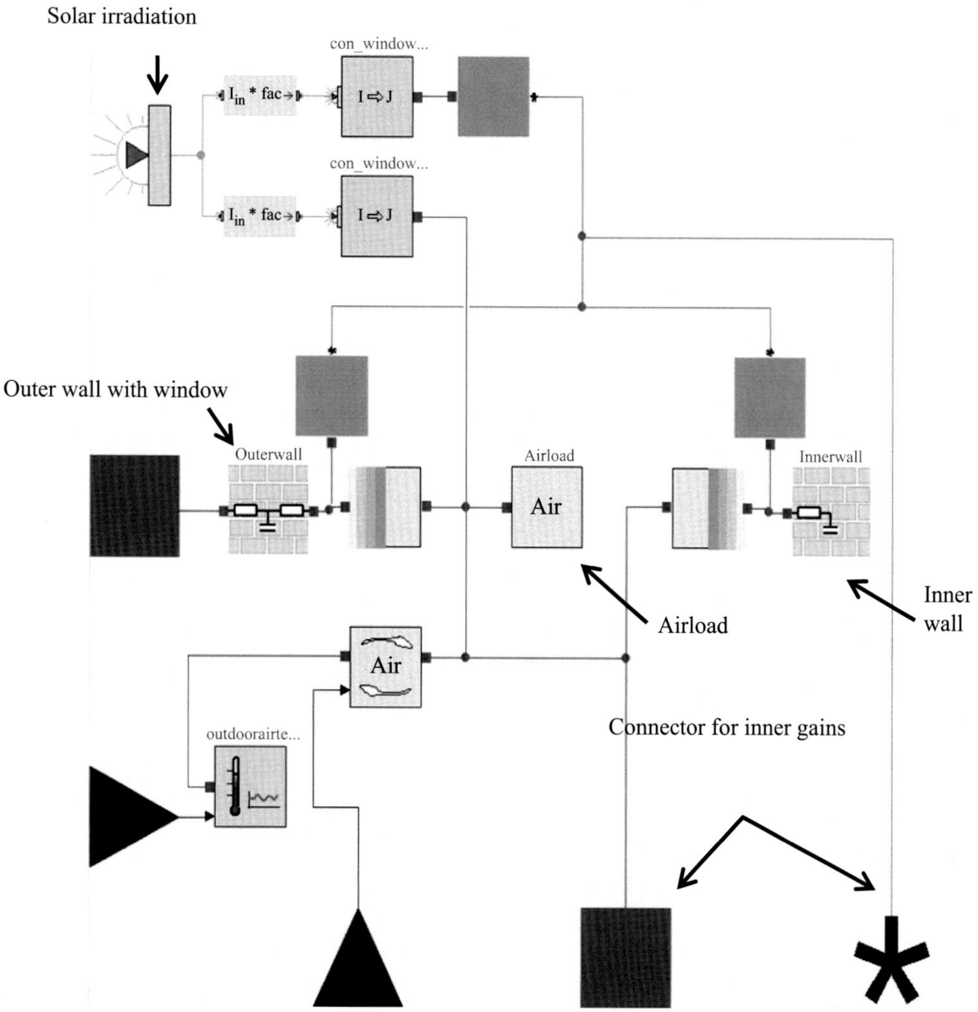

Solar irradiation
con_window...
I_in * fac
I ⇨ J
con_window...
I_in * fac
I ⇨ J
Outer wall with window
Outerwall
Airload
Air
Innerwall
Inner wall
Airload
Air
outdoorairte...
Connector for inner gains

Figure 4.6 ***Schematic of a thermal zone model.***

3.5.1.2 Boiler

Description The Boiler component is a model of a gas or oil boiler, which—depending on the parameterization—can be operated as a conventional or condensing boiler. It increases the enthalpy of the passing medium by raising its temperature and calculates the fuel consumption according to the current boiler efficiency. The model is table based and internally controlled by a PID controller, which controls the flow temperature that is externally prescribed by a heating curve. This boiler model is able to perform a modulated operation in which the range is determined by parameters. These parameters will be taken either from manufacturer specifications or measurement data. The model is able to represent boilers of different kinds and sizes with appropriate parameters.

Parameters
- Initial temperatures (°C)
- Water volume inside the boiler (m^3)
- Nominal heating power (W)
- Modulation rage (%)
- η Efficiency depending on flow temperature and boiler load (−)
- PID controller factors (−)

Interfaces
- Set flow temperature (Real Input)
- On/Off switch (Boolean Input)
- $\dot{E}_{fuel}$ Fuel consumption (Real Output)

Formulation The heating medium which flows into the component via the enthalpy connector is heated up to the prescribed set temperature by an ideal heating element. The thermal output $\dot{Q}_{th}$ of this heating element is determined by a PID controller which controls the flow temperature within the component. By using the efficiency table of the boiler the current fuel consumption is calculated.

$$\dot{E}_{fuel} = \frac{\dot{Q}_{th}}{\eta}.$$

3.5.1.3 Heat Pump

Description The heat pump component is a model suitable for both air and ground source heat pumps. It increases the enthalpy of the passing medium by raising its temperature. Additionally, the power consumption is calculated. Since it is a table-based model, the coefficient of performance (COP) of the heat pump is not influenced by the type of ambient energy source, but its temperature. The source is idealized as an infinite energy source with an externally prescribed temperature (outside air or brine temperature). The heat pump is able to modulate in a defined range. The necessary table data will be taken from the manufacturers' or measurement data. Depending on the parameterization the heat pump represents an air-to-water or brine-to-water heat pump in a system simulation environment. The connection to the electrical system is achieved through the electrical power consumption output which is used to formulate the relation between voltage and current.

Parameters

- Initial temperatures (°C)
- Modulation range (%)
- PID controller factors (−)
- Start-Up and Shut-Down time (s)
- $\dot{Q}_{co, max}$ Condenser heat flow table (W)
- Electric Power table (W)

Interfaces

- Source temperature (Real Input)
- Sink temperature (Real Input)
- N_{target} Target compressor speed (Real Input)
- On/Off switch (Boolean Input)
- COP Current COP (Real Output)
- $P_{el,max}$ Electrical power consumption (Real Output)

Formulation The externally prescribed target compressor speed of the heat pump determines the load state. With the source and sink temperatures as environmental status the maximum condenser heat flow and the corresponding power consumption are delivered by two separate tables. Based on these values the actual COP and thermal output $\dot{Q}_{th}$ can be calculated according to the following equations (Huchtemann & Müller, 2009):

$$COP = \frac{\dot{Q}_{th}}{P_{el,max}},$$

$$P_{el} = P_{el,max} \cdot \left(\frac{1}{N_{target}} \right)^3,$$

$$\dot{Q}_{co} = \dot{Q}_{th} = \dot{Q}_{co,max} \cdot \left(\frac{1}{N_{target}} \right)^3.$$

3.5.1.4 CHP Generator

Description The CHP generator component is a model of a combined heat and power generator with two enthalpy ports, which increases the enthalpy of the passing medium by raising its temperature. Additionally, the generated electrical power is calculated as an output of the model. It can be operated in a heat or power driven mode. Depending on the operation mode a power/heat input signal is used to determine the current generation status. The model is based on empirical data and the produced power and heat are calculated internally with a polynomial formula for which the weight factors must be provided either by the manufacturer or can be determined through the analysis of measurement data. The model's purpose is the representation of a CHP in a multiphysics energy system simulation. The connection to the electrical system is achieved through the electrical power input which is calculated from the voltage and current information at the connection point.

Parameters
- $P_{el,max}$ Maximum power output (W)
- Maximum thermal output (W)
- Modulation range (%)
- a_i,b_i Polynomial weight factors/coefficients
- Control mode (heat or power driven)
- Internal heat capacity (J/K)

Interfaces
- Electrical power (Real Input)
- Thermal power (Real Input)
- Operation mode (Boolean Input)
- On/Off switch (Boolean Input)
- Electrical power (Real Output)
- Thermal output (Real Output)
- Gas/fuel consumption (Real Output)

Formulation
At first the electrical and thermal efficiencies η_{el} and η_{th} are calculated depending on the coefficients of the polynomial, the maximum electrical power output, medium mass flow rate $\dot{m}$, and return temperature T_R:

$$\eta_{el} = a_0 + a_1 \cdot P_{el,\,max}{}^2 + a_2 \cdot P_{el,max} + a_3 \cdot \dot{m}^2 + a_4 \star \dot{m} + a_5 \cdot T_R^2 + a_6 \cdot T_R,$$

$$\eta_{th} = b_0 + b_1 \cdot P_{el,\,max}{}^2 + b_2 \cdot P_{el,max} + b_3 \cdot \dot{m}^2 + b_4 \star \dot{m} + b_5 \cdot T_R^2 + b_6 \cdot T_R.$$

Depending on the generation mode (heat or power driven) the current thermal and electrical output power is calculated using the power/heat demand input signal and the efficiencies. Based on the output flows and the efficiency, the gas consumption $\dot{E}_{fuel}$ is determined by (Rosato & Sibilio, 2012):

$$\dot{E}_{fuel} = \frac{P_{el}}{\eta_{el}}.$$

3.5.1.5 Solar Thermal Panel

Description
The solar thermal panel model has two enthalpy ports and increases the enthalpy of the passing heating medium by raising its temperature. The beam and diffusive irradiation from the environment is collected by the panel and then transferred to the heating medium within the device. The efficiency of this transfer is mainly influenced by the temperature difference of the heating medium and the environmental air.

Parameters
- Reference efficiency of the collector (−)
- Heat exchange coefficient with the environment [W/(m²K)]
- A_{panel} Panel area (m^2)
- Incident angle modifier (−)
- Panel location and orientation
- Ground reflection coefficient (−)

Interfaces
- T_a Environmental air temperature (Real Input)
- Beam and diffusive radiation (Real Input)

Formulation Using the input radiation, orientation, and location data, the total radiation on the tilted panel surface G can be calculated. The thermal efficiency η can be calculated depending on the difference between the mean panel temperature T_m and the environmental air temperature with polynomial coefficients a_i which are either given by the manufacturer or determined empirically. Hereby the total thermal output of the panel $\dot{Q}_{th}$ is determined:

$$\eta = \frac{1}{G}\left[\eta_0 - a_1 \cdot (T_m - T_a) - a_2 \cdot (T_m - T_a)^2\right],$$

$$\dot{Q}_{th} = G \cdot A_{panel}.$$

3.5.1.6 Thermal Storage

Description The thermal storage model represents a stratified water storage tank, where the stratification is achieved by the implementation of n discrete layers. Buoyancy flow within the tank is also considered. The number, position, and type (heating coil/direct flow) of the load and unload cycles can be defined as a parameter. Additionally, the tank volume and insulation is fully adjustable, enabling the model to represent random hot water storage tanks.

Parameters

- Number of discrete layers (−)
- Medium density (kg/m^3)
- Medium specific heat capacity [J/(kgK)]
- Thermal conductivity of medium [W/(mK)]
- Inner and outer heat transfer coefficient [W/(m^{2}K)]
- Tank and coil measures

Interfaces

- Layer temperature (Real Output)

Formulation The enthalpy connectors for loading and unloading the storage tank are, depending on their position, connected to a particular discrete layer of the tank or a coil inside the tank. For the case of tank loading the function of the tank can be described as follows.

The enthalpy of a passing medium is to a certain extent transferred from the medium to the layers through which it flows or to the layers which are connected to the corresponding heating coil. After passing the predefined layers of the tank or the heating coil, the medium leaves the storage tank with a certain enthalpy difference which is transferred to the internal storage medium. The layers itself are also connected to each other. The exchange of enthalpy between these layers occurs on the one hand by simple heat conduction $\dot{Q}_{cond}$ in both directions (to the upper and lower neighboring layer) and on the other hand based on buoyancy induced convection. Those two effects result in the effective heat conduction coefficient λ_{eff} that is used to calculate the effective heat transfer by conduction $\dot{Q}_{cond}$. Additionally each layer is thermally connected with the environment to which heat losses $\dot{Q}_{loss}$ can occur depending on the temperature difference of the medium $T_{layer,n}$ and the outside air T_{env}.

These different heat flows also depend on the height of each discrete volume h_{layer} and their lateral and front surface area $A_{\text{layer,front}}$ and $A_{\text{layer,lateral}}$ respectively:

$$\dot{Q}_{\text{cond}} = \frac{\lambda_{\text{eff}}}{h_{\text{layer}}} \cdot A_{\text{layer,front}} \cdot \left(T_{\text{layer},n} - T_{\text{layer},n+1}\right),$$

$$\dot{Q}_{\text{loss}} = k_{\text{tank,lateral}} \cdot A_{\text{layer,lateral}} \cdot \left(T_{\text{layer},n} - T_{\text{env}}\right).$$

The unloading process of the storage tank is vice versa.

3.5.1.7 Heat Exchanger

Description The heat exchanger component is a model of a plate heat exchanger with $n \times 2$ discrete volumes. It has two enthalpy input and output connectors each and is fully adjustable in size, discretization, and material. The purpose of this component is to enable heat transfer between two different mediums that are not allowed to be in direct contact/mixture, for example, heating water and solar thermal fluid (Fig. 4.7).

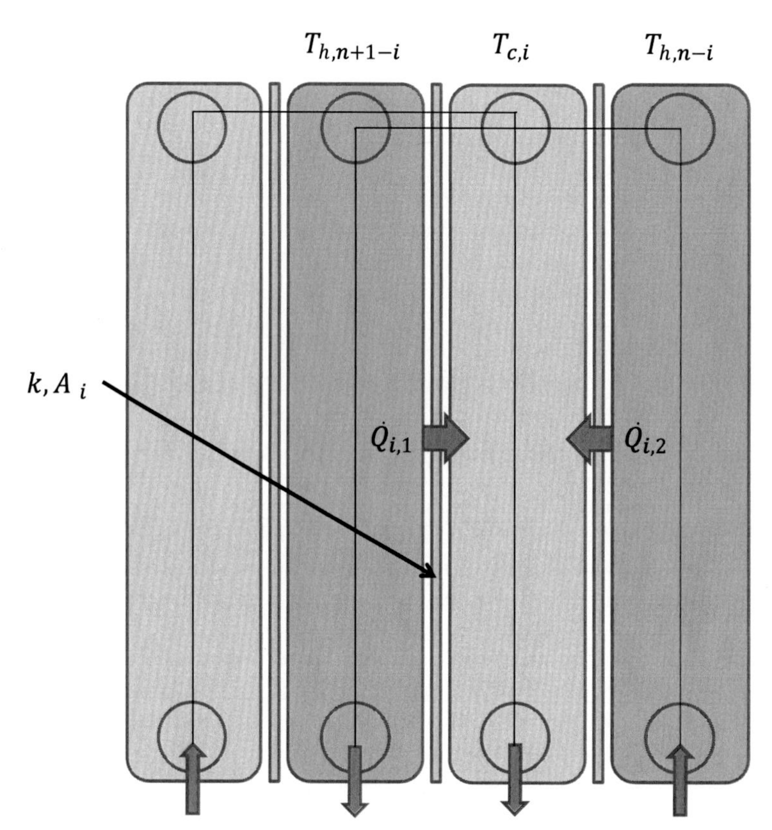

Figure 4.7 ***Heat exchanger discretization scheme.***

Parameters	• Number of discrete volumes (−) • Total heat exchanger volume (m^3) • Total heat exchanger plate area (m^2) • Heat exchanger heat transmission coefficient [W/(m^2K)]
Interfaces	• None
Formulation	The discrete volumes of the heat exchanger are arranged in an alternating manner. Thus, each volume containing the hot medium is neighboring a volume containing the cold medium. Those neighboring discrete volumes are thermally connected to each other by a heat transfer component. Based on the heat flow between the individual volumes $\dot{Q}_i$, which is influenced by the temperatures of hot and cold volumes $T_{h,i}$ and $T_{c,i}$ and the heat transmission coefficient k, the total heat flow within the heat exchanger $\dot{Q}_{th}$ can be calculated:

$$\dot{Q}_i = k \cdot A_i \cdot \left(T_{h,n+1-i} - T_{c,i}\right) + k \cdot A_i \cdot \left(T_{h,n-i} - T_{c,i}\right),$$

$$= k \cdot A_i \cdot \left(T_{h,n+1-i} + T_{h,n-i} - 2T_{c,i}\right),$$

$$\dot{Q}_{th} = \sum_{i=0}^{n} \dot{Q}_i .$$

3.5.1.8 Radiator

Description	The radiator component is a model of a hot water radiator. It has two enthalpy ports and is fully adjustable in size, discretization, and material. The purpose of this component is to enable heat transfer between the circulated heating medium, mostly water, and the room airload described in Section 3.5.1.1.
Parameters	• Nominal heating power (W) • Number of discrete volumes (−) • Geometric measures (m) • Material properties of the radiator
Interfaces	• None
Formulation	The heat $\dot{Q}_{th}$ flowing into the component is transferred to a thermal capacity element representing the mass of the radiator. A radiator wall element calculates the partial energy flows through convection and radiation that are transferred to the surrounding environment. The convective part $\dot{Q}_{conv}$ is determined by the heat transmission coefficient k, the effective radiator area A, and the temperature difference between the radiator wall T_{rad} and the environment T_{env}. The radiative part of the energy transfer $\dot{Q}_{rad}$ is determined by Stefan Boltzmann's law, which uses the emissivity ε of the radiator and the Stefan Boltzmann constant σ. This energy transport is carried out by a conventional thermal connector and a radiation port which will be connected to the thermal zone model (Section 3.5.1.1).

$$\dot{Q}_{th} = \dot{Q}_{conv} + \dot{Q}_{rad}$$

$$\dot{Q}_{conv} = k \cdot A \cdot \left(T_{rad} - T_{env}\right)$$

$$\dot{Q}_{rad} = \sigma \cdot \varepsilon \cdot A \cdot \left(T_{rad}^4 - T_{env}^4\right)$$

3.5.1.9 Pump

Description	The pump component is a model of an ideal pump, which creates a prescribed heating medium mass flow between enthalpy connectors. It is also used for the type declaration of the heating medium. Since the pump can ideally prescribe a mass flow rate, conventional throttle valves are not necessary for a system setup.
Parameters	• Specific heat capacity [J/(kg K)]
Interfaces	• Heating medium mass flow rate (kg/s)
Formulation	The pump component simply prescribes the mass flow $\dot{m}$ of the heating medium in the system. The enthalpy and temperature of the medium stay unchanged: $\dot{m} = \dot{m}_{\text{prescribed}}$, $h_{\text{out}} = h_{\text{in}}$, $T_{\text{out}} = T_{\text{in}}$.

3.5.1.10 Three-Way Valves

Description	The three-way valve model is able to represent either a distribution or a mixing valve. As a distribution valve it splits a mass flow $\dot{m}_{AB}$ into two separate mass flows $\dot{m}_A$ and $\dot{m}_B$ according to the valve opening position. Used as a mixing valve two separate mass flows $\dot{m}_A$ and $\dot{m}_B$ are mixed to one mass flow $\dot{m}_{AB}$ depending on the valve position. Since the pump (Section 3.5.1.9) can ideally prescribe a mass flow rate, conventional throttle valves are not necessary for a system setup.
Parameters	Three-way valve type (distribution/mixing)
Interfaces	• Valve opening position
Formulation	The mass flows and enthalpies of each enthalpy connector for the distribution/mixing valve are calculated as follows: Distribution valve $\dot{m}_A = \dot{m}_{AB} \cdot y$, $\dot{m}_B = \dot{m}_{AB} \cdot (1-y)$, $h_A = h_A$. Mixing valve $\dot{m}_{AB} = y \cdot \dot{m}_A + (1-y) \cdot \dot{m}_B$, $h_{AB} = \frac{\dot{m}_A \cdot h_A + \dot{m}_B \cdot h_B}{\dot{m}_{AB}}$.

3.5.2 Electrical Components

The following list summarizes all electrical components:

- Electrical line
- PV module

- Simplified battery storage system
- Constant power load
- EV charger
- Transformer
- Infinite power bus
- Wind turbine

3.5.2.1 Electrical Line

Description	The electrical line model represents effects which cable properties introduce in electrical systems in low voltage and medium voltage, as opposed to ideal connections in which no losses or capacitive and inductive effects occur. It has three pins: the positive and negative pin as well as a reference pin which may be connected to ground or any other reference point the user would like to choose (Fig. 4.8).
Parameters	• x length of the line (m) • r resistance per meter (Ω/m) • l inductance per meter (H/m) • c capacitance per meter (F/m)
Interfaces	If a specific output is required, electrical quantities can be measured by connecting the corresponding sensor model.
Formulation	The line model is a lumped π line model with a voltage reference pin (Fig. 4.9). The resistance R and the inductance L are the total resistance and inductance of the line, calculated from the parameters, whereas the capacitances C_1 and C_2 share the total line capacitance: $R = r \cdot x,$ $L = l \cdot x,$ $C = c \cdot x,$ $C_1 = C_2 = \frac{1}{2}C.$ The effect of the line capacitance can be neglected in low voltage lines by setting $C = 0$, leading to a purely inductive model.

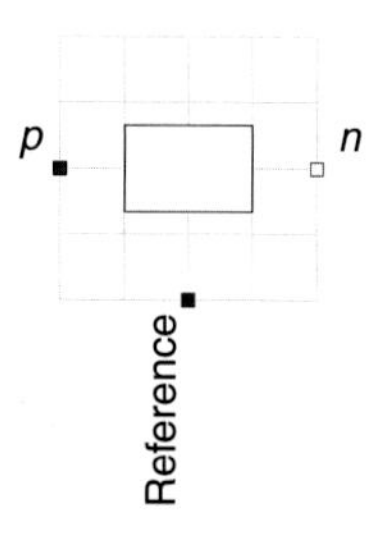

Figure 4.8 ***Electrical line icon.***

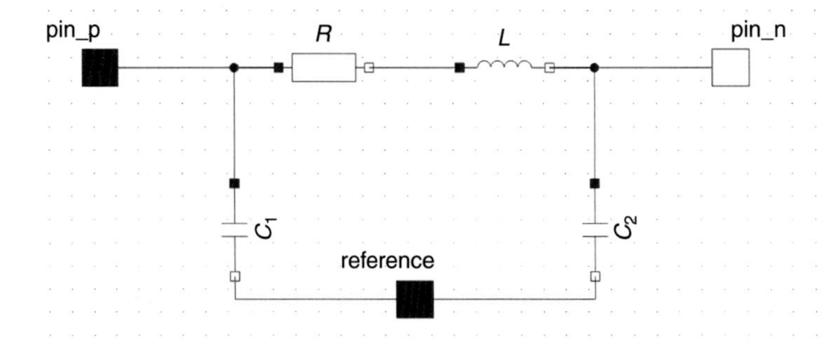

Figure 4.9 ***Line model.***

3.5.2.2 PV Module

Description The photovoltaic module is a simplified model with two pins which converts solar irradiance and ambient temperature data into electrical active power depending on the surface area of the module. This power value is then used to compute the relation between voltage and current. When information on reactive power is not available, its value is set to zero as a default. The nominal operating conditions, such as ambient temperature, cell temperature, and radiation are parameters which may be chosen according to the specific PV panel in use. The efficiency is modeled to decrease at higher ambient temperatures than the nominal ambient temperature, depending on a temperature coefficient which depends on the type of PV panel.
This model is not a component-based model like a diode PV model. Therefore, it is not useful for a detailed analysis of PV panel properties; it should rather be used to represent the behavior of a PV module as a source within an electric system (Fig. 4.10).

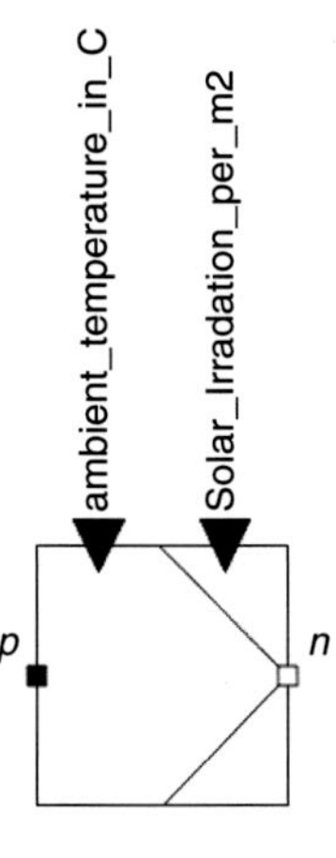

Figure 4.10 ***PV module icon.***

Parameters

- Area (m^2)
- Efficiency at nominal conditions
- Temperature coefficient for efficiency ($°C^{-1}$)
- T_{NOCT} Nominal operating cell temperature (°C)
- $T_{ambient,NOCT}$ NOCT Ambient temperature (°C)
- Irr_{NOCT} NOCT irradiance on cell surface (W/m^2)
- Q Reactive power (default = 0, W)

Interfaces

- Irr Solar irradiance input (W/m^2)
- $T_{ambient}$ Ambient air temperature input (°C)

If a specific output is required, electrical quantities can be measured by connecting the corresponding sensor model.

Formulation

At first, the cell temperature T_{cell} is calculated from the ambient temperature input and the solar irradiance input, depending on the nominal operating conditions:

$$T_{cell} = T_{ambient} + \left(T_{NOCT} - T_{ambient,NOCT}\right) \cdot \frac{\mathrm{Irr}}{\mathrm{Irr}_{NOCT}}.$$

Now, the efficiency η is updated according to the difference between T_{cell} and T_{NOCT}, weighted with the temperature coefficient for efficiency.

Finally, according to Skoplaki and Palyvos (2009) the active power output is given as:

$$P = \mathrm{Irr.\ area.}\ \eta,$$

which specifies the relationship between the dynamic phasor current and voltage across the two pins of the model with the equation:

$$\langle i \rangle = \frac{(P + jQ)^*}{\langle v^* \rangle}.$$

3.5.2.3 Simplified Battery Storage System

Description

The battery storage system is a simple model of a generic battery. It can be charged or discharged up to a maximum capacity with a certain power and efficiency, while updating its state of charge. A charging or discharging logic can be connected to the battery via the charge/discharge status input. When information on the end-of-discharge or end-of-charge voltage are unknown, the default values can be used (Fig. 4.11).

This model does not specify the battery technology and side effects which may occur due to the specific chemical processes some battery types use.

Parameters

- SOC_{max} Storage capacity (Ah)
- SOC_0 Initial state of charge (Ah)
- R_{int} Internal resistance of the battery (Ω)
- v_{nom} Nominal voltage (V)
- v_{min} End-of-discharge voltage (default = $0.v_{nom}$)
- v_{max} End-of-charge voltage (default = $1.v_{nom}$)
- P Charging and discharging power (W)
- η Charging efficiency index (default = 1)

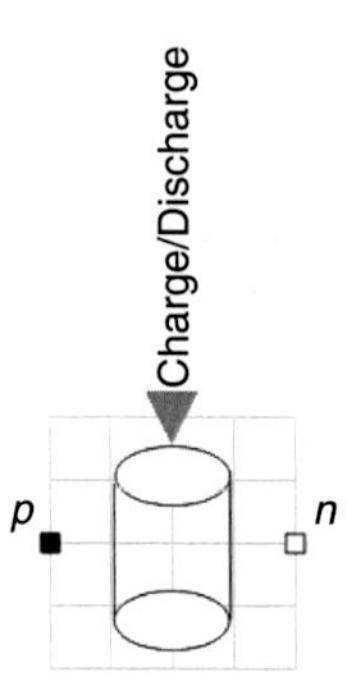

Figure 4.11 ***Simplified battery icon.***

Interfaces • Charge (0) or discharge (1) status input

Formulation The battery is modeled as a voltage source with an internal resistance, whose voltage level depends on the state of charge (Wagner, 2010).

The battery storage is considered to operate in DC, a logic which translates between the dynamic phasor voltage and current in the rest of the system and their root mean square values was included within the battery model (Fig. 4.12).

Assuming that the voltage source depends on the state of charge linearly for simplicity:

$$v_{\mathrm{DCsource}} = v_{\min} + (v_{\max} - v_{\min}) \cdot \frac{\mathrm{SOC}}{\mathrm{SOC}_{\max}}.$$

When the maximal storage capacity is reached, the voltage is limited by the end-of-charge voltage.

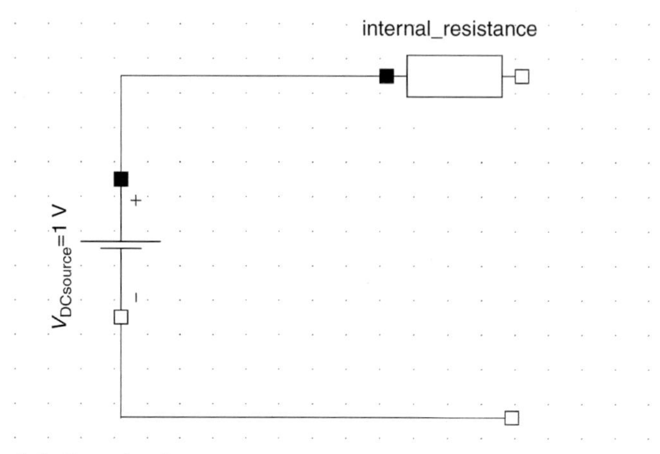

Figure 4.12 ***Simple battery circuit.***

In the first time step, SOC = SOC_0. Through the power parameter, the DC current is computed from:

$$i_{DC} = \frac{P}{v_{source}}.$$

The state of charge is then updated through the equation:

$$SOC = SOC_0 + \int \eta \cdot a \cdot i_{DC} dt,$$

while SOC $\leq SOC_{max}$, and $a = \pm 1$ depending on whether the battery is charging or discharging.

The current and voltage of the DC voltage source are converted back to AC as ideal sinusoidal signals with an adjustable default frequency of 50 Hz, where the amplitude is given by $\sqrt{2}$ times the DC voltage or current.

3.5.2.4 Constant Power Load

Description Due to the difficulty of obtaining all physical load parameters, a simplified model representing the behavior of an electrical consumer with constant power was used. The relation between current and voltage necessary to characterize a two-pin component can be determined from the power values (Fig. 4.13).

Parameters
- P active power (W)
- Q reactive power (var)

Interfaces If a specific output is required, electrical quantities can be measured by connecting the corresponding sensor model.

Formulation The relation between the dynamic phasor current and voltage is given by the complex power:

$$\langle i \rangle = \frac{(P + jQ)^*}{\langle v^* \rangle}.$$

3.5.2.5 EV Charger

Description The EV charger is a type of constant power load. It is an electrical consumer which, depending on the power level of the specific charging station, consumes the same amount of power during the time in which it is in use. By setting the status to "on," the charging of an EV is simulated (Fig. 4.14).

Parameters See constant power load.

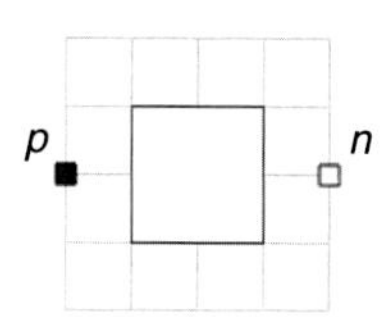

Figure 4.13 ***Constant power load icon.***

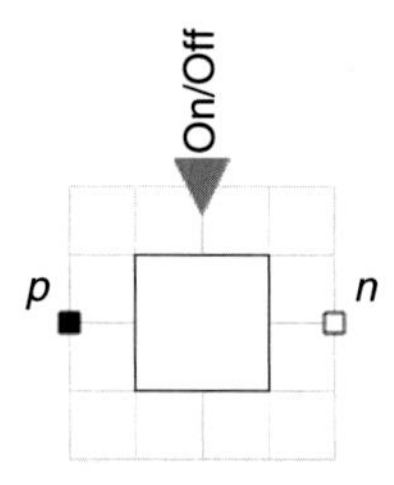

Figure 4.14 ***EV charger icon.***

Interfaces	Off (0) or on (1) status input. If a specific output is required, electrical quantities can be measured by connecting the corresponding sensor model.
Formulation	The formulation of the EV charger is the same as the constant power load except for the additional option of turning it off. This is achieved by setting the current equal to zero when the "off" state is given to the charger as input.

3.5.2.6 Transformer

Description	For the transformer, an ideal model as well as a standard model is available for both single-phase and multiphase applications. The ideal transformer changes the voltage according to the turns ratio with an optional magnetization, whereas the standard transformer is modeled with two inductors and their mutual inductance (Fig. 4.15). Additionally, a prototype of a YY and YD transformer (YD5, YD11) have been modeled.
Parameters	Ideal transformer • n Turns ratio primary/secondary voltage • Option for considering magnetization, default = false • L_m Magnetization inductance Standard transformer: • L_1 Self-inductance of winding1 (H) • L_2 Self-inductance of winding2 (H) • M Mutual inductance (H)

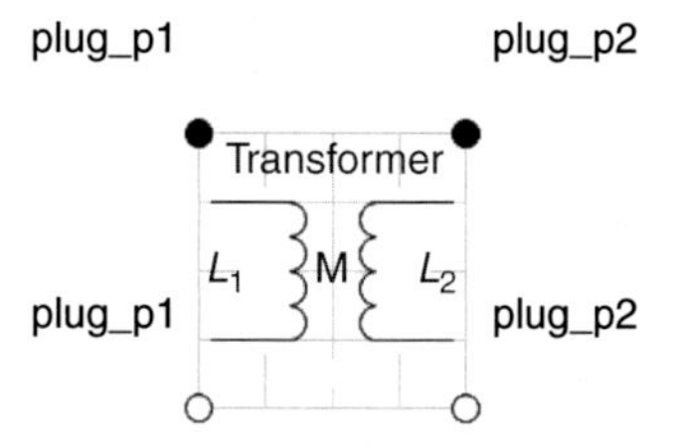

Figure 4.15 ***Transformer icon.***

Interfaces If a specific output is required, electrical quantities can be measured by connecting the corresponding sensor model.

Formulation Ideal transformer

The relation between the primary voltage v_1 and the secondary voltage v_2 is given by:

$$\langle v_1 \rangle = n \cdot \langle v_2 \rangle.$$

If the user chooses to consider the magnetization, the behavior is modified to include the magnetization current i_m (as a function of the primary current i_1 and secondary current i_2) and the magnetic flux ψ_m with respect to the primary side of the transformer:

$$\langle i_m \rangle = \langle i_1 \rangle + \frac{\langle i_2 \rangle}{n},$$

$$\langle \psi_m \rangle = L_m \cdot \langle i_m \rangle.$$

Standard transformer

The standard transformer also includes the primary and secondary inductance as well as their influence on each other. The equations then are:

$$\langle v_1 \rangle = L_1 \cdot \frac{d\langle i_1 \rangle}{dt} + M \cdot \frac{d\langle i_2 \rangle}{dt},$$

$$\langle v_2 \rangle = L_2 \cdot \frac{d\langle i_2 \rangle}{dt} + M \cdot \frac{d\langle i_1 \rangle}{dt}.$$

Three-phase transformers

The three-phase transformers are obtained by connecting single-phase transformers with a wye and/or delta connector.

In case the ideal transformer is used, a resistor and inductor can be added in order to approximate the real behavior. The YY and YD transformers are models comprising the connection logic (wye or delta) and the transformer model (ideal or standard) as well as additional components if the ideal transformer model is used instead of the standard transformer. For example, the connection logic for a YD transformer is obtained by connecting components in series in the following order:

- Wye connector
- Ideal transformer or standard transformer
- Internal resistance (only if the ideal transformer is used)
- Transformer stray inductance (only if the ideal transformer is used)
- Delta connector

3.5.2.7 Infinite Power Bus

Description The infinite power bus is used as a coupling point with the external electrical grid. It does not take the specific grid topology into account (Fig. 4.16). Therefore it is only suitable for studies of a system connected to the grid, but not to analyze the effects a system may have on the external grid.

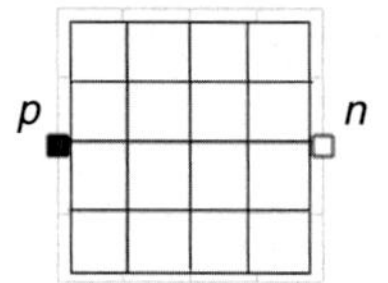

Figure 4.16 ***Infinite power bus icon.***

Parameters	• Amplitude (V) • Frequency (Hz, default = 50 Hz) • Phase
Interfaces	If a specific output is required, electrical quantities of the infinite power bus can be measured by connecting the corresponding sensor model.
Formulation	The infinite power bus is modeled as an ideal sinusoidal voltage source according to the chosen parameters.

3.5.2.8 Wind Turbine

Description	The wind turbine model is a power curve model which, similarly to the solar PV generator, calculates an active power output depending on turbine specifications as well as the wind profile and air density (Fig. 4.17). Voltage and current information are then extracted from the power values. When information on reactive power is not available, it is set to null as a default. Since this is a function-based simplified model without mechanical information or the type of electrical generator, it is not useful for thorough studies of wind turbine properties, but should rather be used to represent a power source within an electrical system.
Parameters	• c_p power coefficient of the turbine, default = 0.593 • A swept area of the rotor (m^2) • ε_g generator efficiency, default = 1 • ε_b gearbox/bearings efficiency, default = 1 • Q reactive power, default = 0

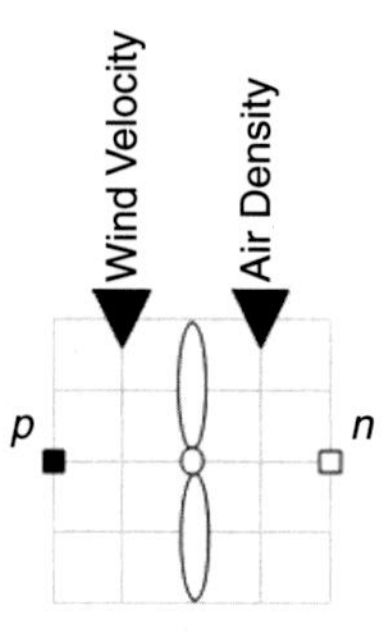

Figure 4.17 ***Wind turbine icon.***

Interfaces

- v_{wind} wind velocity input (m/s)
- ρ air density input (kg/m^3)

If a specific output is required, electrical quantities can be measured by connecting the corresponding sensor model.

Formulation

According to Ghosh and Prelas (2011), the active power curve of a wind turbine is described by the equation:

$$P = c_p \varepsilon_g \varepsilon_b \frac{1}{2} \rho A v^3,$$

where $\frac{1}{2} \rho A v^3$ is the power of the wind acting on a surface A, and the factors $c_p \varepsilon_g \varepsilon_b$ are limiting factors on the actual power exploitation which varies depending on the specific type of wind turbine and is limited by 0.593 according to Betz's law.

From the power output, the dynamic phasor voltage and current can be calculated with the equation:

$$\langle i \rangle = \frac{(P + jQ)^*}{\langle v^* \rangle}.$$

4 NEIGHBORHOOD ENERGY OPTIMIZATION ALGORITHMS

4.1 Mathematical Notations and Definitions

This section highlights the general mathematical notations and definitions used for the formulation of Neighborhood energy optimization algorithms. The notations used in this book are given in Table 4.3.

Table 4.4 presents the definitions to consider the economic criterion in the formulation of optimization algorithms.

Equipment models used for optimization may be simplified with respect to those used for the system-level simulation. In the remaining part of this chapter we review models which can be found in recent literature on energy management and microgrids. First of all, we observe that the optimization formulation can be in general nonlinear therefore nonlinear programming algorithms apply (Bertsekas, 2008). Component models may or may not include system losses. Recent papers report interesting applications of stochastic optimization algorithms, such as particle swarm (Chung, Wenxin, Cartes, & Schoder, 2008; Peng & Duo, 2014) and genetic (Deng, Gao, Zhou, & Hu, 2011) which can handle nonlinearities. However, it is observed that by approximating efficiencies of CHP (electrical efficiency η_{CHPel} and thermal efficiency η_{CHPth}) and boiler efficiency η_{bo} with constant terms, the optimization problem can be conveniently linearized. In that case, deterministic optimization algorithms for linear programming can be applied. The mathematical equations used to describe equipment, such as CHP, Fossil-Fuel generation; RES and Storage are discussed later in the chapter (Moghaddam, Saniei, & Mashhour, 2016).

Table 4.3 Centralized Optimization algorithm—notations and definitions

General

$k \in \{1, N\} \subset \mathbb{N}$ time step
N: prediction horizon
$i, j \in \{1, B\} \subset \mathbb{N}$ index of actors
B: number of actors in the Neighborhood
$S \in \{1, N\} \subset \mathbb{N}$: number of schedules per building

Generation

Combined Heat and Power (CHP)	Fossil Fuel Generators	Renewable generation
$P_{CHP,el,i}$: Electrical CHP output power P_{CHP,el,i_max}: Max electrical CHP output power P_{CHP,el,i_min}: Min electrical CHP output power $P_{CHP,th,i}$: Thermal CHP output power $\eta_{CHP,el,i}$: Electrical CHP efficiency $\eta_{CHP,th,i}$: Thermal CHP efficiency $s_{CHP,on,i} \in \{0,1\}$: CHP ON/OFF state variable $s_{CHP,up,i} \in \{0,1\}$: Start command $s_{CHP,down,i} \in \{0,1\}$: Stop command $n_i \in \{1 \ldots 8\}$: Minimum running time (time steps)	$P_{boiler,i}$: Boiler output power $P_{boiler,i,min}$: Min boiler output power $P_{boiler,i,max}$: Max boiler output power $\eta_{boiler,th,i}$: Boiler efficiency $s_{boiler,on,i} \in \{0,1\}$: Boiler ON/OFF state variable $P_{Diesel,i}$: Power produced by Diesel generator	$P_{renew,i}$: Power produced by renewables

Storage

$W_{th,i}$: Thermal energy stored in kWh
W_{th,i_max}: Max thermal energy stored in kWh
$\eta_{sto,th,i}$: Thermal storage efficiency
$P_{sto,in,i}$: Storage thermal charging power
$P_{sto,out,i}$: Storage thermal discharging power
P_{sto,in,i_max}: Max storage charging rate
P_{sto,out,i_max}: Max storage discharging rate
$W_{el,i}$: Electrical energy stored
W_{el,i_max}: Max electrical energy stored
$P_{C,i}$: Electrical storage charging power
$P_{D,i}$: Electrical storage discharging power
P_{C,i_max}: Max electrical storage charging power
P_{D,i_max}: Min electrical storage charging power
$\eta_{C,i}$: Electrical storage charging efficiency

Consumption

$L_{th,i}$: Thermal load of actor i
$L_{el,i}$: Electrical load of actor i

Table 4.4 Definitions—power purchases/sell, prices and costs

Power purchases/sells

Outside grid (infinite power bus)	Neighborhood grid	Neighborhood energy exchange
$P_{gb,i}$: Power purchased from the grid $P_{gs,i}$: Power sold to the grid	$P_{nb,i}$: Power purchased from the Neighborhood $P_{ns,i}$: Power sold to the Neighborhood	B_i: Profit of actor i when acting as an energy seller [€/kWh] B_j: Savings of actor j when buying energy from the Neighborhood [€/kWh] $P_{n,i\rightarrow j}$: Power exchanged from actor i to actor j (kW) $C_{n,i\rightarrow j}$: Price of the energy exchanged from actor i to actor j [€/kWh] $x_{i\rightarrow j}$: Ratio of power exchanged from actor i to j

Prices/costs

Fuel cost	Grid prices	Greenhouse emissions penalties
$C_{\mathrm{gas},i}$: Cost of gas $C_{\mathrm{Diesel},i}$: Cost of diesel	$C_{gb,i}$: Cost of electrical energy imported from the grid $C_{gs,i}$: Reward for electrical energy grid exports $C_{nb,i}$: Cost of electrical energy imported from the Neighborhood $C_{ns,i}$: Reward for electrical energy Neighborhood exports	$\mu_{\mathrm{CO_2,CHP},i}$: Penalty for CO_2 emissions of CHP $\mu_{\mathrm{CO_2,Fuel},i}$: Penalty for CO_2 emissions of fuel-based generators $\mu_{\mathrm{CO_2},g,i}$: Penalty for CO_2 emissions content of the energy bought from grid E_{bs}: The emissions of building b when the set of schedules S is selected

4.1.1 Combined Heat and Power Mathematical Description

A general model of CHP suitable for energy optimization purposes is described in Moghaddam et al. (2016). The mathematical description for a CHP starts with the power constraint of this equipment. In a CHP system, electrical and thermal powers are not generated independently, but the output level of one of them also determines the other one. This dependency determines a feasible region wherein the generation of electricity and heat can be controlled. The total output power is given by:

$$s_{\mathrm{CHP,on},i}(k)\cdot P_{\mathrm{CHP},i\_\min} \leq P_{\mathrm{CHP},i} \leq s_{\mathrm{CHP,on},i}(k)\cdot P_{\mathrm{CHP},i\_\max}$$

Where the CHP electrical power output is represented as:

$$P_{\mathrm{CHP,el},i}(k) = \eta_{\mathrm{CHP,el},i}\cdot P_{\mathrm{CHP},i}(k)$$

And the thermal power output by:

$$P_{\mathrm{CHP,th},i}(k) = \eta_{\mathrm{CHP,th},i} \cdot P_{\mathrm{CHP},i}(k)$$

The efficiencies relating the above electrical and thermal output powers to the total output power can be considered constant to linearize the optimization formulation, even though Mixed Integer Non-Linear Programming formulations are recently being considered and evaluated (Moghaddam et al., 2016). The minimum CHP running time is expressed by the following constraints:

$$s_{\mathrm{CHP,up},i}(k) + s_{\mathrm{CHP,down},i}(k) \leq 1$$

$$s_{\mathrm{CHP,on},i}(k) + s_{\mathrm{CHP,on},i}(k-1) \leq s_{\mathrm{CHP,up},i}(k) + s_{\mathrm{CHP,down},i}(k)$$

$$s_{\mathrm{CHP,on},i}(k+\ell) \geq s_{\mathrm{CHP,up},i}(k) \text{ where } \ell \in \{1, N\}$$

4.1.2 Boiler Mathematical Description

The operating range of a boiler can be represented as (Moghaddam et al., 2016):

$$s_{\mathrm{Boiler,on},i}(k) \cdot P_{\mathrm{Boiler},i\_\min}(k) \leq P_{\mathrm{Boiler},i}(k) \leq s_{\mathrm{Boiler,on},i}(k) \cdot P_{\mathrm{Boiler},i\_\max}(k)$$

4.1.3 Thermal Storage Mathematical Description

Similar to the previous subsections, the thermal energy stored can be represented in terms of (Moghaddam et al., 2016):

$$W_{\mathrm{th},i}(k+1) = \eta_{\mathrm{sto,th},i} W_{\mathrm{th},i}(k) + P_{\mathrm{sto,in},i}(k)\Delta t - P_{\mathrm{sto,out},i}(k)\Delta t$$

Constrained to:

$$0 \leq W_{\mathrm{th},i}(k) \leq W_{\mathrm{th},i\_\max} \qquad \text{Maximum capacity}$$

$$0 \leq P_{\mathrm{sto,in},i}(k) \leq P_{\mathrm{sto,in},i\_\max} \qquad \text{Maximum input power}$$

$$0 \leq P_{\mathrm{sto,out},i}(k) \leq P_{\mathrm{sto,out},i\_\max} \qquad \text{Maximum output power}$$

4.1.4 Electrical Storage Mathematical Description

The electrical energy stored in, for example, a battery can be defined by (Moghaddam et al., 2016):

$$W_{\mathrm{el},i}(k+1) = W_{\mathrm{el},i}(k) + P_{C,i}(k)\Delta t - P_{D,i}(k)\Delta t$$

Constrained to:

$$0 \leq W_{\mathrm{el},i}(k) \leq W_{\mathrm{el},i\_\max} \qquad \text{Maximum capacity}$$

$$0 \le P_{C,i}(k) \le P_{C,i\_\max} \qquad \text{Maximum input power}$$

$$0 \le P_{D,i}(k) \le P_{D,i\_\max} \qquad \text{Maximum output power}$$

4.2 Centralized Optimization—Single Ownership Neighborhood

In the single-owner case, a single subject owns multiple buildings of the Neighborhood. The energy optimization architecture is of centralized type.

Recent published work on optimization of microgrids and heating systems includes several subsystems for energy generation and storage. A comprehensive list of components and related models is included in Moghaddam et al. (2016). Some recent papers (Rahbar, Xu, & Zhang, 2015; Parisio, Rikos, & Glielmo, 2014; Gamez Urias, Sanchez, & Ricalde, 2015; Kanchev, Colas, Lazarov, & Francois, 2014; Chaouachi, Kamel, Andoulsi, & Nagasaka, 2013) consider the optimization of the electrical microgrid systems. Potential of microgrids in enhancing the energy efficiency in buildings is already emerging from recent studies (Xiaohong, Zhanbo, & Qing-Shan, 2010).

A (centralized) Neighborhood energy optimization algorithm:

- Minimizes Neighborhood running energy cost.
- Considers all buildings electrical and thermal loads, local generation sources.
- Accounts for both the economic criterion and environmental impact (energy cost and CO_2 emissions).
- Keeps the total Neighborhood pollution emissions below a given threshold.
- Can simultaneously handle Renewable Energy Sources (RES), fossil fuel generators, cogeneration units, as well as electrical and thermal storages.
- Is implemented as energy optimization service in the NEM Platform as a prototype enabling the implementation of optimized schedules by actuating the relevant components (e.g., adjustable generation and storage equipment) with the optimized set-points.

With the definitions corresponding to different generation and storage equipment from Section 4.1, objective functions for a centralized algorithm in a single-ownership Neighborhood can be formulated, respectively, for the minimization of total energy operating:

$$\min\left\{\sum_{j=1}^{B}\sum_{n=1}^{N}\overbrace{C_{\text{gas}}(P_{\text{CHP},j}(n)+P_{\text{bo},j}(n)\Delta t)}^{\text{CHP+Boiler}}+\overbrace{C_{\text{Diesel}}P_{\text{Disel},j}(n)\Delta t}^{\text{Diesel generator}}+\overbrace{C_{gb}P_{gb,j}(n)\Delta t}^{\text{Energy purchased from Grid}}-\overbrace{C_{gs}P_{gs,j}(n)\Delta t}^{\text{Energy sold to grid}}\right\}$$

and:

$$\min\left\{\sum_{j=1}^{B}\sum_{n=1}^{N}\left[\mu CO_{2,gb}P_{gb.j}(n)-\mu CO_{2,gs}P_{gs,j}(n)+\mu CO_{2,\text{CHP},j}P_{\text{CHP},j}(n)+\mu CO_{2,\text{bo},j}(n)+\mu CO_{2,\text{Diesel},j}P_{\text{Diesel},j}(n)\right]\Delta t\right\}$$

A combination of the two through a weighted sum of the two objectives can also be adopted to account for energy cost and environmental impact at the same time. However, in the next section a more general method based on two levels of optimization exploiting multiple equipment profiles will be introduced.

The electrical and thermal load balances are given by the following equations [see also (Moghaddam et al., 2016)]:

$$\sum_{j=1}^{B}\left[P_{\mathrm{CHP\_el},j}(n)+P_{\mathrm{renew},j}(n)+P_{\mathrm{grid\_buy},j}(n)\right]=\sum_{j=1}^{B}\left[L_{\mathrm{el},j}(n)+P_{\mathrm{grid\_sold},j}(n)\right]$$

$$\sum_{j=1}^{B}\left[P_{\mathrm{CHP\_th},j}(n)+P_{\mathrm{bo},j}(n)\right]=\sum_{j=1}^{B}\left[L_{\mathrm{th},j}(n)+P_{\mathrm{sto},j}(n)+P_{\mathrm{waste},j}(n)\right]$$

Fig. 4.18 shows the architecture and interactions for the centralized energy optimization algorithm.

The inputs and outputs of the centralized energy optimization algorithm are illustrated in Fig. 4.18. The day-ahead energy price is used together with the thermal and electrical building loads to determine the set-points of local generation and (electrical) storage. Depending on the level of power generated and stored, the amount of grid demand is also determined.

Examples of the application of the proposed optimization formulation are shown in Figs. 4.19–4.22 for two buildings of the Cork Institute of Technology Neighborhood. They refer to the energy cost minimization case, where the optimizer computes

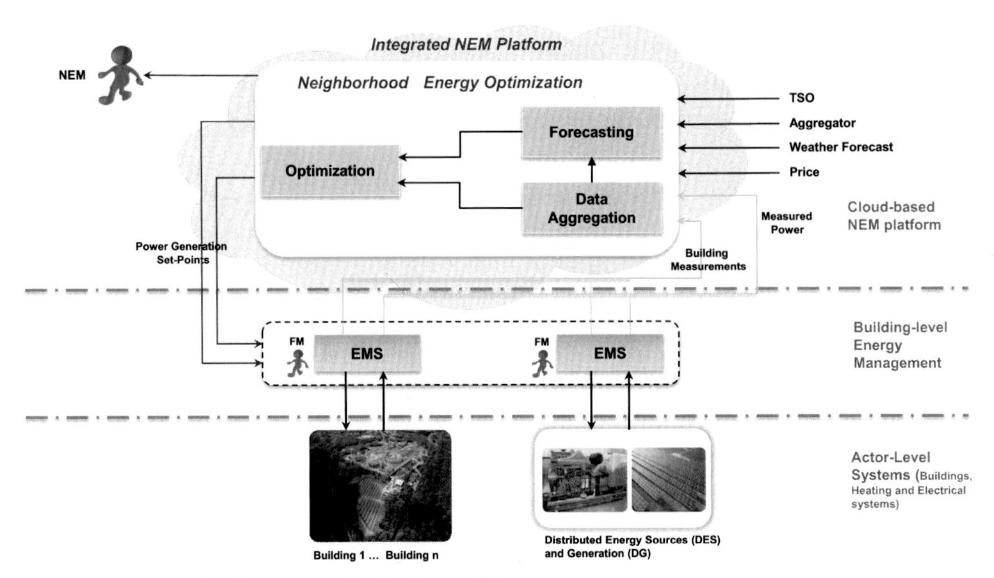

Figure 4.18 ***Centralized energy optimization architecture.***

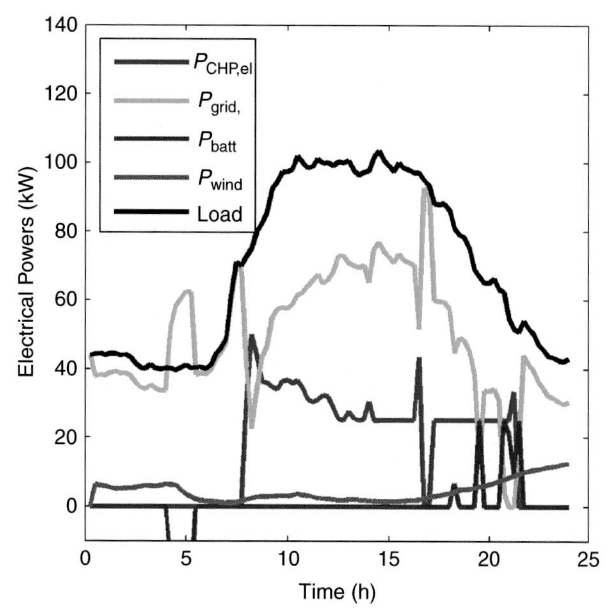

Figure 4.19 ***Nimbus Electrical Supply Optimization.***

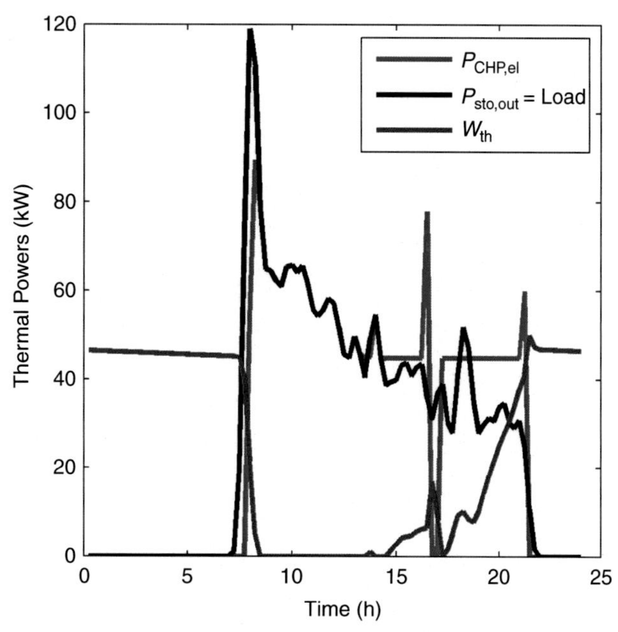

Figure 4.20 ***Nimbus Thermal Supply Optimization.***

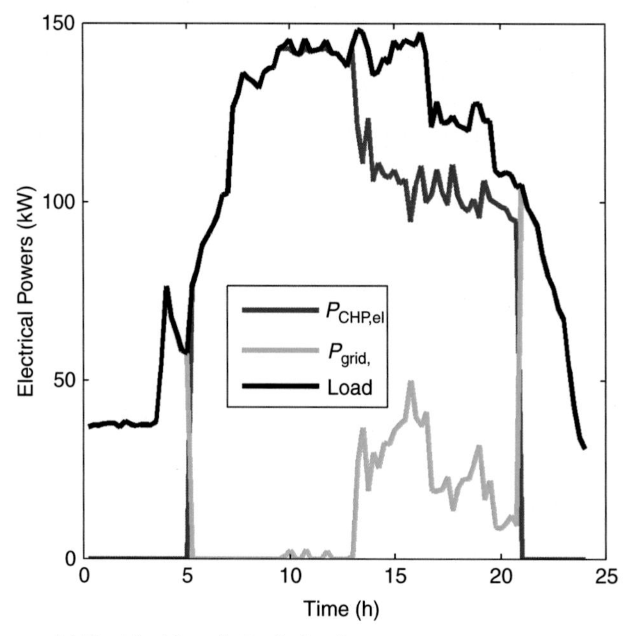

Figure 4.21 ***Leisureworld Electrical Supply Optimization.***

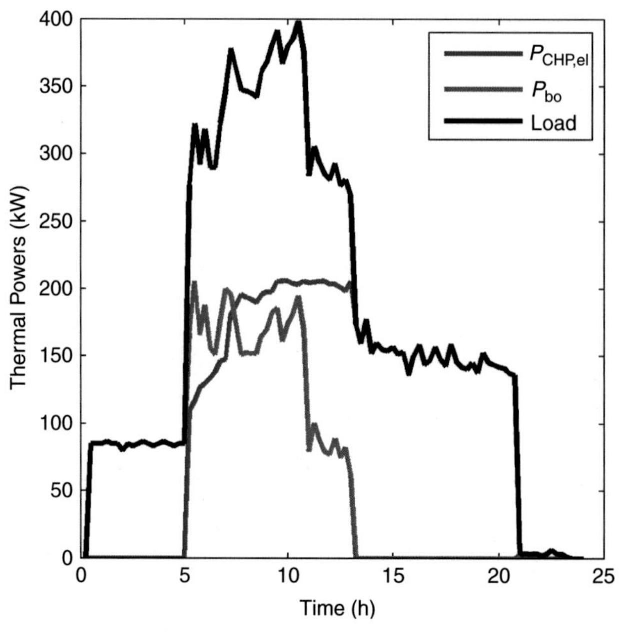

Figure 4.22 ***Leisureworld Thermal Supply Optimization.***

schedules of CHP, boiler, and battery accounting for electrical and thermal load forecasts as well as wind power forecasts. More results from the algorithm demonstration will be discussed in Chapter 7.

4.3 Hierarchical Two-Level Optimization

In the multiowner case, different subjects own the Neighborhood buildings and each of them wants to minimize their own energy cost. In addition, the building owners agree to offer flexibility to the whole Neighborhood and operate their own local generation and storage units in a coordinate manner such that Neighborhood-level objectives are fulfilled. The energy optimization architecture is of hierarchical type (Fig. 4.23).

With the multiowner type of Neighborhood, an algorithm based on the schedule flexibility concept was designed that does not necessarily require sharing the consumption or generation information with the Neighborhood controller.

Recent papers about microgrid optimization (Kanchev et al., 2014; Ma, Kelman, Daly, & Borrelli, 2012; Hui, Anwei, Weijun, & Rongjia, 2012; Parisio & Glielmo, 2012; Afshar, Moravej, & Niasati, 2013; Mohamed & Koivo, 2007) also consider the microgrid pollution emissions. In particular, Parisio and Glielmo (2012) report the application of multiobjective optimization for simultaneous optimization of energy cost and pollution emissions. Two formulations are proposed:

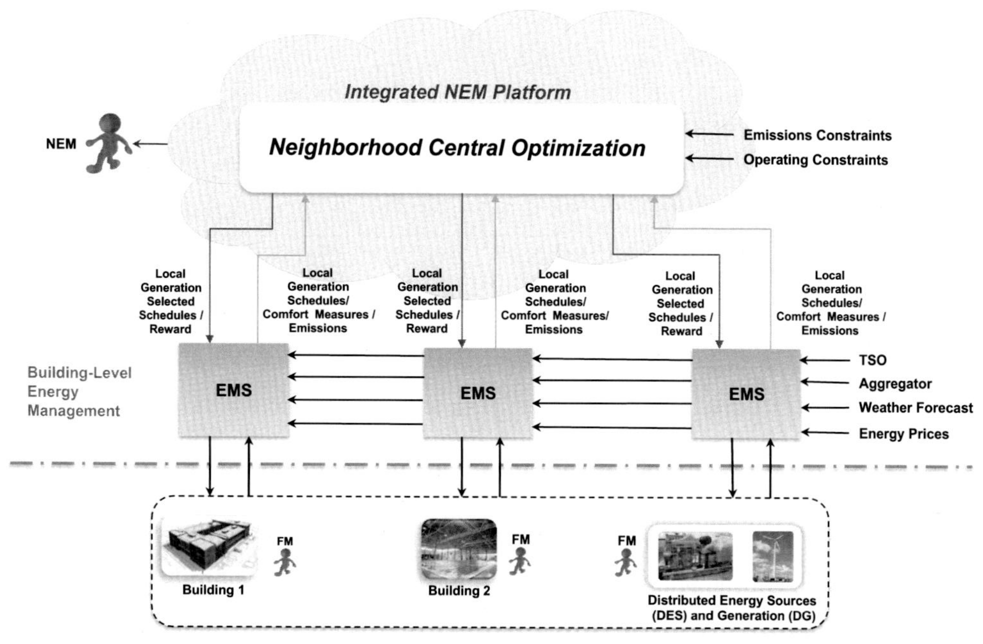

Figure 4.23 ***Hierarchical two-level energy optimization architecture.***

- The first one is the minimization of a weighted sum of the economic objective and emission objective.
- The second one is the minimization of the economic objective under an emission constraint.

We argue that with the first criterion the objective functions may recede their economic meaning, whereas with the second criterion the optimization lacks flexibility and relies on a so-called "hard constraint."

A general energy optimization algorithm applicable to a Neighborhood:

- Features a decentralized optimization for single actors according to their own goals (first layer) that minimizes the single actor's operating cost as long as they can offer some flexibility to the centralized optimizer (second layer).
- Implements a centralized optimization (second layer) that uses this flexibility as well as all other known parameters to target the Neighborhood objective (e.g., maximize the Neighborhood average comfort level and fulfillment of a CO_2 constraint).
- Selects equipment schedules within the flexibility offered by the decentralized optimizations and communicates them back to the decentralized level.
- Accounts for both the economic criterion and environmental impact (energy cost and CO_2 emissions).
- Is implemented as energy allocation optimization service in the NEM Platform as a prototype, it can provide a recommendation to each building owner on which set of schedules to implement in order to fulfill top-level goals.

Using the mathematical notations and descriptions from Section 4.1 the centralized optimization performed at building level (first layer) is formulated to minimize the single actor's operating energy cost and determine suboptimal sets of local generation schedules to offer a flexibility to the Neighborhood-level optimizer (second layer), the first layer objective function is:

$$\min \left\{ \sum_{n=1}^{N} \overbrace{\left(C_{\text{gas}}(P_{\text{CHP},j}(n) + P_{\text{bo},j}(n)\right)\Delta t}^{\text{CHP+Boiler}} + \overbrace{C_{\text{Diesel}} P_{\text{Diesel},j}(n)\Delta t}^{\text{Diesel generator}} + \overbrace{C_{gb} P_{gb,j}(n)\Delta t}^{\text{Energy purchased from grid}} + \overbrace{C_{gs} P_{gs,j}(n)\Delta t}^{\text{Energy sold to grid}} \right\}$$

The electrical and thermal load balances per building are given by the following equations:

$$P_{\text{CHP\_el},j}(n) + P_{\text{renew},j}(n) + P_{gb,j}(n) + P_{nb,j}(n) = L_{\text{el},j}(n) + P_{gs,j}(n) + P_{ns,j}(n)$$
$$P_{\text{CHP\_th},j}(n) + P_{\text{bo},j}(n) = L_{\text{th},j}(n) + P_{\text{sto},j}(n) + P_{\text{waste},j}(n)$$

As an example of the application of the proposed framework, we consider the case of gas-fired heating systems where the whole building's thermal consumption is supplied by CHP and Boiler. The building owners agree to participate in the Neighborhood CO_2 emission reduction program by offering flexibility in their thermal loads. This means

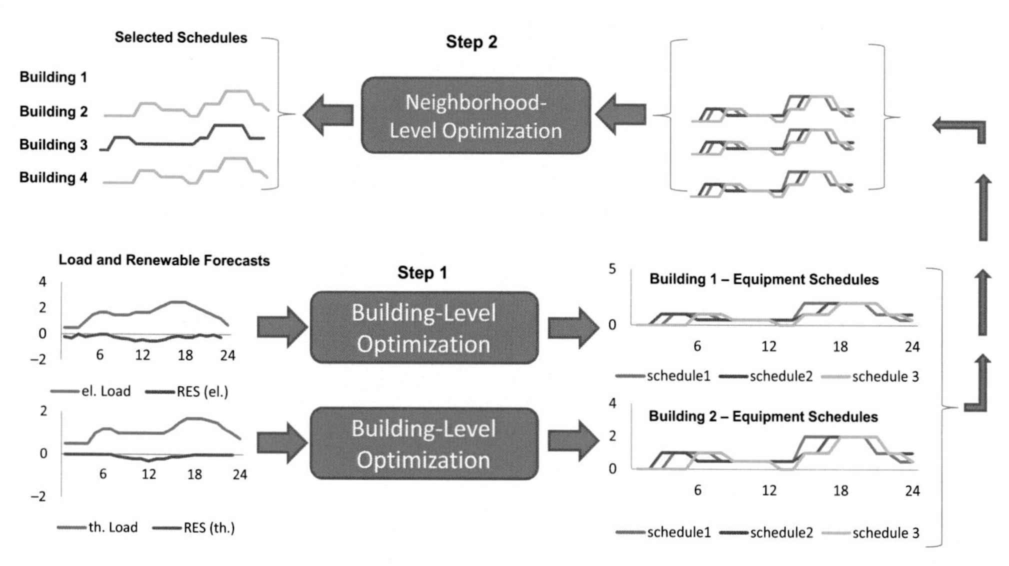

Figure 4.24 ***Two-level Neighborhood energy optimization algorithm.***

that relaxing thermal comfort requirements, such as the maximum temperature deviation around the zone temperature set-points, different thermal load profiles can be computed given the environmental conditions (outdoor temperature profile and building internal gain). Hence, each single actor can compute suboptimal CHP/Boiler schedules by using the slightly relaxed thermal load profiles to help the Neighborhood to achieve its CO_2 emission target. In order to facilitate the participation of each building, the Neighborhood CO_2 emission reduction program is rewarded. This general concept is illustrated in Fig. 4.24.

Each single actor suboptimal CHP/Boiler set of schedules can be associated with a measure of comfort C_{bs} (the level of comfort achieved in building b when the set of schedules s is selected) and CO_2 emissions E_{bs} (the emissions of building b when the set of schedules s is selected).

The optimization problem can be formulated as a combinatorial problem, the maximization of average building thermal comfort is given by:

$$C_{\text{neigh}} = \frac{1}{\sum \gamma_b} \max\left[\sum_{b=1}^{N_b} \gamma_b \left(\sum_{s=1}^{N_S} C_{bs} x_{bs}\right)\right]$$

Subject to:

$$\sum_{b=1}^{N_b}\sum_{s=1}^{N_S} E_{bs} x_{bs} < E_{\text{neigh,max}} \qquad \text{Neighborhood } CO_2 \text{ emissions constraint}$$

$$\sum_{s=1}^{N_S} x_{bs} = 1, \; b = 1 \ldots N_b \qquad \text{Schedule selection constraint}$$

The binary variables x_{bs} are introduced to select a certain set of schedules. In particular, when $x_{bs} = 1$ the set of schedules s is selected in the building b, whereas $x_{bs} = 0$ means that the set s has not been selected and a different one will be.

For completeness, building-dependent weighting factors γ_b were introduced in the objective function because buildings of a Neighborhood have typically different sizes and number of occupants, therefore one may want to weight the comfort therein accordingly.

If the pollution limits are sufficiently high, the optimizer will select the nominal load profiles such that the comfort is not penalized at all. As the limits become stricter, lower and lower profiles are selected. If the limits are too low, the optimization problem will be infeasible and the lowest profiles will be selected (maximum curtailment).

The concept of building schedule flexibility was first introduced in Molitor, Marin, Hernandez, and Monti (2013) to minimize power fluctuations between renewable sources and residential demand. In this work, an optimization framework is used to solve an energy management problem account for environmental constraints. In addition, the coordination algorithm selecting the equipment schedules was formulated as a binary integer programming, whereas in Molitor et al. (2013) it was formulated as brute-force search. This means that all the possible combinations were tested to obtain the solution of the optimization problem.

Concerning the data-privacy issue, the solution of the optimization only requires the knowledge of the comfort measures C_{bs} and corresponding emissions E_{bs} associated to the set of schedules S_i. The actual sets of building schedules do not have to be necessarily shared with the Neighborhood central optimization engine.

The flow of the hierarchical two-level optimization algorithm is illustrated in Fig. 4.25. The centralized, Neighborhood-level optimization is the selection of optimal schedules from the decentralized optimization outputs (building level). If the Neighborhood global goal cannot be fulfilled exploiting the first set of schedules, the optimization procedure can be iterated relaxing some constraints (e.g., allowing further curtailment of thermal load to reach a CO_2 Neighborhood emission target).

An example of the application of the proposed hierarchical algorithm is shown in Figs. 4.26–4.28 and Table 4.5. Different optimized profiles are generated optimizing the system with full and curtailed thermal load. Schedules blue, green, and red correspond respectively to 100, 95, and 90% thermal load. In Table 4.5 the selected schedules for different levels of the emission constraint limit are shown. It can be seen that the lower the limits the greater the curtailment of the load.

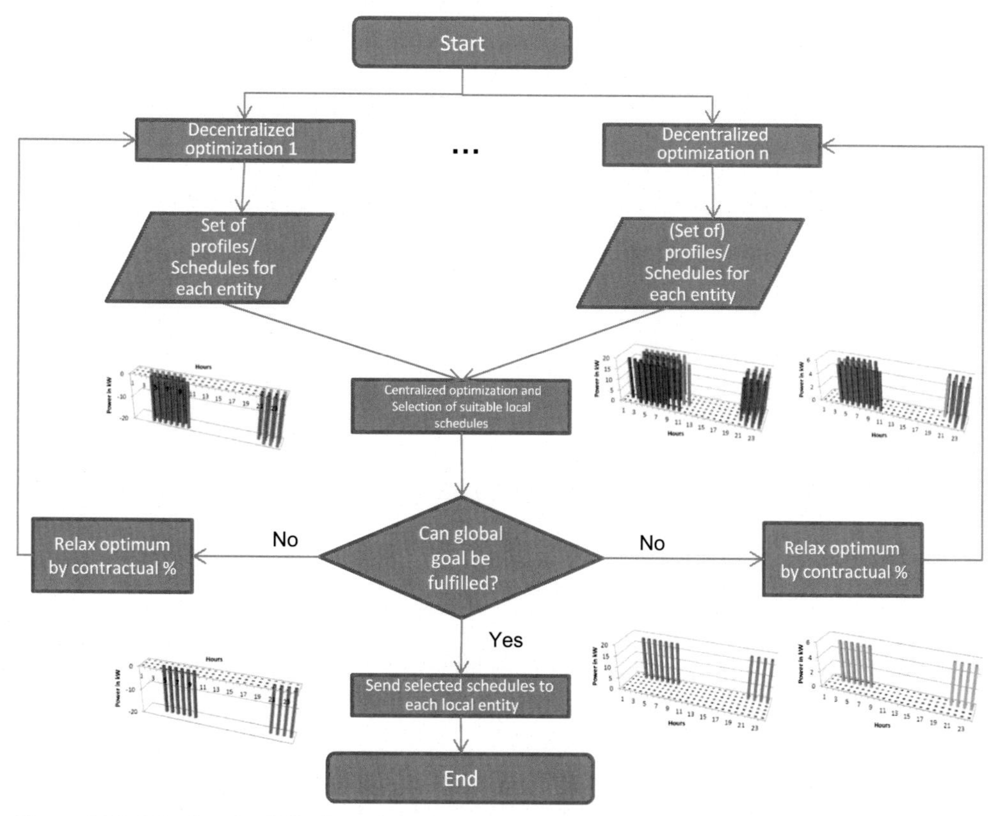

Figure 4.25 *Schedule flexibility based optimization algorithm.*

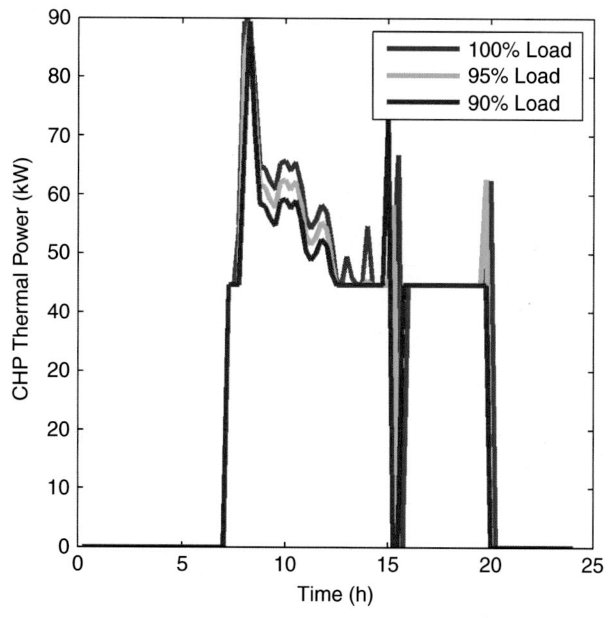

Figure 4.26 *Nimbus CHP Schedules.*

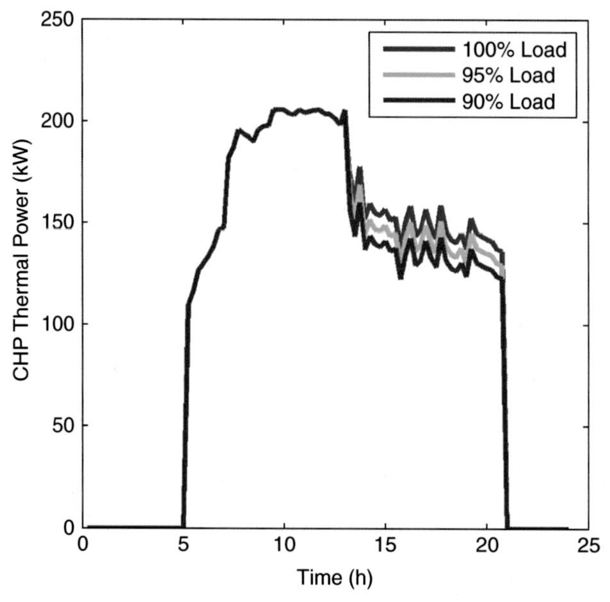

Figure 4.27 ***Leisureworld CHP Schedules.***

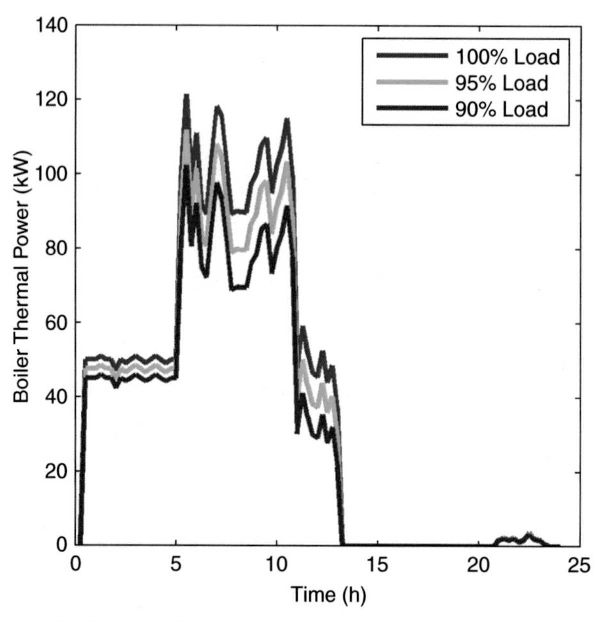

Figure 4.28 ***Leisureworld Boiler Schedules.***

Table 4.5 Neighborhood-Level Optimization using Building Schedules Flexibility

Neighborhood Total CO_2 Emission-Level Constraint		Selected Schedules	
		Building A CHP	Building B CHP/Boiler
1	<3000 kg/day		
2	<2300 kg/day		
3	<2000 kg/day		

5 SUMMARY

This chapter has introduced different schemas for energy optimization in Neighborhoods and a Modelica-based library for simulating Neighborhoods' energy systems. Section 3 presented a Modelica-based library which accounts for the different nature of numerous components at the Neighborhood energy system (thermal and electrical); this multiphysics library allows efficient simulation and testing of the optimization actions in Neighborhoods before actually deploying them on-site. Thus, it supports the field tests with more extensive assessments.

In Section 4 different optimization algorithms have been formulated to account for different levels of transparency in Neighborhoods: the centralized optimization for a single-ownership Neighborhood with a high level of transparency and the hierarchical two-level optimization for a multiple-ownership Neighborhood with a lower level of transparency. Moreover, the Neighborhood energy optimization algorithms perform an energy price driven scheduling of the generation and storage equipment to reduce the Neighborhood net-load seen from the grid side while keeping the Neighborhood pollution emissions below a given threshold. It has been shown that the usage of flexible resources, such as electrical storage or thermal storage (related to thermal comfort levels) aids to pursue an economic objective, while leveraging on the Neighborhood energy flexibility.

ACKNOWLEDGMENTS

The authors wish to acknowledge Dr. Kanali Togawa for the valuable work and contributions to this topic and the EU FP7 Project COOPERaTE.

REFERENCES

Afshar, H., Moravej, Z., & Niasati, M. (2013). Modeling and optimization of microgrid considering emissions. *Smart Grid Conference (SGC)*, *1*, 225–229.

Bertsekas, D. P. (2008). *Nonlinear programming* (2nd ed.). Belmonth, Massachusetts: Athena Scientific.

Chaouachi, A., Kamel, R. M., Andoulsi, R., & Nagasaka, K. (2013). Multiobjective intelligent energy management for a microgrid. *IEEE Transactions on Industrial Electronics*, *60*(4), 1688–1699.

Chung, I. -Y., Wenxin, L., Cartes, D. A., & Schoder, K. (2008). Control parameter optimization for a microgrid system using particle swarm optimization. *IEEE International Conference on Sustainable Energy Technologies*. ICSET 2008, pp. 837–842.

Demiray, T. (2008). Simulation of power system dynamics using dynamic phasor model. *Dissertation submitted to the Swiss Federal Institute of Technology Zurich, Department of Information Technology and Electrical Engineering., ETH Zurich* , Zurich, Switzerland.

Deng, Q., Gao, X., Zhou, H., & Hu, W. (2011). System modeling and optimization of microgrid using genetic algorithm. *Second International Conference on Intelligent Control and Information Processing (ICICIP), Vol. 1*, pp. 540–544.

Gamez Urias, M. E., Sanchez, E. N., & Ricalde, L. J. (2015). Electrical microgrid optimization via a new recurrent neural network. *IEEE Systems Journal, 9*(3), 945–953.

Ghosh, T. K., & Prelas, M. A. (2011). *Energy resources and systems, Vol. 2: Renewable resources*. Dordrecht, Heidelberg, London, New York: Springer.

Huchtemann, K., & Müller, D. (2009). Advanced simulation methods for heat pump systems. In: *Proceedings of seventh Modelica conference*. pp. 798–803, Como, September 2009.

Hui, R., Anwei, X., Weijun, T., & Rongjia, C. (2012). Economic optimization with environmental cost for a microgrid. *2012 IEEE Power and Energy Society General Meeting*, 1–6.

Kanchev, H., Colas, F., Lazarov, V., & Francois, B. (2014). Emission reduction and economical optimization of an urban microgrid operation including dispatched PV-based active generators. *IEEE Transactions on Sustainable Energy, 5*(4), 1397–1405.

Lauster, M., Streblow, R., & Müller, D. (2012). Modelica-Bibliothek und Gebäudemodelle. Tagungsband Symposium: Integrale Planung und Simulation in Bauphysik und Gebaeudetechnik, Dresden.

Ma, Y., Kelman, A., Daly, A., & Borrelli, F. (February 2012). Predictive control for energy efficient buildings with thermal storage. Modeling, simulation and experiments. *IEEE Control Systems Magazine, 32*(1), 44–64.

Mattavelli, P., Verghese, G., & Stankovic, A. (1997). Phasor dynamics of thyristor-controlled series capacitor systems. *IEEE Transactions on Power Systems, 12*(3), 1259–1267.

Moghaddam, I. G., Saniei, M., & Mashhour, E. (2016). A comprehensive model for self-scheduling an energy hub to supply cooling, heating and electrical demands of a building. *Energy, 94*, 157–170.

Mohamed, F. A., & Koivo, H. N. (2007). Online Management of Micro-Grid with Battery Storage Using Multiobjective Optimization. *International Conference on Power Engineering, Energy and Electrical Drives*. POWERENG 2007, pp. 231–236.

Molitor, C., Marin, M., Hernandez, L., & Monti, A. (2013). Decentralized coordination of the operation of residential heating units. *Fourth IEEE PES Innovative Smart Grid Technologies Europe (ISGT Europe)*, Lyngby, pp. 1–5.

Parisio, A., & Glielmo, L. (2012). Multi-objective optimization for environmental/economic microgrid scheduling. *IEEE International Cyber Technology in Automation, Control, and Intelligent Systems (CYBER)*, Bangkok, pp. 17–22.

Parisio, A., Rikos, E., & Glielmo, L. (2014). A model predictive control approach to microgrid operation optimization. *IEEE Transactions on Control Systems Technology, 22*(5), 1813–1827.

Peng, L., & Duo, X. (2014). Optimal operation of microgrid based on improved binary particle swarm optimization algorithm with double-structure coding. *2014 International Conference on Power System Technology (POWERCON)*, pp. 3141–3146.

Rahbar, K., Xu, J., & Zhang, R. (2015). Real-time energy storage management for renewable integration in microgrid: an off-line optimization approach. *IEEE Transactions on Smart Grid, 6*(1), 124–134.

Rosato, A., & Sibilio, S. (2012). Calibration and validation of a model for simulating thermal and electric performance of an internal combustion engine-based micro-cogeneration device". *Applied Thermal Engineering, 45–46*, 79–98.

Skoplaki, E., & Palyvos, J. A. (2009). *On the temperature dependence of photovoltaic module electrical performance: A review of efficiency/power correlations*. Solar Energy, Elsevier. Available from http://www.journals.elsevier.com/solar-energy/.

Stinner, S., Schumacher, M., Finkbeiner, K., Streblow, R., & Müller, D. (2015). FastHVAC : a Library for Fast Composition and Simulation of Building Energy Systems. *The Eleventh International Modelica Conference*, France, 2015.

Wagner, A. (2010). *Photovoltaik Engeneering: Handbuch für Planung, Entwicklung und Anwendung* (3rd ed.). Dordrecht, London, New York: Springer.

Xiaohong, G., Zhanbo, X., & Qing-Shan, J. (2010). Energy-efficient buildings facilitated by microgrid. *IEEE Transactions on Smart Grid, 1*, 243–252.

CHAPTER FIVE

Information and Communication Technology for EPN

K.A. Ellis*, D. Pesch**, M. Klepal**, M. Look†, T. Greifenberg†, M. Boudon‡, D. Kelly*, C. Upton*
*IoT Systems Research Lab, Intel Labs, Intel Corporation, Ireland
**NIMBUS Centre for Embedded Systems Research, Cork Institute of Technology, Cork, Ireland
†Software Engineering, RWTH Aachen University, Aachen, Germany
‡EMBIX, Issy-les-Moulineaux, Paris, France

Contents

1 INTRODUCTION

Our urban infrastructure, for example homes, commercial and industrial buildings, outdoor spaces, streets, transport infrastructure, and also our energy grids are heavily instrumented with a variety of monitoring and control systems. These systems have been embedded into our environment to manage energy flows and usage, control buildings and our homes, to control traffic and lighting in our streets and so on, thus helping to improve our daily life's. These monitoring and control systems have traditionally been standalone and vendor specific, for example, a building monitoring and control system manages just the building it is installed in, the street lighting system controls the light along just one street, traffic lights are controlled along one street or even only at a single junction, a parking control system manages one car park. There are many more examples of these soiled system that manage our environment. However, our increasingly digitalized world now offers opportunities to connect these individual systems via the Internet into a much larger system. The emergence of the Internet of Things (IoT) (Yan, Zhang, Yang, & Ning, 2008) and the continued digitalization trend, provide opportunities to create a system-of-systems (SoS) (Boardman & Sauser, 2006) beyond individual, isolated monitoring and control systems. The opportunity of networking individual systems into a bigger, SoS offers new business cases, that is opportunities to gather data, analyze it and to develop new services based on the availability of data from a broad spectrum of systems. In particular the opportunity to develop new energy management services based on a virtual neighborhood management platform that connects individual systems within a neighborhood and shares data motivated the EU funded FP7 COOPERaTE project. The project's focus was to develop concepts and services for Energy Positive Neighborhoods (EPN) as outlined in Chapter 2. COOPERaTE, like other initiatives focused on making urban neighborhoods more energy efficient and energy flexible, envisioned new services that would monitor energy use across the neighborhood and aggregate the information, services that manage energy consumption in individual buildings, forecast energy usage, optimize energy purchase between grid based energy supply and on-site generation, and demand response (DR) services that trade energy between buildings and the grid (Pesch, Ellis, Kouramas, & Assef, 2013). Value add services were also envisioned that link the building infrastructure to the outside spaces within an urban neighborhood to optimize traffic, parking, street lighting, and security.

However, the main road blocks to achieving an integrated IoT based SoS are the many different standards that are used for machine-to-machine (M2M) communication technologies, radio frequencies, networking protocols, data formats, management platforms, service formats, and security protocols. The plethora of technologies and formats has hindered integration across existing systems at both the system and the data layer. Techniques to overcome such issues are the focus of much current research and invariably the COOPERaTE project experienced some key challenges.

The first key challenge identified by the project was the issue of "Interoperability". Interoperability whether technical, syntactic, semantic, or organizational is arguably the number one barrier to IoT and SoS adoption. Section 2 focuses on identifying technology and standards to consider with respect to the connectivity and interoperability of resources and services within the built environment, smart city, smart grid sector, and more generally the IoT, which as Fig. 5.1 (Postscapes, 2015) and Fig. 5.2 (Carrez et al., 2013; Cisco, 2013; ITU-T, 2012; Intel, 2015; Open Interconnect Consortium, 2015; Industrial

Figure 5.1 ***IoT alliance round-up (Postscapes, 2015).***

Internet Consortium, 2015) suggests is a bourgeoning landscape of multiple standards, technologies, and alliances.

Heterogeneous data, low level communication protocols, and networks all pose interoperability challenges to be navigated. Uncertainty regarding which technology and standards should be applied act as a barrier to IoT adoption and any envisaged domain services. Once a data format, a communication protocol, and a transfer medium have been chosen, the next interoperability question arises, semantics. Like before there are many existing standards with different advantages and disadvantages that has to be decided upon. The COOPERaTE approach is to integrate these heterogeneous standards while minimizing changes to existing systems.

The second key challenge identified by the project relates to "Trust, Security, and Privacy". Trust, Security, and Privacy has become a huge challenge in the digitalized world. This is due to the increasing amount of available data containing personal information, and the cloud computing techniques used to cope with same. Fig. 5.3, is taken from an IDC published report (Bradshaw, Folco, Cattaneo, & Kolding, 2012), "estimates of the demand for cloud computing in Europe and the likely barriers to take up", while Fig. 5.4 (Bonagura, Folco, Kolding, & Laurini, 2012) is a follow on analysis of the same report. The data shown in Fig. 5.3 suggests that amongst those stakeholders surveyed, "data location" and "security" were the highest ranking barriers to cloud adoption with almost half of all stakeholders rating "security" as the biggest barrier. Approaches to increase the trust in cloud computing are described in (Helmut Krcmar, Ralf Reussner, & Bernhard Rumpe, 2014). Interestingly "interoperability" and "standardization" where

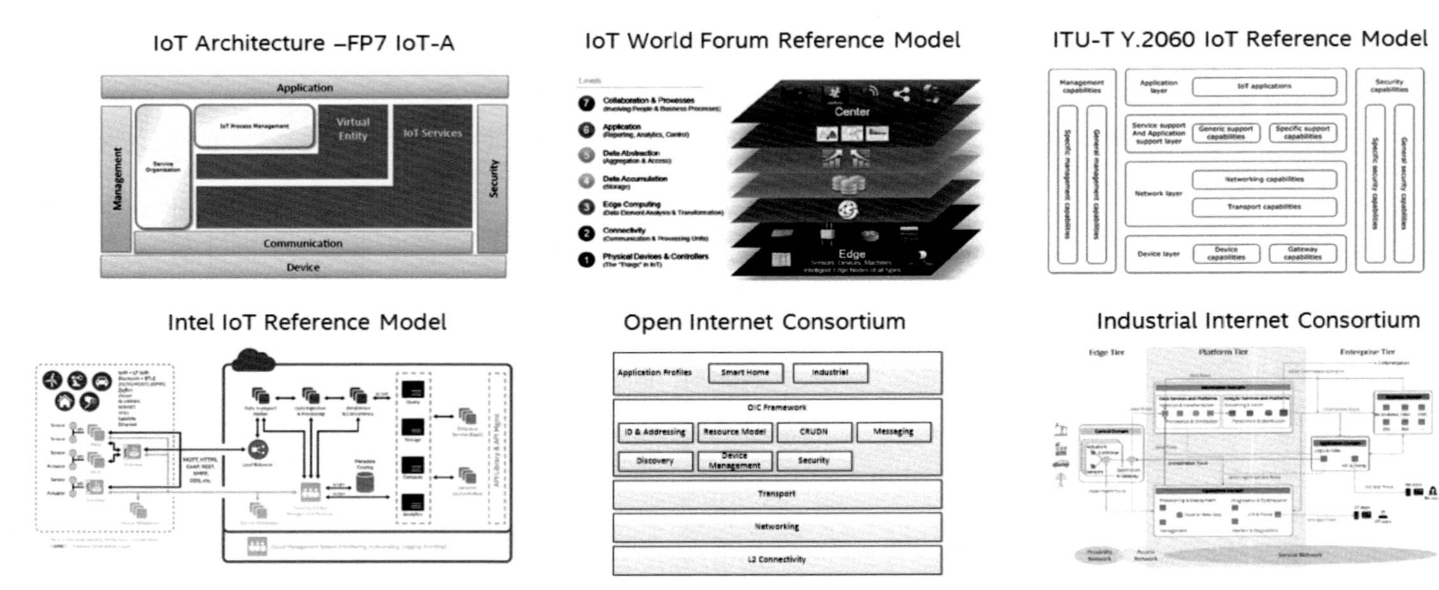

Figure 5.2 ***IoT reference architecture exemplars (Carrez et al., 2013; Cisco, 2013; ITU-T, 2012; Intel, 2015; Open Interconnect Consortium, 2015; Industrial Internet Consortium, 2015).***

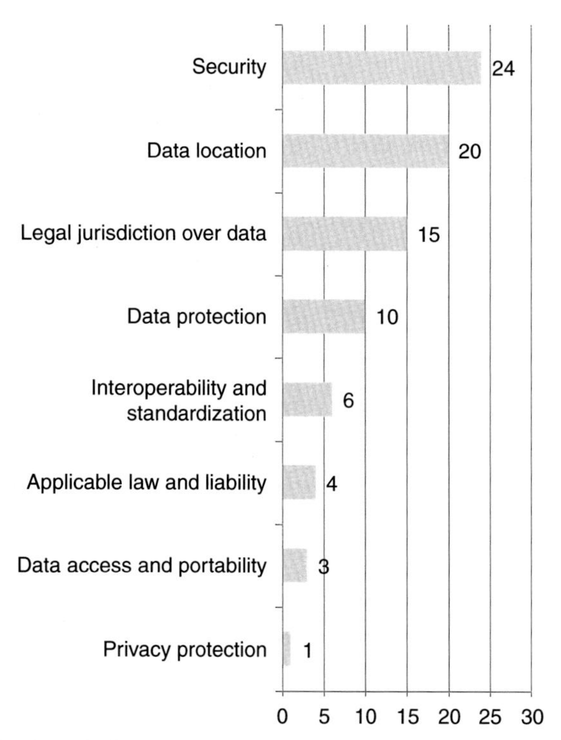

Figure 5.3 ***Stakeholder identified main barriers to adoption (Open Interconnect Consortium, 2015).***

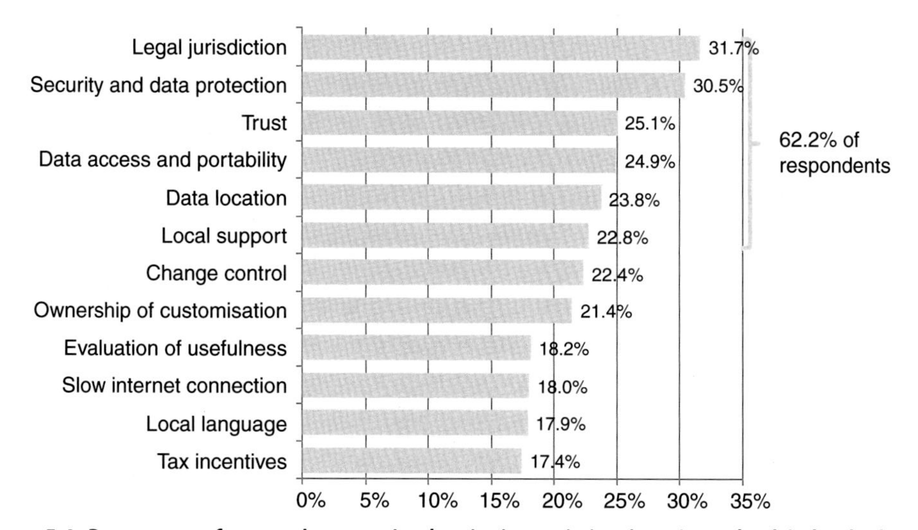

Figure 5.4 ***Percentage of respondents stating barrier is restricting (very/completely) cloud adoption (Bonagura et al., 2012).***

also identified in the IDC report. Fig. 5.4 illustrates that 62.2% of survey respondents mentioned at least one of the top six concerns, which indicates quite strongly that trust, security, and privacy are significant barriers to adoption.

While the reports (Bradshaw et al., 2012; Bonagura et al., 2012) are cloud specific they can arguably be taken as a reasonable proxy as to the importance of security, privacy, and interoperability to the ICT solutions investigated in the context of EPN. Whereby the solutions investigated are a combination of traditional embedded monitoring and control systems and cloud computing based data services, that is, solutions increasingly being described as IoT systems. Within the COOPERaTE project this challenge is set against the context of an energy sector that tends to be conservative in terms of ICT technology adoption.

The third key challenge identified by the project relates to "Complexity" and the impact that has on federation of data. This challenge essentially goes hand-in-hand with the interoperability challenges. The respondents of (Bradshaw et al., 2012; Bonagura et al., 2012) are in the main considering adoption from an organizational perspective. Delivery of an Energy Positive Neighborhood (EPN) is a far more complex brownfield scenario with heterogeneous systems and multiownership. The technologies described in Section 2 are a mixture of well-established and increasingly adopted technologies that can form part of any proposed EPN solution. However, in the main, adoption to date has been within silos. The issue faced in the context of EPN is that scale is measured at the neighborhood level not at an organizational level. Information needs to flow between systems and actors and this compounds the trust, security, and privacy issues and critically, the interoperability challenge. Cloud technology is arguably fit for purpose today, in terms of delivering the scale, security, and flexibility required at the neighborhood level, but the challenge is more one of ownership and interoperability with legacy in situ systems and new increasingly distributed resources.

To tackle these challenges the COOPERaTE project adopted an interoperable data driven approach that mitigates some of these issues without forcing major changes to existing systems. However, before discussing the specifics of the approach, its necessary to first review the standards and technologies to consider when developing IoT/SoS solutions.

2 STANDARDS AND TECHNOLOGIES OF INTEREST

This section describes some of the technologies and standards to be considered when developing ICT/IoT/SoS solutions in general and also for the delivery of energy services at the building and neighborhood level in particular. Any given solutions will most likely be composed of technologies and standards at different conceptual levels. This analysis uses the Industrial Internet Consortium (IIC) (Industrial Internet Consortium, 2015) "Edge, Platform and Enterprise tiers" (Fig. 5.5) as a structural guide

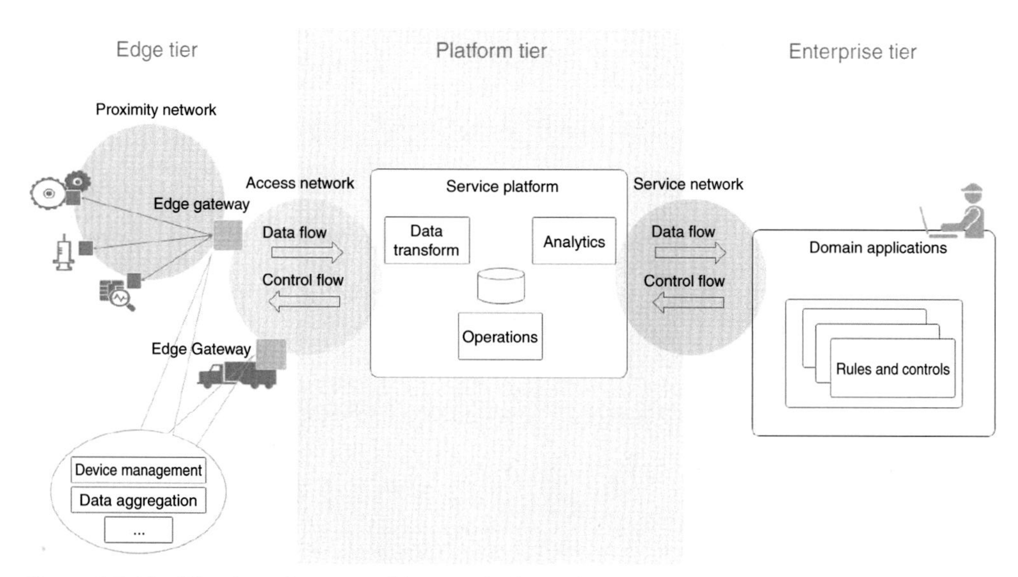

Figure 5.5 ***The IIC 3-tier reference architecture (Industrial Internet Consortium, 2015).***

to the conceptual position of different technologies. Additionally, the layers of the "IoT World Forum reference model" (Postscapes, 2015) were loosely used as a more granular structure/taxonomy as they cover all required functional elements.

The proximity network (Fig. 5.5) connects the sensors, actuators, devices, control systems, and assets, collectively called "*edge nodes*". It typically connects these nodes, as one or more clusters related to a gateway that bridges to other networks. The access network enables connectivity for data and control flows between the edge and the platform tiers. It may be a corporate network, or an overlay private network over the public Internet, or a 3G/4G network. The access network bridges edge compute/gateway functions with greater compute capacity, storage, and analytical tools. The architecture depicted in Fig. 5.5 largely assumes a *gateway-centric view* of the IoT. What this means is that edge nodes need not be "smart" as the gateway provides much of the required functionality and compute capacity. There are other views, for example a *cloud-centric view* whereby, individual edge nodes communicate directly into a remote cloud backend. An onpremises *or fog view* sees both backend compute and edge nodes geographically localized, perhaps also integrating with remote cloud compute. While a *distributed edge-centric view* effectively sees increased intelligence, control, and peer-to-peer edge communications. (http://www.wi-next.com/2015/03/living-cloud-gateway-edge-iots-fragmented-future/)

2.1 Edge Compute and Communication Technologies

The IoT involves "physical things/devices" for example, sensors, motors, fridges, fans, locks, lights, boilers, controllers, and so on. Such edge nodes are varied and voluminous.

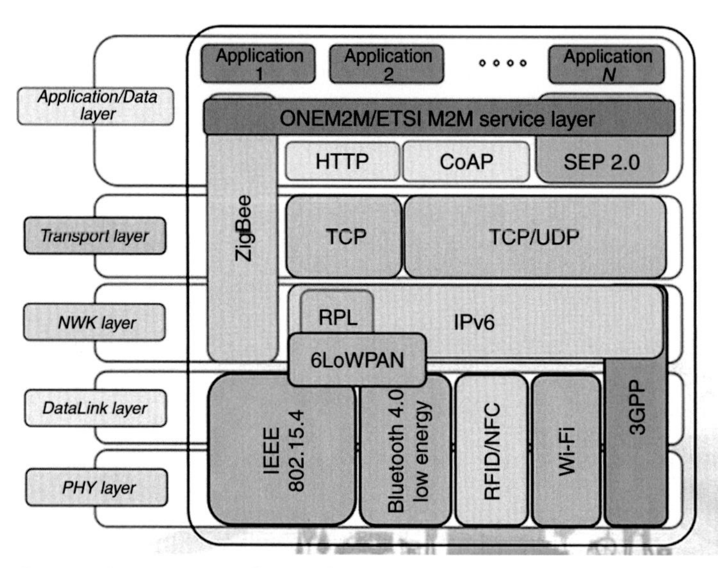

Figure 5.6 ***Communication protocol exemplars Source the Butler project (http://www.slideshare.net/butler-iot/butler-project-overview-13603599).***

As per Fig. 5.1 there are numerous alliance all of which utilize various M2M communication protocols in connecting the compute functions associated with such nodes. Examples of wired and wireless communications include Ethernet, Wi-Fi, IEEE 802.15.4 based (Z-wave, Zigbee, WirelessHart, 6LoWPAN), LoRa, DASH 7, Modbus, Profibus, RS232, and RS485, and many others. Fig. 5.6 (http://www.slideshare.net/butler-iot/butler-project-overview-13603599) outlines example protocols utilized by the alliances/architectures of Fig. 5.1 and 5.2 superimposed over the OSI 7 layer protocol stack model (Zimmermann, 1980). Some M2M communications like Zigbee are full protocol stacks which in this case is built on the IEEE 802.15.4 standard, but are commonly understood to relate to the connection of edge nodes.

Often within the built environment M2M communications are based on more industrial oriented protocols, such as PROFIBUS, PROFINET, IO-Link, LON, BACNet, Modbus, OPC-UA, FDI, ISA100.11a, HART, WirelessHART. Traditional IP technologies, such as IP, IPv6, or Ethernet (IEEE 802.3), and other specific communication technologies like Near Field Communication (NFC) and ultrawide bandwidth (UWB) have also been used more recently. Figs. 5.6 and 5.7 (https://entrepreneurshiptalk.wordpress.com/2014/01/29/the-internet-of-thing-protocol-stack-from-sensors-to-business-value/) highlight just how onerous a task it is to select an IoT solution given the number of options for communicating with physical things is reflective of the multitude of things to be connected, monitored, and controlled.

To date, the approach to deal with complexity has typically been to adopt a leading standard, or propose one, but this can be a labored process. Additionally, IoT type services

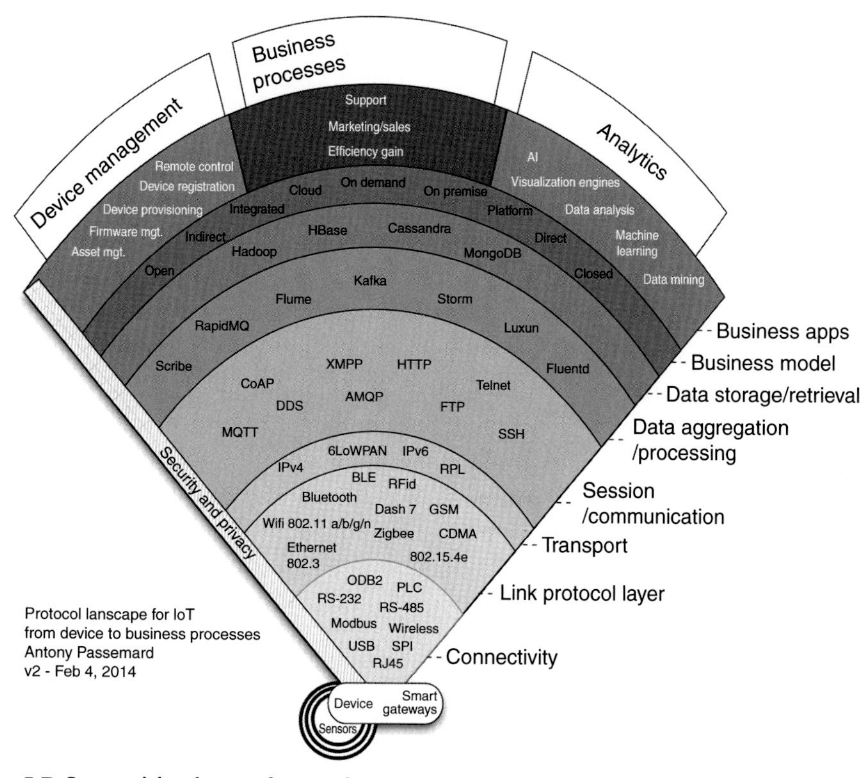

Figure 5.7 ***Protocol landscape for IoT from device to business processes, Antony Passemard v2, Feb.4,2014.(https://entrepreneurshiptalk.wordpress.com/2014/01/29/the-internet-of-thing-protocol-stack-from-sensors-to-business-value/).***

increasingly look to leverage multiple data sources to infer knowledge and insights. One could adopt a cloud-centric, gateway-centric, or distributed edge-centric approach as introduced previously, but invariably there will be a requirement to enable interoperability, that is, data exchange/translation needs to happen at some point. Whether that happens in a data center, more locally or in a truly distributed fashion is at the crux of the various approaches. The description of "edge gateways" and "abstraction layer technologies" that follow envisage a combined gateway and cloud-centric view, but increased computational power in the embedded space coupled with improvements in power envelope and communications means some aspect of distributed approaches are on the increase.

An *edge gateway* is a device which provides an entry point into enterprise or service provider core networks. Examples include routers, routing switches, integrated access devices (IADs), multiplexes, and a variety of metropolitan area network (MAN) and wide area network (WAN) access devices (https://en.wikipedia.org/wiki/Edge_device). Put simply, an edge gateway is a device that bridges two networks. It is considered to be at the edge of the networks because data must flow through it before entering either network.

Table 5.1 Edge hardware exemplars

x86 micro-PC	Raspberry Pi
Intel Galileo, Edison,	Building Automation Gateways
Intel DK series	Alcatel–Lucent HSG(Home Sensor Gateway)
Honeywell XYR300G	Lantiq GRX family

Edge gateways in this context can be data-center hosted or in-the-field devices, which (Fig. 5.5) connect the proximity network of the lower level wired/wireless edge nodes and the access network linking to larger scale compute, storage, and analytics services. Based on the latter description Table 5.1 lists examples of hardware that can act as gateways in the built environment.

The gateway itself is tasked with providing the requisite compute, memory, communication, and other interfaces required to link edge nodes (sensors and actuators) and the internet. It is a combination of hardware and software that enables what is essentially a translation/abstraction functionality, whereby the gateway links the data protocols of edge nodes and higher order compute nodes. As such, data abstraction technologies form an integral element of the gateway function.

As discussed previously, in IoT-enabled scenarios, invariably there is the requirement to enable interoperability between heterogeneous low level communication protocols. For example, in the built environment one might want to convert from Modbus RTU and say BACnet IP whereas in residential settings one might have Zigbee, Z-Wave, and Bluetooth sensors. Ideally one would want a gateway that could communicate/connect all three. A combination of hardware and software is required to do so. The gateway requires the communication hardware, but from a software perspective, some means of parsing the various data formats is required, as is some logic to decide what to do with the data. This can, of course, be done on a case by case basis, but that is not ideal. One means of addressing the issue is to utilize a "devise abstraction layer" type technology which acts as an extensible standard and handles data translation/exchange from different connecting protocols. Table 5.2 identifies some examples, many of which are OSGI (https://www.osgi.org/) Java based.

Table 5.2 Examples of device abstraction layer software frameworks

OIC—IoTivity	https://www.iotivity.org/
OpenHab Eclipse SmartHome	http://www.openhab.org/ http://www.eclipse.org/smarthome/
Thread	http://threadgroup.org/
ProSyst	http://www.prosyst.com/what-we-do/
Kura gateway	http://www.eclipse.org/kura/
HGI	http://www.homegatewayinitiative.org/
OPC-UA	https://opcfoundation.org/about/opc-technologies/opc-ua/
oBIX	http://www.obix.org/

Some proximity network communication protocols that is, Zigbee are meshed technologies and can cover a reasonable range within a building, however, the "access network", linking the edge tier, and the platform tier, typically utilizes fiber or cellular based technologies, such as GPRS, EDGE, HSPA, LTE, LTE+, or wireless metropolitan area network (WMAN) technologies such as WiFi (wireless mesh) and WiMAX. However, increasingly low-power wide-area network (LPWAN) technologies, for example, LoRA, SigFox, Neul, and WEIGHTLESS, are being seen as alternatives for IoT M2M based communications, both within the proximity network and to some extend at the access network level. LPWAN technologies are aimed at addressing high coverage and very low power. The technology is particularly suited to M2M use cases, where data rates can be low and infrequent. While cellular based solutions play an important role in today's M2M networks, they often have significant power requirements and are chatty. This type of overhead is unsuitable for many M2M applications. Moreover the current capacity limitations reduces the number of concurrent connections available at macro cells (Guibene, Nolan, & Kelly, 2015). LPWAN networks generally operate in the Sub-1GHz unlicensed spectrum, enabling longer range and building penetration and allowing installation without having to acquire a spectrum license. Moreover, LPWAN solutions which operate at lower frequencies can cover significant geographic distances when compared to high-bandwidth cellular technologies. Fig. 5.8 (Guibene et al., 2015) shows how LPWAN technologies (SigFox, Neul, and LoRa) compare to classical cellular technologies (3G/4G/5G) and to low power short range mesh technologies (IEEE802.15.4, ZigBee) in terms of latency, cost, power consumption, and geographical coverage (Guibene et al., 2015). The number of options at the access network hamper

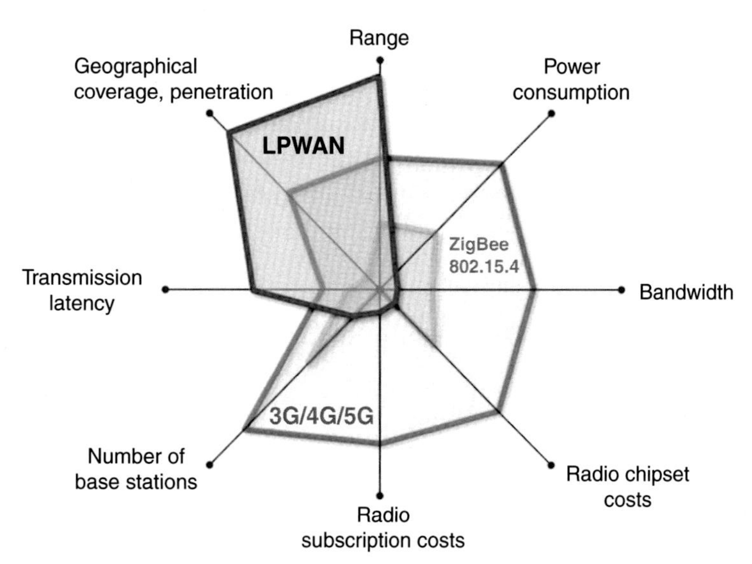

Figure 5.8 ***LPWAN target characteristics (Guibene et al., 2015).***

solution selection for domain actors as much needs to be considered in identifying the best technological fit.

2.2 Data Analytics Technologies, Frameworks, and Applications

Technologies and standards in the Platform tier are essentially tasked with providing access to what is typically described as "data storage", "data analytics", or "big data analytics" infrastructure and frameworks. As can be seen from Table 5.3, this tier is a myriad of some well-established and some emerging technologies and standards. Similar to the networking technologies, the spectrum of available technologies and solutions creates a challenge in identifying the best fit for a particular requirement. Therefore, a careful analysis of the requirements is required to identify which technology is the best fit for a particular application need.

2.3 Enterprise Technologies

In the pre-IoT era, Supervisory Control and Data Acquisition (SCADA) systems were the predominant approach taken to build automated monitoring and control systems. Traditional SCADA involves Programmable Logic Controllers, Telemetry Systems, Remote Terminal Units, and Human Machine Interfaces. These systems require custom design and deployment to a specific domain and the Human Machine Interfaces tended to be desktop native applications. Traditional SCADA is still widely used in industrial processes and facilities management, however, the growth of IoT as well as advances in predictive analytics and cloud computing has created demand for more flexible solutions. Modern commercial IoT based SCADA systems make better use of data modeling techniques to map sensed data back to control interfaces and take advantage of scalable cloud-hosted software-as-a-service (SaaS) to run more complex control algorithms. At the same time there has been growing demand for frameworks and software solutions that can merge traditional Business Intelligence (BI) functionality with that of SCADA systems. New trends are starting to emerge including:

Broader device support: Most reporting tools have been designed as native desktop applications but there is an increasing demand for mobile and touch screen support. While it seems unlikely that complex queries will be developed on mobile phone or tablet screens, the consumption of reports on mobile platforms is a common request. Furthermore, trends around increasing self-service BI and Visual improvements open up possibilities for touchscreen and gestural query composition.

Improved Visualization: Many solutions include charting components that allow users to generate basic charts from their query results, however users often struggle to get the views they required due to limited chart libraries and clumsily configured GUI's. Extended charting libraries with improved composition capabilities are expected.

Table 5.3 Big data analytical infrastructure and frameworks

Hadoop	
HDFS	For storing large datasets. Hadoop utilizes inexpensive hard drives in a very large cluster of servers. While one can expect failure on these drives, the Mean time to Failure (MTTF) is well understood. HDFS divides data into blocks and copies these blocks of data across nodes in the cluster, thus embedding built-in fault-tolerance and fault compensation within Hadoop
MapReduce	For processing large data sets. MapReduce is a model of programming for processing and generating large data sets utilizing a parallel, distributed algorithm on a cluster. The MapReduce framework marshals the distributed servers, running the various tasks in parallel, manages all communications and data transfers between the various parts of the system, and provides for redundancy and fault tolerance.
Pig	For analyzing large data sets. Is a platform for analyzing large data sets. It consists of a high-level language for expressing data analysis programs, coupled with infrastructure for evaluating these programs. The salient property of Pig programs is that their structure is amenable to substantial parallelization, which in turns enables them to handle very large data sets. Pig also allows the user to define their own user defined functions (UDFs).
Yet Another Resource Negotiator (YARN)	Supporting multiple processing models in addition to Map Reduce. Designed to address the tendency of MapReduce to be I/O intensive, with high latency not suitable for interactive analysis. Additionally, MapReduce was constrained in support for graph, machine learning (ML) and other memory intensive algorithms.
Zookeeper	Is a centralized service for maintaining configuration information, naming, providing distributed synchronization, and providing group services. Distributed applications utilize these kinds of services and they are typically difficult to implement. Zookeeper combines these services into a interface to a centralized coordination service. The coordinated service itself is distributed and highly reliable.
Apache Mesos	Is a cluster manager that abstracts CPU, memory, storage, and compute resources away from machines this enables fault-tolerant and elastic distributed systems to be managed. It is built similarly to the Linux kernel, but at a different level of abstraction. The Mesos kernel runs on every machine and provides applications with APIs for resource management and scheduling across cloud environments. It can be used by Hadoop, Spark, Kafka, and Elastic Search.
Cloudera Enterprise, IBM big Insights, EMS/Pivotal HD, Hotonworks	Enterprise distributions of Hadoop.

(*Continued*)

Table 5.3 Big data analytical infrastructure and frameworks (*cont.*)

Big data storage	
HBase, Cassandra	NOSQL Big table stores
CouchDB, MongoDB	NOSQL Document Based
Riak, Redis, HANA RDBMS, VoltDB RDBMS, OpenTSDB, KairosDB	Key Value and In-Memory Databases (both RDMS & NOSQL)
Neo4j	Graph Databases
Big Data Processing & Querying	
Hive	Is a data warehouse software which supports querying and management of large datasets residing in distributed storage. Hive provides a mechanism to project structure onto this data and query the data using a SQL-like language called HiveQL (HQL).
Apache Shark	Is a port of Apache Hive designed to run on Spark. It is still compatible with existing Hive data, megastores, and queries like HiveQL. The reason for the port is that MapReduce has simplified big data analysis but users want more complex analysis capabilities and multistage applications.
Apache Tajo	Data warehousing system on top of HDFS. Designed for low-latency and scalable ad-hoc queries, online aggregation, and ETL (extract-transform-load process) on large-data sets stored on HDFS and other data sources.
Apache Drill	Is a low latency SQL query engine for Hadoop and NoSQL. Drill provides direct queries on self-describing and semistructured data in files (such as JSON, Parquet) and HBase tables without needing to define and maintain schemas in a centralized store, such as Hive metastore.
Cloudera Impala	Is an open-source interactive SQL query engine for Hadoop. Built by Cloudera, it provides a way to write SQL queries against your existing Hadoop data. It does not use Map-Reduce to execute the queries, but instead uses its own set of execution daemons which need to be installed alongside your data nodes.
Apache Phoenix (for HBase)	Provides a relational database layer over HBase for low latency applications via an embeddable JDBC driver. It offers both read and write operations on HBase data.
Presto by Facebook	Is a distributed SQL query engine optimized for ad-hoc analysis. It supports standard ANSI SQL, including complex queries, aggregations, joins, and window functions.
Big Data Acquisition and Distributed Stream Processing	
Apache Samza	Is a distributed stream processing framework. It uses Apache Kafka for messaging, and apache Hadoop Yarn which provides fault tolerance, processor isolation, security, and resource management.
Apache Storm	Is a distributed real-time computation system for processing large volumes of high-velocity data. Storm on YARN provides real-time analytics, machine learning, and continuous monitoring of operations.

Table 5.3 Big data analytical infrastructure and frameworks (*cont.*)

Apache Spark Streaming	Uses the core Apache Spark API which provides data consistency, a programming API, and fault tolerance. Spark treats streaming as a series of deterministic batch operations. It groups the stream into batches of a fixed duration called a Resilient Distributed Dataset (RDD). This continuous stream of RDDs is referred to as Discretized Stream (DStream).
Apache Spark Bagel	Is a Spark implementation of Google's Prgel A System for Large-Scale Graph Processing. Bagel currently supports basic graph computation, combiners, and aggregators.
Typesafe ConductR and Akka Stream processing	ConductR is a Reactive Application Manager that lets Operations conveniently deploy and manage distributed systems. It utilizes Reactive Akka stream processing which is an open source implementation of the reactive streams draft specification. Reactive Streams provides a standard for asynchronous stream processing with nonblocking back pressure on the Java Virtual Machine (JVM).
Big Data Analytics Frameworks and Tools	
Apache Spark	Is a fast and general engine for large-scale processing. It provides in-memory processing for efficient data streaming applications while retaining the Hadoop's MapReduce capabilities. It has built-in modules for machine learning, graph processing, streaming, and SQL. Spark needs a distributed storage system and a cluster manager. Spark is quick and runs programs up to 100× faster than Hadoop MapReduce in memory, or 10× faster on disk
Apache Flink	Is an open source system for data analytics in clusters. It supports batch and streaming analytics, in one system. Analytical programs can be written in APIs in Java and Scala. It has native support for iterations, incremental iterations, and programs consisting of large Directed acyclic graphs (DAG) operations.
H2O	Is an open source big data analysis offering. Using the increased power of large data sets, analytical algorithms like the generalized linear model (GLM), or K-means clustering, are available and utilize parallel computing power, rather than by truncating data. Efficiency is achieved by dividing data into subsets and then analyzing each subset simultaneously using the same algorithm. Iteratively results from these independent analysis are compared, eventually convergence produces the estimated statistical parameters of interest.
Weka	This is a collection of data mining tasks algorithms that provide machine learning. It has tools for visualization, for data preprocessing, for classification, for regression, for clustering and for association rules. It facilitates the development of new machine learning schemes.
Massive Online Analysis (MOA)	Branched from Weka, designed for data streams and concept drift. It has APIs to interact with Scala and R.
RapidMiner and RapidMiner Radoop	The Rapidminer platform provides an integrated environment for machine learning, data mining, text mining, predictive analytics and business analytics. Rapidminer Radoop is a big data analytics system it provides visualization, analysis, scripting, and advanced predictive analytics of big data. It is integrated into RapidMiner on top of apache Hadoop.

(*Continued*)

Table 5.3 Big data analytical infrastructure and frameworks (*cont.*)

Apache SAMOA	Enables development of new machine learning (ML) by abstracting from the complexity of underlying distributed stream processing engines (DSPE). Development of distributed streaming ML algorithms can be done once and then can be executed on DSPEs. Such as Apache Storm, Apache S4, and Apache Samza.
Apache Spark Mlib	Is a scalable machine learning library. It consists of common learning algorithms and utilities, such as classification, dimensionality reduction, regression, clustering, collaborative filtering, and optimization primitives.
Apache Spark SparkR	SparkR is an R package that provides a light-weight frontend to use Apache Spark from R. Through the RDD class SparkR exposes the Spark API. Users can interactively run jobs from the R shell on a cluster.
Amazon, AWS ML, Microsoft Azure ML, Google prediction ML	Commercial machine Learning (ML) solutions.

Agile self-service Reporting: Traditionally data querying and analysis was carried out by the IT department or more recently by data scientists. There is an increasing demand from business users to be able to query data directly. Many advanced analytics platforms include visual query builders that support drag and drop query workflow building and charting. This codeless development will become more widespread. A range of new reporting and visualization solutions have appeared over the last decade. These range from programming libraries to dash boarding technologies to off the shelf applications. While off the shelf applications tend to provide simpler development and deployment, lower level programming libraries provide greater customization and integration capabilities. Solution providers should consider their choices accordingly.

Programming Libraries: In order to satisfy the trends of broad device support and self-service graphics many browser based java script frameworks have emerged, for example:

D3.js	http://d3js.org/
Processing.js	http://processingjs.org/
AM charts	https://www.amcharts.com/

Dashboard Solutions: A number of dashboard building frameworks exist that enable even inexperienced programmers or general users to build effective dashboards, examples are:

Freeboard.io	https://freeboard.io/
Dashing.io	http://dashing.io/
Finalboard.io	http://finalboard.com/

Commercial Applications: Finally, a number of online services exist that allow users to build online, browser based dashboards through drag and drop techniques, examples are:

Tableau	http://www.tableau.com/
Birst	https://www.birst.com/
Finalboard.io	http://finalboard.com/

2.4 Security and Privacy Considerations

Security and privacy is of increasing concern in the deployment of IoT systems as they connect the physical world to the cyber world. We are all familiar with security issues in the cyber world and the thought of cyber security issues translating into the physical world is worrying. Gathering data from the physical world also creates privacy issues in terms of linking data to people and their activities, data ownership, access to data, and so on. Current security and privacy issues around IoT are due to their:

- physically distributed nature
- mixture of very small to very large devices
- use of open or untrusted networks
- large scale deployments, which may extend to tens of thousands of components

They also have complex attributes due to their system complexity and SoS nature, such as

- different parts of the system may be created by different vendors (multiownership and multitenancy)
- use and functionality changes over the duration of the system's lifecycle

Table 5.4 outlines elements that should be incorporated at the development stage that is, security by design. Security needs a holistic approach from the physical up to network level and then service level authentication.

Table 5.5 outlines some of the many security standards that are used in wireless and wired communications and their applications. Any solution needs to be based on best

Table 5.4 EPN security considerations

Network aspects	Other aspects
• Firewall • Virtual private networks • Authentication	• Key management • Device attestation • Runtime controls • Stack simplification • Integrity measurement • Data encryption • Data authentication
Physical aspects	
• Device-specific cert • Trusted Platform Module Platform Configuration Registers • Secure boot • Physical access	

Table 5.5 Communication/Data Security standards

Encryption standards	
Triple-DES data encryption standard	Symmetric-key block cipher, which applies the original Data Encryption Standard (DES), which is now obsolete, cipher algorithm three times to each data block.
Advanced Encryption Standard (AES)	AES also known as Rijndael, is a specification for the encryption of electronic data established by the US National Institute of Standards and Technology (NIST) in 2001. AES is based on the Rijndael cipher developed by two Belgian cryptographers, Joan Daemen and Vincent Rijmen, who proposed it to NIST. Rijndael is a family of ciphers with different key and block sizes.
RSA	Named after Ron Rivest, Adi Shamir, & Leonard Adleman at MIT, RSA s one of the first practical public-key cryptosystems and is widely used for secure data transmission. In such a cryptosystem, the encryption key is public and differs from the decryption key which is kept secret.
OpenPGP	Pretty Good Privacy (PGP) is a data encryption and decryption computer program that provides cryptographic privacy and authentication for data communication. PGP is often used for signing, encrypting, and decrypting texts, e-mails, files, directories, and whole disk partitions and to increase the security of e-mail communications. It was created by Phil Zimmermann in 1991. PGP and similar software follow the OpenPGP standard (RFC 4880) for encrypting and decrypting data.
Wireless standards	
Wi-Fi Protected Access (WPA) WPA2/802.11i uses AES	Wi-Fi Protected Access (WPA) and Wi-Fi Protected Access II (WPA2) are two security protocols & security certification programs developed by the Wi-Fi Alliance to secure wireless computer networks. The Alliance defined these in response to serious weaknesses researchers had found in the previous system, Wired Equivalent Privacy (WEP). WPA2 became available in 2004 & is a common shorthand for the full IEEE 802.11i-2004 standard.
A5/1 cell phone encryption for GSM	A5/1 is a stream cipher used to provide over-the-air communication privacy in the GSM cellular telephone standard. It is one of seven algorithms which were specified for GSM use. It was initially kept secret, but became public knowledge through leaks and reverse engineering. A number of serious weaknesses in the cipher have been identified.
Transport Security	
Secure Socket layer	Cryptographic protocol designed to provide communications security over a computer.
Transport Layer Security	Evolved from SSL cryptographic protocol used to provide privacy and data integrity between two communicating computer applications. Symmetric cryptography is used to encrypt the data transmitted

practice incorporating aspects outlined in Table 5.4 and standards listed in Table 5.5. It can be arduous for domain actors in understanding the best approach when deploying an IoT system for their urban neighborhood.

Aside from technical security, "privacy by design" should also be considered as a standardized practice. Incorporating privacy aspects at the data model level should help with appropriate data lifecycle management. Additionally, developing interfaces and mechanism for allowing users to tag data in intuitive ways could offer a means of mitigating uncertainty regarding privacy legislation which can be a barrier to investment in IoT infrastructure. While good practice in terms of data minimization, that is, collecting, storing, and processing only data strictly required for a task, and anonymity, that is, removing individually identifiable data where not required, should be considered. Also the level of granularity should be questioned for example, The City of Amsetrdam's Energy Atlas project uses clustering as a means of protecting privacy will giving a suitable level of data granularity in order to be useful for services (http://amsterdamsmartcity.com/news/detail/id/186/slug/the-launch-of-the-energy-atlas-in-amsterdam-arena). Some relevant standards/initiatives (with varied degrees of adoption and/or recency) include:

- P3P—is a protocol allowing websites to declare their intended use of information they collect about web browser users. Designed to give users more control of their personal information when browsing, P3P was developed by the World Wide Web Consortium (W3C). However, it is not widely adopted. (https://www.w3.org/P3P/)
- XACML—the Extensible Access Control Mark-up Language together with its Privacy Profile is a standard for expressing privacy policies in a machine-readable language which a software system can use to enforce the policy in enterprise IT systems. (http://docs.oasis-open.org/xacml/3.0/xacml-3.0-core-spec-os-en.html)
- EPAL—the Enterprise Privacy Authorization Language is very similar to XACML. It is a formal language for writing enterprise privacy policies to govern data handling practices in IT systems according to fine-grained positive and negative authorization rights. It has been submitted by IBM to the World Wide Web Consortium (W3C) to be considered for recommendation. (https://www.w3.org/Submission/2003/SUBM-EPAL-20031110/)
- Open Authorization (OAuth)—OAuth is an authentication protocol that allows users to approve application to act on their behalf without sharing their password. (http://oauth.net/)
- Open Web Application Security Project: top 10 privacy project—The OWASP Top 10 Privacy Risks Project provides a top 10 list for privacy risks in web applications and related countermeasures. It covers technological and organizational aspects that focus on real-life risks, not just legal issues. The Project provides tips on how to implement privacy by design in web applications with the aim of helping developers and web application providers to better understand and improve privacy (https://www.owasp.org/index.php/OWASP_Top_10_Privacy_Risks_Project)

Within the COOPERaTE project the information regarding these issues are foreseen within the NIM data model. The data model offers storing information about authorization and authentication. The actual control is left to the overall architecture which may reuse the aforementioned described standards.

2.5 Semantics, Data Models, and Ontologies of interest

Relevant semantic and ontology standards can be divided into those which support the modeling and deployment of knowledge in cloud-based systems, and those which capture knowledge regarding the built environment, energy and intelligent sensing domains. Within the former category the most relevant standards include—the web ontology language (OWL) (OWL, 2015), the resource description framework (RDF) (RDF, 2016), the SPARQL protocol, and RDF query language (SPARQL, 2015) (SPARQL) and the semantic web rule language (SWRL) (SWRL, 2016). These standards are ICT specific, they to not pertain to any given domain.

The existing standards which capture knowledge in the built environment/smart grid and intelligent sensing domains form a plethora of languages, data models, and taxonomies, towards a variety of different purposes, and within different parts of a broad domain. Semantic models of interest include the common information model (CIM) (International Electrochemical Commission, In Press), the industry foundation classes (IFC) (BuildingSmart, 2016a), the city geography mark-up language (CityGML) (CityGML, 2016), the smart city concept model (SSCM) (BSI, 2014) and the semantic senor network (SSN) ontology (Compton et al., 2012).

The CIM has been adopted by the International Electrotechnical Commission (IEC) (IEC, 2016), and aims to allow the interoperation of software in electrical networks by facilitating data exchange. The CIM is natively expressed as a universal mark-up language (UML) model, but further IEC standards define extensible mark-up language (XML) serializations of the model, to allow their federation into an RDF format (BuildingSmart, 2016b). This is relevant as it acts as a benchmark for information models within utility companies, and whilst it is arguably not sufficient for next generation smart grids, much can be learnt from the CIM.

The industrial foundation class (IFC) is the data model which facilitates open building information modeling (open BIM), and has been accepted as an ISO standard (BuildingSmart, 2016a). IFC was primarily developed to facilitate information exchange between the design and construction phase of buildings. It is based on the EXPRESS schema, but can also be expressed in XML (BuildingSmart, 2016b), and ongoing research is enriching the IFC to an OWL model (BuildingSmart, 2016c). This model is also undergoing development to describe data and concepts from the broader role of connected buildings within urban environments.

CityGML has foundations in the geospatial field, and facilitates the visualization and exchange of data at the city level (CityGML, 2016). CityGML is an extension of the geography mark-up language (GML) for the purposes of specifically modeling cities to various levels of detail, where GML is an extension of XML for the purposes of modeling geospatial data. Several domain extensions are under development for CityGML, to allow standardized descriptions of data related to various "city domains", including utility networks. The smart city concept model (SCCM) is a conceptual model presented in BSI:PAS 182 (BSI, 2014), and outlines the concepts and relationships deemed most critical to the smart city field. The SCCM is somewhat domain neutral, in that it aims to remain a middle-level conceptual model, suitable for further development as the smart city modeling field matures. The SCCM also is not serialized in a normative manner; it is officially termed a "guide" rather than an actual model. Finally, the SSN ontology is an OWL model formalizing a language for the description of intelligent sensor networks, and utilizes significant abstraction and domain neutrality.

The energy efficient Building Data Model (eeBDM) community (https://webgate.ec.europa.eu/fpfis/wikis/display/eeSemantics/Home), is a European Commission (EC) supported community focused on semantic ontology development. The premise being that R&D focused on ICT integration and standardization will ultimately address interoperability issues between building subsystems, building to buildings and between buildings and other sectors, particularly smart utility grids (not just electric but also water, waste, heat/cool, etc...). It is envisaged that the introduction of ICT interoperability standards, models, and tools, that can consider all levels of complexity in energy management and optimization, will enable greater adoption of solutions.

Within the COOPERaTE project a meta model based approach is used in order to not enforce such a concrete standard. Moreover the ability to integrate the different data model standards is facilitated by the NIM.

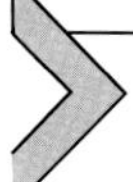

3 THE COOPERaTE APPROACH

3.1 An End-to-End Reference Model

As was discussed in the inroduction and illustrated in Fig. 5.2 there are multiple IoT architectural options. All offer a reference in terms of end-to-end implementation and all to some degree focus on enabling data acquisition, aggregation, persistence, analytics, presentation, or visualization. What follows describes a logical view of the COOPERaTE reference architecture and its logical components. Fig. 5.9 shows an overview of the COOPERaTE architecture components and logical scopes. The architecture is based on a service-oriented system's approach. From a technical perspective it could be a standalone platform offering services that can scale to the neighborhood level. In practice it is more likely to be a combination of existing systems that act in a transactional nature to deliver the range of functionality outlined in the architecture and required for neighborhood level services, the components outlined can be considered broadly reflective of those required within any architecture for delivering services at scale.

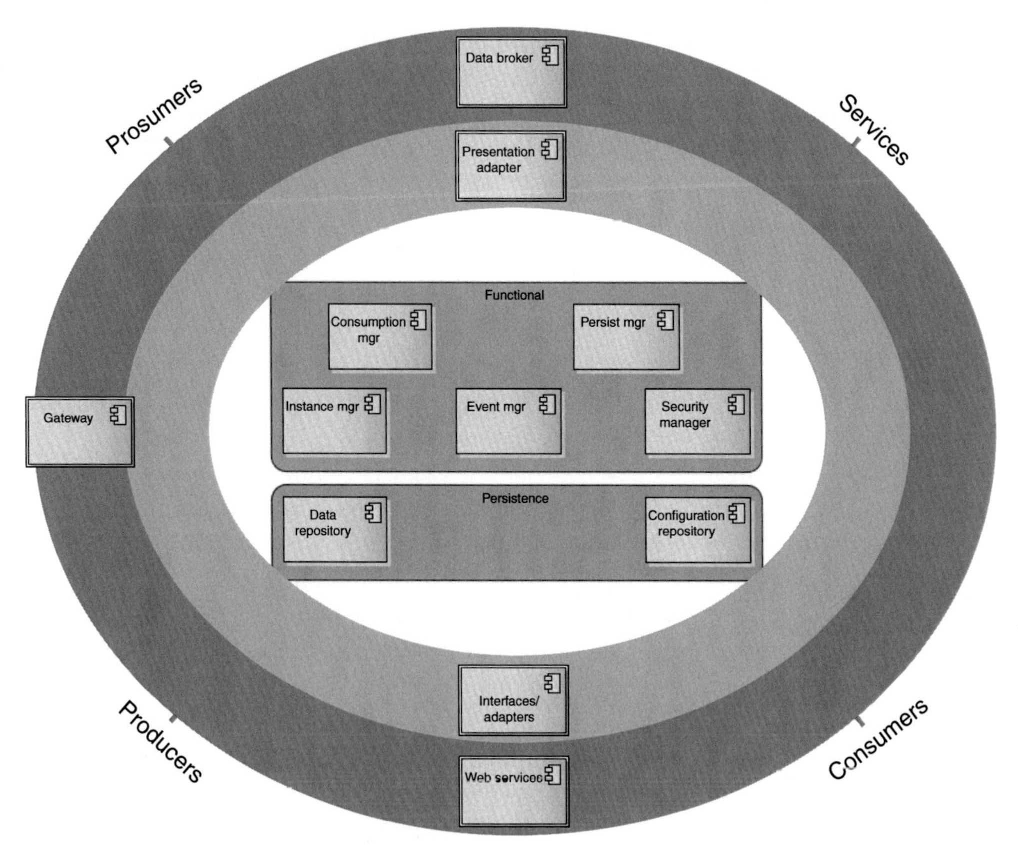

Figure 5.9 ***COOPERaTE Reference Architecture.***

Logically the components can be grouped into several scopes as follows:

- The Communication scope (blue ring) contains a broker, a web service component and a gateway focusing on message transport and management. This just means that some form of M2M communication is required whether that be publish/subscribe, RESTful or web-sockets based.
- The Integration and presentation scope (grey ring) contains a presentation component and an interface/adapter component focused on integration of external systems and on ensuring consumers, producers, and services can interact with the COOPERaTE services. For example, the gateway component straddles this scope and contains the required adapters and interfaces to communicate with the underling physical systems, in the case of a home environment this could be Z-Wave or Zigbee based or in a commercial building system it may be OPC based. Essentially the blue and grey rings mean that services, prosumers, producers, or consumers connect via a gateway, web service, or broker to the presentation layer or adapters, to access any of the components in the circle. This is consistent with the proximity and access network description of Section 2.

- The Functional scope focuses on logic and service and management components that is, platform tier of Fig. 5.5.
- The Persistence scope focuses on efficient data storage and retrieval, again the platform tier of Fig. 5.5.
- The security manager is listed as a functional component, but of course it is in practice a horizontal that is managed by an instance in each cooperating system of the SoS.

Each externally facing service defines the message types it can accept. In the case of COOPERaTE such services will conform to the NIM and/or subscribe to a NIM service (Section 4). If one service wishes to communicate with another service it must send a message which meets the requirements of the service which will receive the message or utilize a NIM or similar service that is cognizant of the underline systems data model.

Components are references and in practice the cooperating systems of the neighborhood host the required functionality while participating in NIM (or similar) enabled neighborhood wide services. Table 5.6 lists the COOPERaTE architecture components.

Table 5.6 COOPERaTE architecture components

Component	**Presentation Adapter PADT**
Functionality	The presentation Adapter allows targeted adapters for specific consumption clients. There may be several types of PADT and each PADT may have several Instances configured differently.
Component	**Consumption Mgr CMgr**
Functionality	The CMgr provides the initial default access point to the system for Actors/consumers. The consumption Mgr will manage and ensure best fit for consumers and presentation adapters.
Component	**Event Mgr EMgr**
Functionality	The EMgr facilitates the communication between *components via* messaging mechanisms negotiating service contracts. In essence it is responsible for event propagation, initial message route configuration and queuing. The EMgr is not involved in the actual communication between components but rather routing.
Component	**Security Mgr SMgr**
Functionality	The SMgr is responsible for ensuring authentication and authorization and integrating security provided by individual components within each layer.
Component	**Instance Mgr IMgr**
Functionality	The *Instance Mgr* is a component to manage the life-cycle of logic/adapters / interfaces instances and their configuration. It handles the initialization of adapters and logic components and the component instances with their corresponding configurations.

(*Continued*)

Table 5.6 COOPERaTE architecture components (*cont.*)

Component	**Persistence Mgr PMgr**
Functionality	The *PMgr* is responsible for storage, long-time archiving, retrieving and deleting of any kind of data. Accordingly, it allows access to all persisted configuration and historical data, which are stored respectively in the configuration and data repositories.
Component	**Data Repository DR**
Functionality	The database component provides historical data. Historical data is collected data in the past. This data could be results from the event processing or raw data (but not limited to these). Basically, the DR is used for pattern mining and other analysis
Component	**Configuration Repository CR**
Functionality	This component maintains Configuration data, which includes all data necessary to configure the internal components of the COOPERaTE system including COOPERaTE consumer profile information.
Component	**Interfaces/Adapters IA**
Functionality	Adapters support receiving notifications, events or data from external systems devices as well as querying of same. It is possible to forward data and trigger actions to external systems. Each adapter provides an interface that allows another Layer to query for data from external systems. Several types of adapter and several instance of each adapter type may exist.
Component	**Data Broker Br**
Functionality	The purpose of the Br is to deliver content/messages to the consumers that is actors or systems components.
Component	**Gateway G**
Functionality	The G component is responsible for providing a subset of the COOPERaTE system functionality but on a compute constrained device. The G will be utilized with respect to edge v cloud processing or rather a compute-centric approach whereby compute processing occurs where most appropriate.

3.2 A System-of-Systems Approach

An EPN is challenged with integrating the many local monitoring and control functions prevalent in urban neighborhoods with the power, flexibility, and scalability of cloud computing platforms in order to delivery energy and other valued services at the district level. This level of scale, Heterogeneity and multiownership is significant and is why end-to-end system approaches, such as those referenced in Section 3.1 are unlikely to be adopted. An EPN must acquire, aggregate, analyze, and act on data from disparate and dispersed sources, but must always do so on the basis of agreed cooperation from system owners. Hence an EPN must have ICT systems that allow for federation of information set against a hybrid-cloud landscape whereby public and private clouds, remote hosted and on-premise, can exchange data, while at the same time ensuring data privacy and security is maintained. In this regard, the overarching premise posited by COOPERaTE is that by

> *"allowing for interoperability at the data level and by leveraging existing solutions via common communication interfaces, one can produce a loosely coupled integrated solution that goes beyond the current state of the art and which is likely to be adopted given an existing brownfield reality".*

This is defined as a "System-of-Systems" (SoS) approach. This approach allows systems to both act independently and to interoperate based on an agreed semantic meaning. The COOPERaTE Neighborhood Information Model (NIM) (Look & Greifenberg, 2013a) is an example for achieving such common semantic meaning. The SoS approach is pragmatic in addressing barriers to adoption due to ownership, control, security, and privacy concerns within and across districts. The approach ensures flexibility whilst minimizing impact as individual downstream systems retain decision making capability that is, these systems may or may not take action based on suggestions provided upstream by NIM enabled logic. This approach is effectively a recommender system whereby recommendations are subject to associated business/service level agreements. Fig. 5.10 illustrates the SoS scenario whereby different systems control different physical assets within a neighborhood. Information regarding these different assets maybe required for EPN level services. However, there is often a reluctance to share information and more understandably an averseness to allowing any external system control those assets. Therefore the approach proposed here is to allow for federation of data and information whereby the existing systems continue to execute their established control and have the degrees of freedom to implement any aspect of the technologies discussed in Section 2. However, more importantly from an EPN perspective an approach that offers such systems the opportunity to cooperate in a neighborhood level process, by opting to carry out recommendations. The "red dashed line" of Fig. 5.10 represents the NIM type service, which is a proposed means of permitting this cooperation to happen. The NIM is presented in the following section.

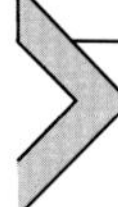

4 A NIM ENABLED SYSTEM-OF-SYSTEMS

4.1 Overview

The approach adopted in the COOPERaTE project in realizing a SoS was to focus on semantic interoperability at the data and service level, rather than insisting on a single standardized system architecture with selected M2M communication standards. The concept of the Neinghborhood Information Model (NIM), the integration of

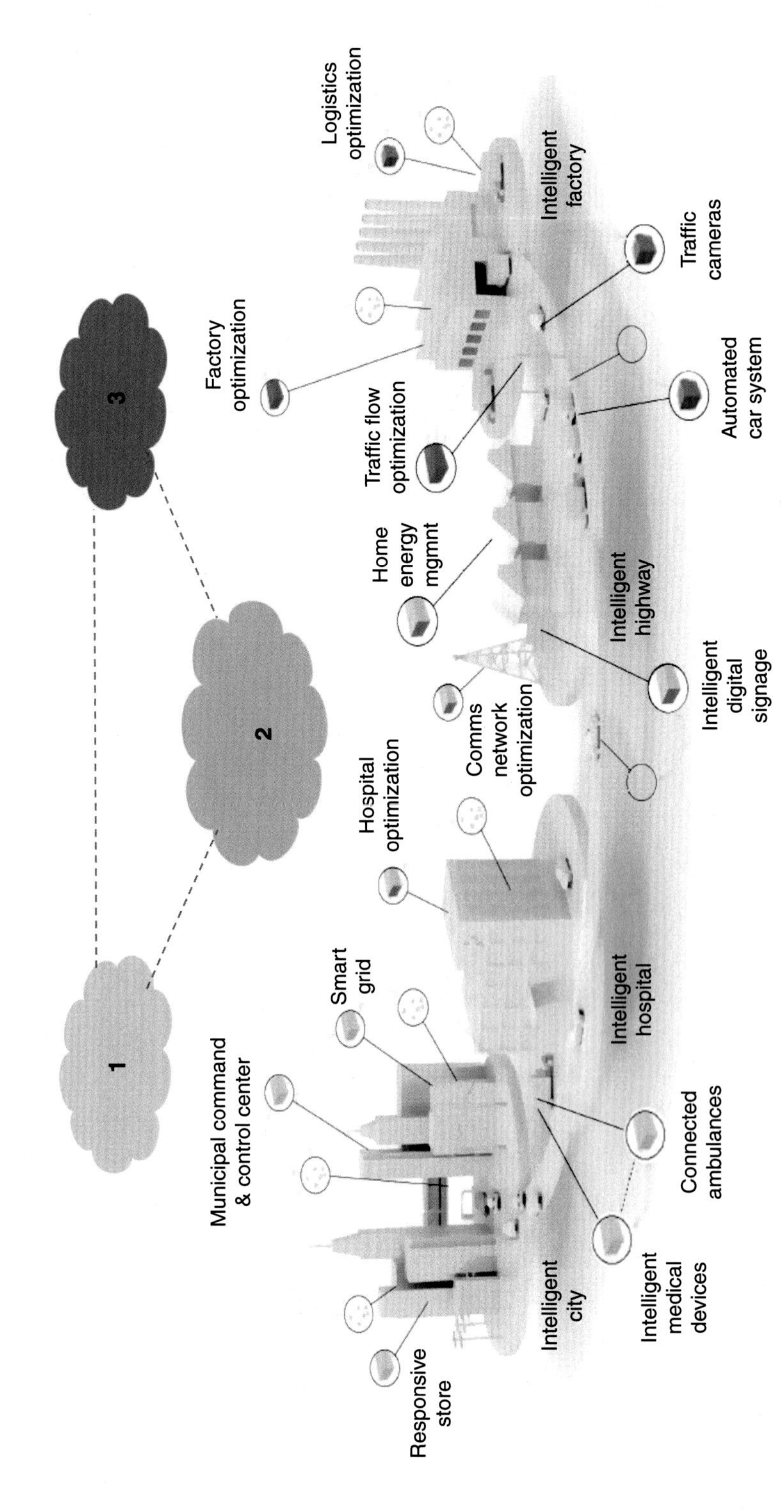

Figure 5.10 ***System-of-systems (SoS) approach to EPN.***

heterogeneous data models, the security and privacy issues as well as a prototype implementation has already been discussed in (Greifenberg, Look, Rumpe, & Ellis, 2014).

In the previous chapters the existing technologies, data models, and protocols for devices, neighborhoods, buildings, or M2M communication have been presented. Here the focus is on the integration of existing heterogeneous data models into a common data model that can be used by all participants, such as users, services, or sensors. This integration of heterogeneous models is necessary, since a holistic approach to model every aspect of a neighborhood is unfeasible. There has been put a lot of effort into the creation of ontologies for different parts and components of buildings and neighborhoods within the eeSemantics Community. (eeSemantics (Login required), 2016). These ontologies can be used as specific data models. If the features of the data model do not suffice, that is, due to new requirements, it has to be extended to a new version of the ontology. This leads to a plethora of different data models and extensions that cannot be integrated into a unified data model since this would also require extensibility and would therefore have the same problems.

The neighborhood information model (NIM) (Look & Greifenberg, 2013a,b,c) aims at solving these issues. It has been developed following a metamodel based approach. Relying on a metamodel makes use of its generic character and is easily extendible at runtime and is able to integrate heterogeneous data models. For the integration a Domain Specific Language (DSL) is used to define a mapping between the NIM elements and the data model to be integrated. Furthermore, we make use of generative engineering techniques in order to generate adapters between different data models. Thus, the NIM allows for semantic interoperability of different systems via data model integration. Albeit being rather generic, the NIM makes a distinction between different types of data that is stored and exchanged.

These data types include:

- Measured sensor data that changes frequently, close to real time, stored as time series data.
- Infrequently changing data, that provides information on systems in a building or a neighborhood.
- Non-changing, constant data, that does not change, such as latitude and longitude of a neighborhood.

Apart from the different types the data may be historical, real-time or forecasted. The sources of the data include measurements, third-party services or simulation services. Metadata, such as the source of the data and different value ranges are also allowed. By using the DSL, defined by a context free grammar, small simple chunks of the different data models are easily described. This description can be reused by other models. Overall, data of the described data models, used in a neighborhood can be integrated into the overall NIM. The concept of DSLs is a well established instrument in model based software engineering to enable the expression of domain knowledge by experts.

The DSL puts domain specific concepts into the foreground while simultaneously creating a formal description of each used data model within a neighborhood. Thus, the complete data of the neighborhood can be integrated. This enables the development and implementation of value-added services since they are able to access the complete neighborhood data. As said before, the models are used as input for code generators generating concrete adapters from the more abstract mapping descriptions. The extension at runtime is dome, by simply adding a new model to the set of existing models and by generating a new adapter between the involved data models. Thus, the new data can be transformed into already existing and vice versa. Introducing a NIM to a new neighborhood can be done quickly, by describing the existing data models, automatically inferring them from different standards or reusing already existing models of a different neighborhood. By packaging the existing models into libraries this be enhanced even more. Apart from the integration of existing data models the implementation of new value-added services at the SoS level follows a similar approach as shown in Section 5.

The employed model based and generative software engineering techniques, such as the aforementioned DSLs (Fowler, 2010) are created with the *MontiCore framework* (Krahn, Rumpe, & Völkel, 2010). MontiCore processes context free grammars as language definitions and creates a DSL out of them. To achieve this, MontiCore generates necessary infrastructure, such as a parser, a prettyprinter, editors, context condition checking support and code generation capabilities (Völkel, 2011). For implementing the code generators, generating the adapters the template engine *freemarker* is used. While the DSL provides an abstraction from a technical to a domain problem space the code generator provides the transformation back from the domain to the problem domain space. The designed data model DSL within the NIM approach follows a set of other languages provided by MontiCore, namely the UML/P (Rumpe, 2011; Rumpe, 2012; Schindler, 2012), a slightly modified derivate of the UML (OMG, 2011).

As stated before, the code generation transforms the domain problem space back into the technical problem space by creating technology specific source code out of the model. As target platform in the NIM Java is used to provide a prototype implementation for the COOPERaTE project. While creating a code generator and defining a DSL typically requires a lot of effort it is worth it, since the DSL can be used in different neighborhoods with different requirements and scenarios. The effort for creating a DSL and code generators can be reduced if certain guidelines are followed (Karsai et al., 2009).

4.2 Technical NIM Description

The design of the NIM makes use of two different information sources. In (Pesch et al., 2013) requirements of an EPN based on identified use cases were presented and taken into account during the NIM design. Furthermore, the design of existing data models within the building domain (Corrado & Ballarini, 2012) has been analyzed and the outcome included in the NIM design. The NIM design aims at providing a common data understanding of the complete neighborhood to enable value-added services to be implemented inside an EPN. The prototype NIM implementation uses this flexible approach going beyond simply using a fixed integration model since extensibility of the platform and the therefore the service landscape of the EPN is important and necessary. The concrete data model providing available data fields is detailed in the following. Consequently the abstraction of the aforementioned concrete data model used as COOPERaTE domain model is shown. After that, the DSL and the code generators are introduced.

The concrete data model is an extension of the SEMANCO data model (Corrado & Ballarini, 2012). The SEMANCO data model is a basic ontology for newly implemented services. It was extended by additional categories. These categories, as well as additional entries and data model details can be found in the COOPERaTE project Deliverable D1.2 (Look & Greifenberg, 2013a). Fig. 5.11 shows that the neighborhood element is a composite element that consists of several other elements.

The information persons, geographical information, energy grid connections, reports, and traffic have been added due to the requirements shown in Deliverable D1.1 (Pesch et al., 2013). This information is modeled as a subclass hierarchy of the Elements

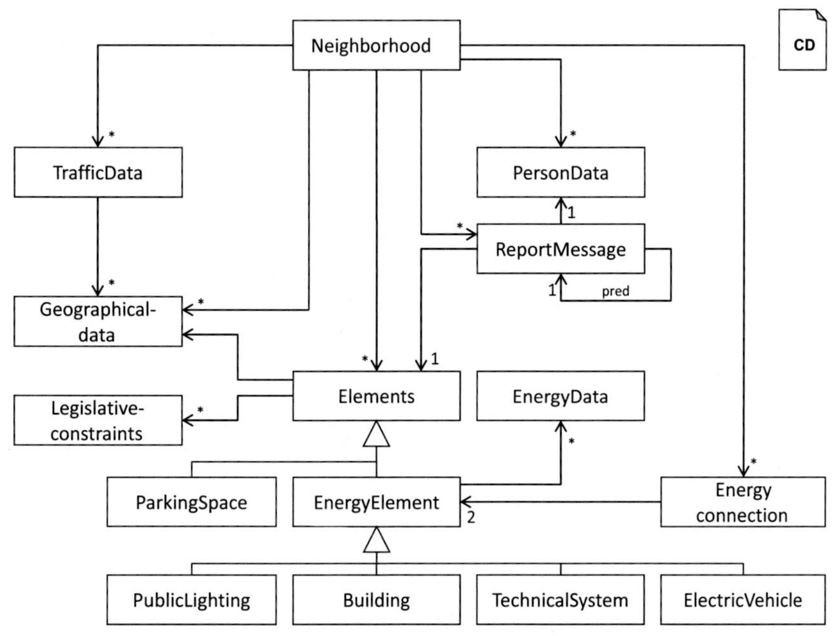

Figure 5.11 ***Concrete data model.*** *(Greifenberg, T., Look, M., Rumpe, B., & Ellis, K. (2014). Integrating Heterogeneous Building and Periphery Data Models at the District Level: The NIM Approach. In: Proc. 5th Workshop on eeBuilding Data Models (eeBDM) (part of 10th European Conference on Product & Process Modelling (ECPPM) 2014). Vienna, Austria).*

class. Further information, such as energy data, public lighting, building, technical systems, parking spaces, energy elements, and electric vehicles are contained in the model. The energy grid connection links to two energy elements in order to store necessary information on energy connections in the neighborhood. Such a connection information is needed when thinking about scenarios where there is no underlying grid but direct connections.

In the background a meta model is used to integrate the heterogeneous data entries from different existing data models into the NIM. This generic model is the underlying basis to the aforementioned concrete NIM. The generic NIM, as shown in Fig. 5.12 consists if several NIM-Components which make use of a composite design pattern (Gamma, Helm, Johnson, & Vlissides, 1994). These components are divided into entries and categories, where the latter can contain other components. A link relation can be established between categories to support cross referencing between categories. Additionally, privacy enabling files and metadata, such as a unit and a name are included. An

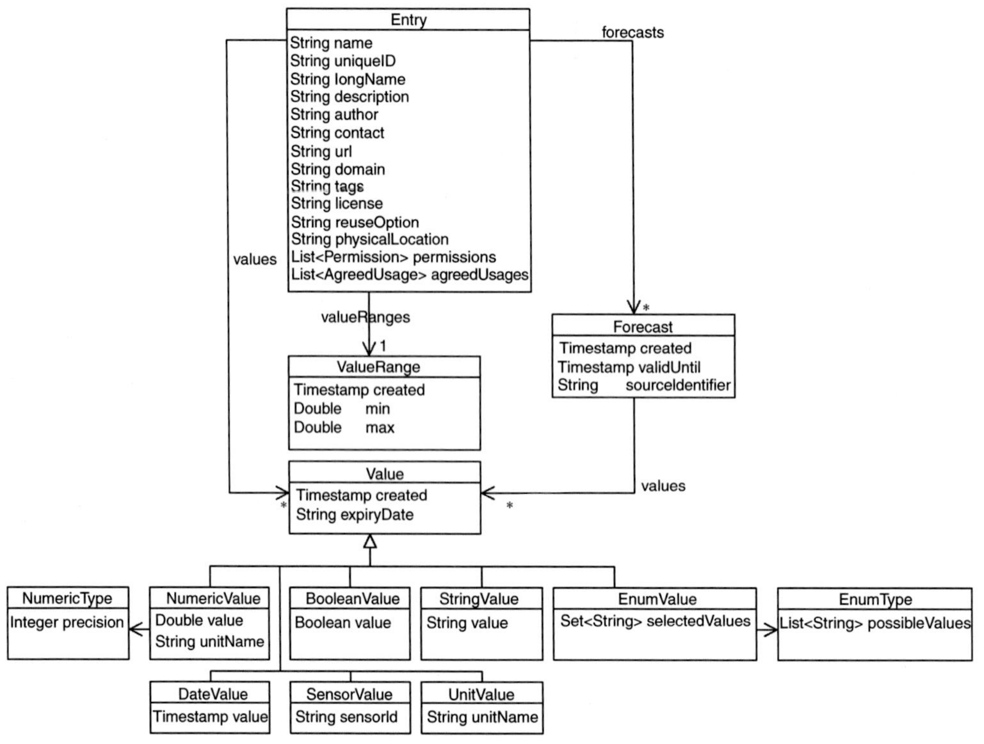

Figure 5.12 ***A UML class diagram detailing the generic NIM.*** *(Greifenberg, T., Look, M., Rumpe, B., & Ellis, K. (2014). Integrating Heterogeneous Building and Periphery Data Models at the District Level: The NIM Approach. In Proc. 5th Workshop on eeBuilding Data Models (eeBDM) (part of 10th European Conference on Product & Process Modelling (ECPPM) 2014). Vienna, Austria).*

entry may have different information on values can be stored: values, forecast data, meta information, value ranges, and historical data.

As mentioned earlier a value might be calculated, measured, or manual inserted into an entry. Timestamps are used for the values where the actual value of an entry shows the most recent timestamp. Older values are available too and are considered as historical data available for an entry. The value field in the value element stores the value itself. Data, that a said to be present in the future at a given timestamp is called Forecast data modeled as an explicit element in the NIM. This is different from the historical data that is inferred by the timestamps. For Forecasts this is not possible since there should be multiple forecasts regarding the same point in time available. This enables that each forecast is stored separately together with its source that differentiates different forecast sources by an identifier. Upper and Lower bound of valid values are stored by value ranges.

Security aspects concerning the data are directly built into the model. Such aspects include that only authorized users or systems may access or store other users or systems data. Furthermore, forgetting a value is enabled by introducing an "expiry date" for each value. This field should be used to delete expired data. This is especially important since historical data is stored. In order to not keep the data forever users and systems can mark the data as expired in order to specify the lifetime. Additionally the "agreed usage" field has been introduced as additional information. This field describes the entities that might use the data. Therefore, it should be checked, if an accessing person, service identifier or role is contained in the field. A list of all possible users is not possible in a neighborhood setting. Thus, role based access control (RBAC) is used to assign the same role to several users. The roles can be used within the "agreed usage" field.

The "physical location" field is also embedded into the data model as a security mechanism. It specifies the geographical location where the data entry might be stored. Due to legislative constraints this information may become necessary in order to be compliant. This field can be used by, for example, data centers to determine the actual physical storage place. Apart from providing necessary information to storage provider it can also be used by users to decide which information should be uploaded to a platform which uses the NIM as data format.

The claimed adaptability and extensibility required a mechanism that enables the overall system to store data unknown beforehand. Due to its generic nature the data model supports this by transforming the data into the meta model. This leads to the drawback that the data model, developers have to use for their implementation is quite generic, does not provide type safety and is in general very inconvenient. Also data from other data models has to be transformed into the meta model leading to a huge manual implementation effort, if done for each data model within the neighborhood. Additionally, this extension would not be possible at runtime of the system. As a solution a plugin based architecture has been used within the NIM implementation. This architecture allows the addition of new plugins at runtime. Furthermore, the plugins are generated

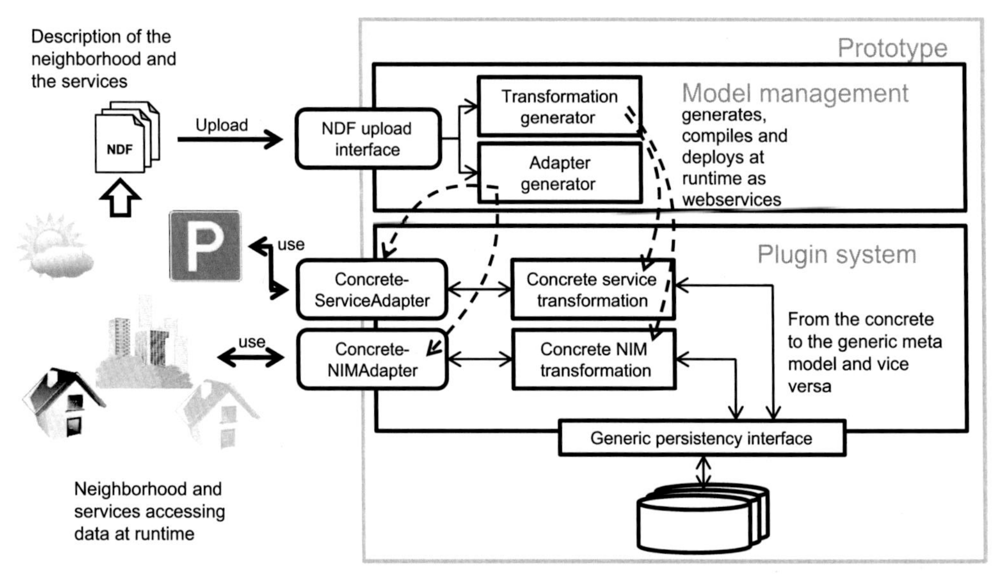

Figure 5.13 ***Plugin based implementation of the NIM.*** *(Greifenberg, T., Look, M., Rumpe, B., & Ellis, K. (2014). Integrating Heterogeneous Building and Periphery Data Models at the District Level: The NIM Approach. In Proc. 5th Workshop on eeBuilding Data Models (eeBDM) (part of 10th European Conference on Product & Process Modelling (ECPPM) 2014). Vienna, Austria).*

by a code generator using models of the previously mentioned DSL as input. The code generation, as well as compilation and plugin instantiation is done at runtime. The transformation from a data model into the meta model is put into the generated code. The DSL is used to create models that describe concrete NIM data models in a NIM data format (NDF). This NDF describes a data format not actual data. Fig. 5.13 shows the overall approach and methodology to configuring the running prototype.

The prototype implementation consists of a Model Management component and the Plugin System. The Model Management component processes the NDF models and generates new adapters containing new transformation code. The Plugin System incorporates the overall system that is used by services and neighborhood elements for accessing the data. Three steps are necessary to connect new participants, such as a new neighborhood or a new service to the platform(Greifenberg et al., 2014). First the NDF model has to be created in its textual syntax and uploaded to the system. Analysis and context condition checking is done by the Model Management component. If it is valid, it is passed on to the transformation and the adapter generator which in turn generate the necessary transformation and plugin code. After generation the plugin is automatically deployed and made accessible. In a second step the newly connection service or neighborhood may use the adapter for operating on the defined data Model. In the third consecutive step the data is transformed between the concrete data model and the generic data model. Again, this is threefold: first the transformation from a concrete to the generic data model is considered which is relatively straight forward. A "Room" element would become a category with name "room" in the generic data model. The name of the room would be represented by an entry contained in the category. This entry has the name "roomName". The actual name of the room, not present in the NDF since it defines a data format, but available in the uploaded data, becomes the value of the entry. The NDF allows the definition of hierarchical data types within other data types which can be reflected via the composition pattern used inside the categories. The second case is the transformation from the generic data model back to the concrete data model. If the meta model data belongs to the concrete model it should be transformed, this is straight forward and simply the inverse of the previously described first case. The third case is the most complex one and reflects the transformation of data from different concrete models. This is probably the most common case since this reflects interoperability between services. For a service requiring data about buildings from two existing services, there are two possibilities:

- Making use of the already existing adapters and collect the data separately.
- Describing the desired data in its own NDF containing a mapping.

Figs. 5.14 and 5.15 provide a high level depiction of the role the NIM plays in integrating cooperating neighborhood management platforms within the COOPERaTE project prototypes and their associated services. The NIM is implemented here as a service, providing a demonstrator platform and a common data format, which is applied to data from the 3 COOPERaTE demonstration sites (neighborhoods). Alternatively, a platform could implement a NIM compliant adapter/parser, which simply means that data made available via their API would be made available in NIM format. Here the NIM acts as a standard data exchange service.

The data made available via the NIM was used in what was essentially an abstraction service (Fig. 5.16) whereby the NIM was used to translate between different cooperating systems data models/formats. As such, one can appreciate how energy services developed and implemented within the three cooperating platforms (Section 5) can by complying with the NIM enable the project use-cases, namely:

- near-real-time (NRT) monitoring
- run day ahead forecasts of electrical/thermal consumption
- run day ahead forecasts of renewable sources, such as PV and wind power
- optimize the power purchase versus on-site generation
- use the neighborhood flexibility to participate in DR programs

5 EXAMPLE NEIGHBORHOOD SERVICES

A number of example energy management and decision support services were adapted and/or developed in the COOPERaTE project based on a prototype implementation of the COOPERaTE architecture and the NIM service. The services

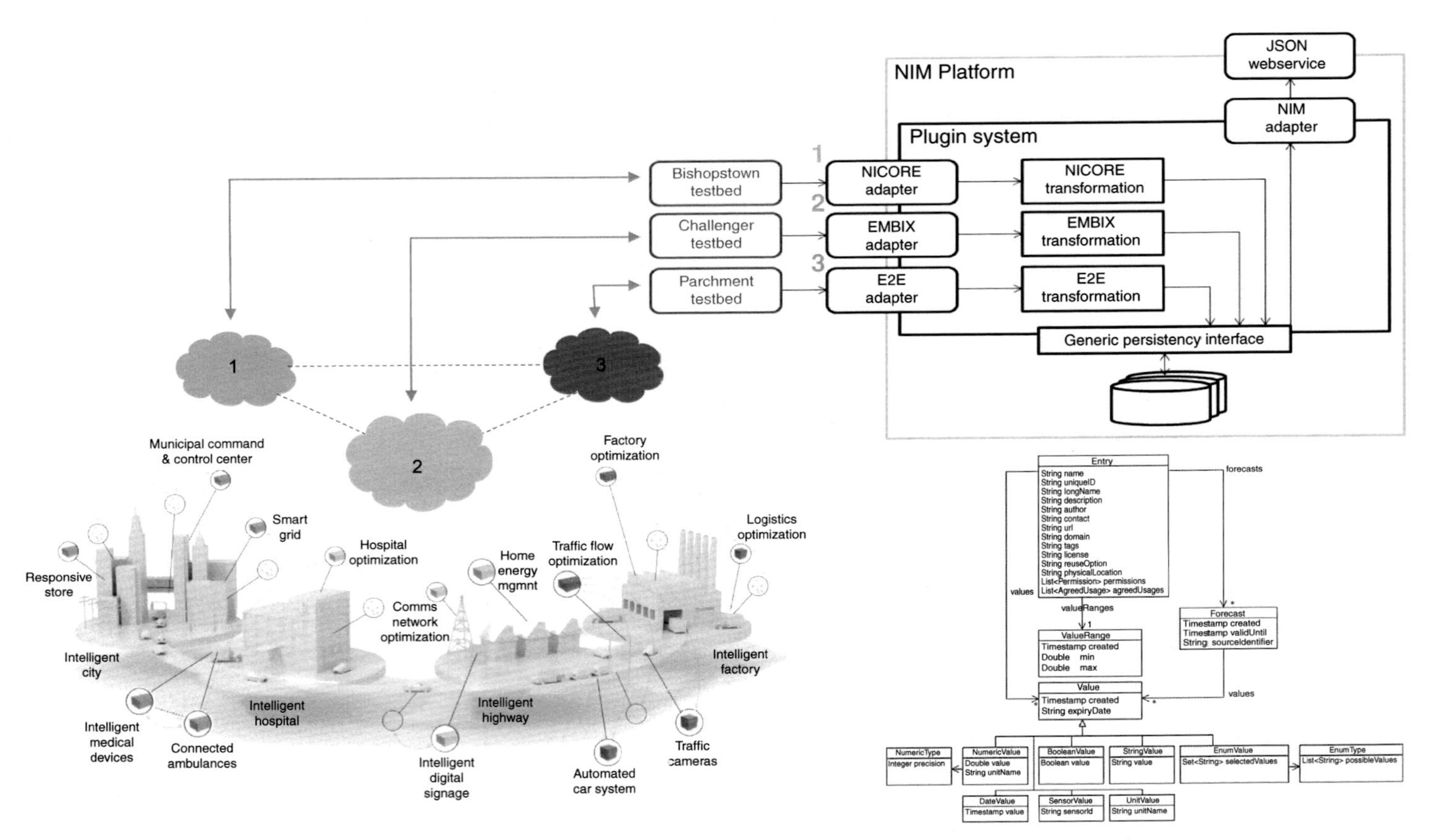

Figure 5.14 ***NIM acting as a translation service for 3 cloud based neighborhood management systems.***

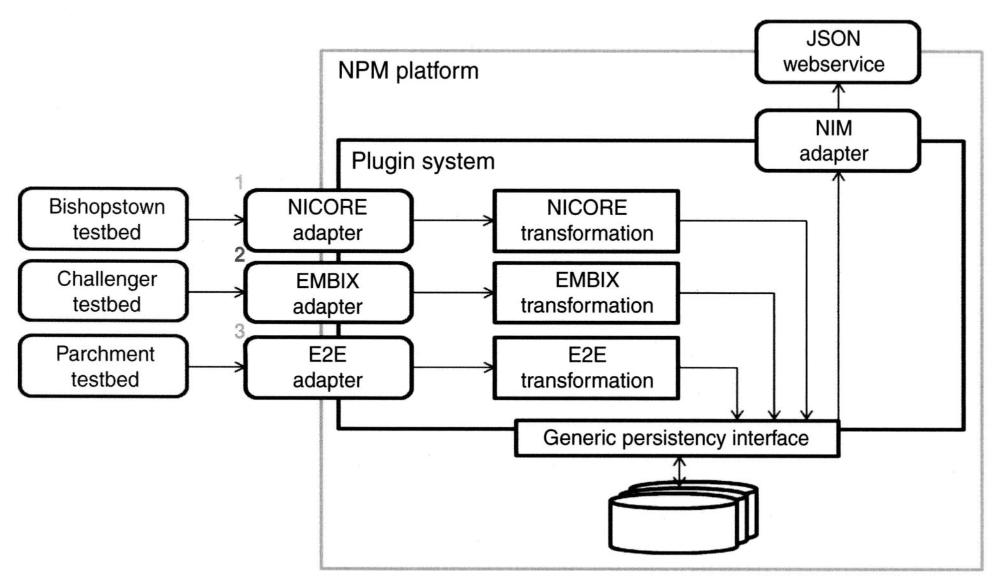

Figure 5.15 ***Integrating heterogeneous platforms and services.***

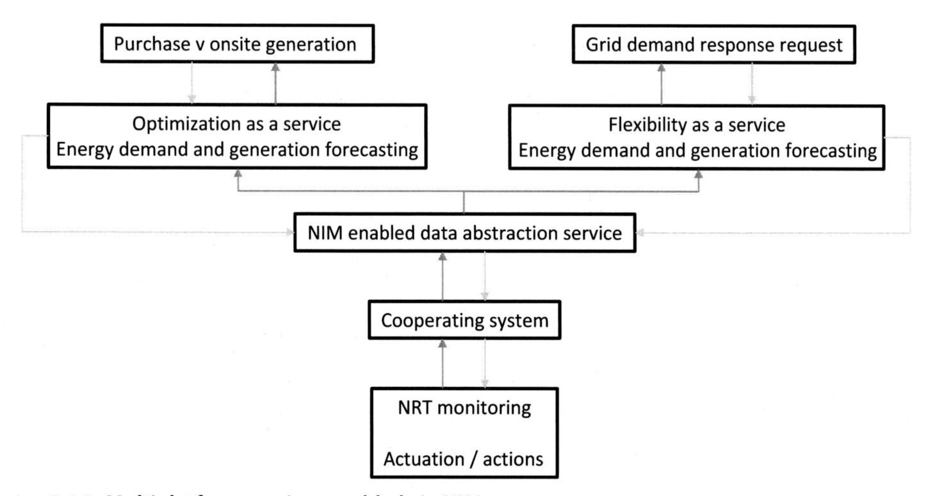

Figure 5.16 ***Multiplatform services enabled via NIM.***

were implemented within the three test-bed environments available in the project, the CIT Bishopstown Campus (commercial neighborhood), the Parchment Square student village (residential neighborhood), and the Bouguyes Challenger Campus (industrial neighborhood).

5.1 Energy Services—Commercial

The COOPERaTE energy services deployed for the CIT Bishopstown Campus neighborhood were implemented on the CIT NICORE IoT application enablement platform (McGibney, Beder, & Klepal, 2012). These services are a set of reusable software components that are run in NICORE. These services facilitate implementation and execution of neighborhood energy management and forecasting algorithms, usage scheduling, and setpoint profiles, as well as forecasting and estimation algorithms. The energy services address the neighborhood energy management services that were identified by the COOPERaTE consortium (Pesch et al., 2013). Three main enabling prototype services supporting real-time monitoring and actuation have been developed within NICORE, namely:

- *a Building Management System (BMS) Connector service*
- *a Core Control service*
- *a Data Aggregator service*

Figs. 5.17 and 5.18 show the NICORE energy services for the two main usage scenarios for example, the single-owner and multiowner case. Fig. 5.19 illustrates the Command module.

The *BMS Connectors* service acts as an interface between each embedded building monitoring and control system and the NICORE platform. In the Bishopstown testbed, there are two interfacing technologies used by the respective BMS, (OLE for Process Control) OPC, or web services. Both technologies provide read/write capabilities thus

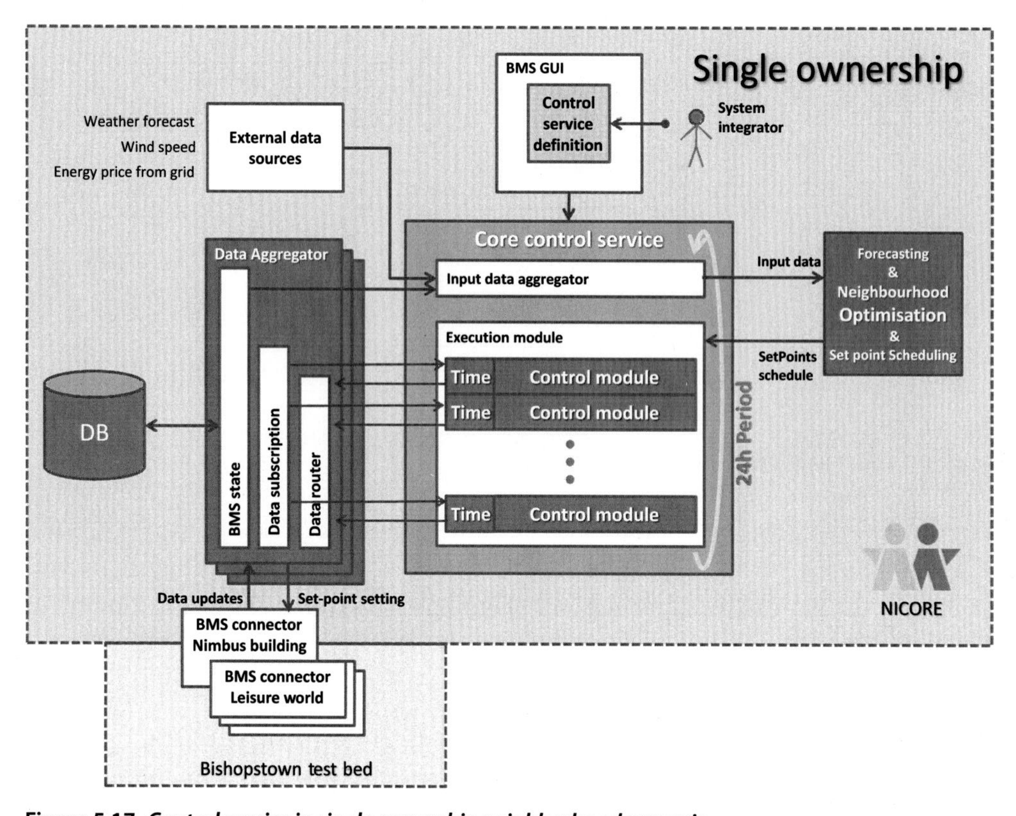

Figure 5.17 ***Control service in single ownership neighborhood scenario.***

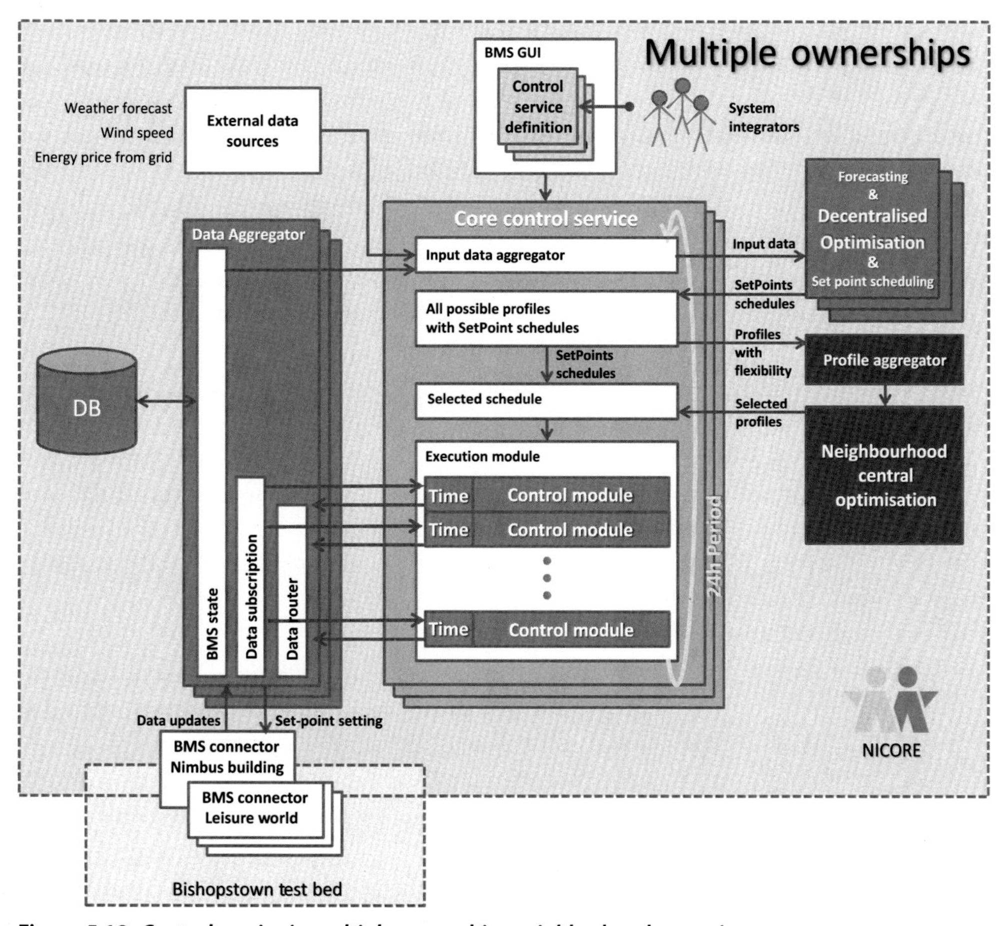

Figure 5.18 *Control service in multiple ownerships neighborhood scenario.*

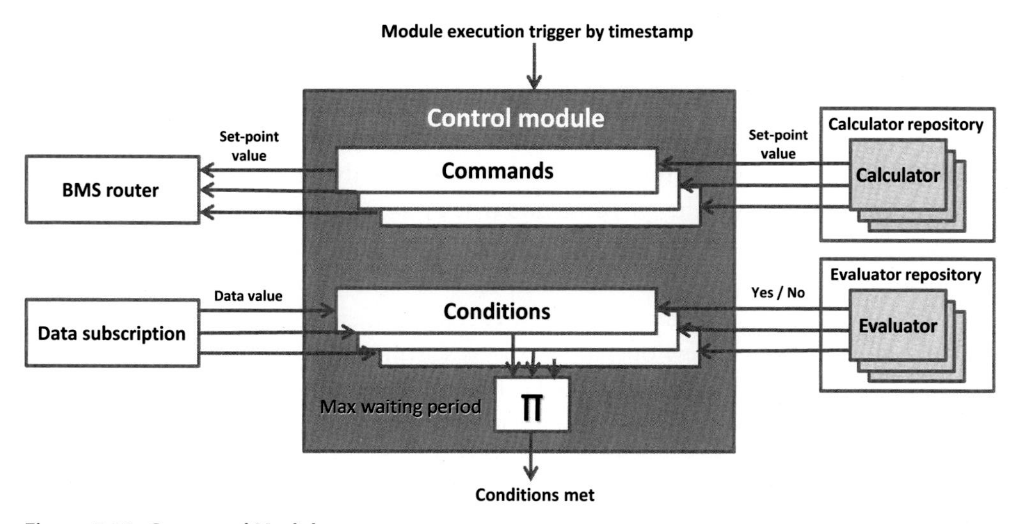

Figure 5.19 *Command Module.*

enabling the reading of and controlling of BMS states. When a *BMS Connector* receives a BMS state from a BMS, OPC, or web service interface, it requests provisioning of a *Data Aggregator* instance from the NICORE platform manager in the cloud. The *Data Aggregator* caches the BMS state, forwards the state updates to the subscribed *Control Modules* of the Control Services and store changes in a database.

In addition, a number of *Forecasting and Optimization Services* were developed and provided by neighborhood and building system integrators. These services are uploaded to the NICORE platform *Service Repository* and instantiated on the request of a *Core Control Service* in the cloud. *The Core Control Service is the central component* of the NICORE energy services. There can be multiple *Control Services* in a neighborhood system. For example, if two BMS subsystems in a building were to be optimized independently, two separate *Core Control Service* instances would be created to execute their control tasks independently. More often, however, in the case of the single ownership neighborhood only single *Core Control Service* is required to run and act on the results of the *Centralized Neighborhood Optimization*. In the case of the multiownership neighborhood, a separate *Core Control Service* is required for each of the *Decentralized Optimizations* additionally to the *Neighborhood Central Optimization*, Fig. 5.17.

The operation process is the following:

1. Based on a *Control Service Definition* provided by the *Neighborhood System Integrator*, the *Core Control Service* interrogates the *Data Aggregator* and *External* Resources for input data for the *Neighborhood Forecasting and Optimization* component.
2. Optimization returns set point schedule for each building equipment (for the next 24 h with 15 min interval, for example), upon which the *Core Control Service* instantiates a new *Execution Module*.
3. The *Execution Module* processes the received schedule into a sorted sequential list of *Control Modules*, which are sequentially executed at scheduled time. Each *Control Module* (Fig. 5.19) contains a list of *Commands* and list of *Conditions*.
4. Each of the commands contains single set point specification and a reference to a *Calculator* which calculates the set point value based on the current BMS state. The calculated set point value is then sent to the *BMS Router* which forward it to the right a *BMS Connector*.
5. The set of *Conditions* in the *Control Module* insures that the certain BMS state is achieved after applying the set points and that all the *Conditions* are met before moving to the next *Command Module* in the sequence. The *Conditions* are evaluated in an *Evaluator* and they have to be met within a predefined maximum waiting period.

When all *Control Modules* are successfully executed (after 24 h, for example), the *Core Control Service* repeats the new operation process starting from the point 1 again.

In the case of a multiownership neighborhood there is a *Core Control Service* running at every separately owned premise of the neighborhood. The overall system architecture and the *Core Control Service* were designed with reusability in mind when the same *Core*

Control Service together with all other services can operate in both single and multiple ownership scenarios. In the multiple ownership case, the *Core Control Service* requests a full list of possible set point schedules from the *Decentralized Optimization*. It then anonymizes schedule profiles and forwards them to the neighborhood *Profile Aggregator*, which collects all schedules from the neighborhood before the *Neighborhood Central Optimization* selects the best schedule profile for each owner. The *Core Control Service* discards all schedules but the selected one and the operation process continue the same way as in the single ownership neighborhood scenario.

5.1.1 Forecasting Services

For the forecasting services a number of key energy performance related variables, for example, thermal and electrical loads, indoor temperatures, microgrid power generation and so on, can be forecasted based on available historical and weather data ranging from a few hours to days or years. In the COOPERaTE project the following variables were considered for example:

- Thermal loads
- Electrical loads
- Building Indoor temperatures

The forecasts can be obtained by the building and neighborhood energy managers on request or implemented in real-time in the NICORE platform to support energy optimization services based on the architecture given in the previous section. For the NIMBUS and Leisure World buildings of the COOPERaTE Bishopstown demo-site, the day-ahead forecasts are obtained by using previous day historical data of the electricity and thermal power consumption, indoor temperatures, and outdoor weather data (such as outdoor temperatures and wind speed). The forecasting algorithms allow for, the period of the prediction, as well as the range and period of historical data, to be selected by the system integrator. The forecasts can also be utilized by the energy optimization and DR services (described later) for the coordination and management of the optimal profiles, schedules and setpoints of the neighborhood and building electrical and thermal assets (thermal storage, batteries, energy microgrid).

5.1.2 Optimization Service

The Optimization service developed addresses both types of single-owner and multiowner neighborhoods. The objectives of the optimization service can vary between:

- Whole neighborhood energy cost optimization (electricity and thermal bill)
- Whole neighborhood CO2 emissions optimization
- Combination of the aforementioned objectives

It can also be used to address both the building-level energy optimization and neighborhood energy optimization scenarios. The optimization service aims to optimize the neighborhood power purchase versus on-site generation, taking into account prices, current measurements (from the real-time monitoring service), load and weather forecasts (from the forecasting service), and provide predictions of the on-site neighborhood generation over a period of time that can meet the neighborhood objectives aforementioned.

For the single-owner case, the service uses single-owner optimization to collectively optimize the optimal profiles, schedules, and setpoints of the microgrid and thermal system assets of the neighborhood. In this prototype service, the objective of the optimization service has been set to optimize the overall neighborhood energy cost. The service can be used in two ways:

- as a service to the neighborhood or building energy managers to determine the optimal power and energy generation of the neighborhood for a day or longer period time, or
- as a real-time optimization service that automatically sets the optimal profile and setpoints of the neighborhood assets. It can also be used to determine the neighborhood or buildings flexibilities for example batteries and microgird generation profiles.

For the multiowner neighborhood type where data privacy issues between the various owners of buildings and distributed energy sources (including district renewables, district-level generation etc.), the energy service utilizes multiowner neighborhood optimization. It receives the flexibilities (a number of generation, storage, and renewable profiles) from each building or district power generation owner/user and determines the optimal selection of power generation and consumption profiles for the neighborhood that both optimizes the global neighborhood and individual actors benefits simultaneously. The objective chosen for this prototype example was the whole neighborhood CO2 emissions reduction but equally the overall neighborhood energy cost could be selected. The service is offered to the neighborhood energy manager and building facilities managers to determine the optimal power and energy generation of the neighborhood and buildings respectively. Finally, the energy service can be offered for the energy optimization of each individual building. Currently the energy optimization service has been implemented and demonstrated in the CIT Bishopstown demo-site.

5.1.3 Demand Response Service

A DR service was created by using an *external data sources* interface. The external data sources interfaces to a number of external date relevant to neighborhood energy management including DR, such as real-time energy prices from an aggregator or grid operator and weather forecasts including wind speed and solar irradiation measurements and forecasts. Typical external data sources for the Bishopstown test-site are the Single electricity Market Operator (SEMO, http://www.sem-o.com) web-site sources and the Met Eireann (www.met.ie) and Weather Analytics services (www.weatheranalytics.com) for weather data and forecasts. In this case the control service interrogates the external

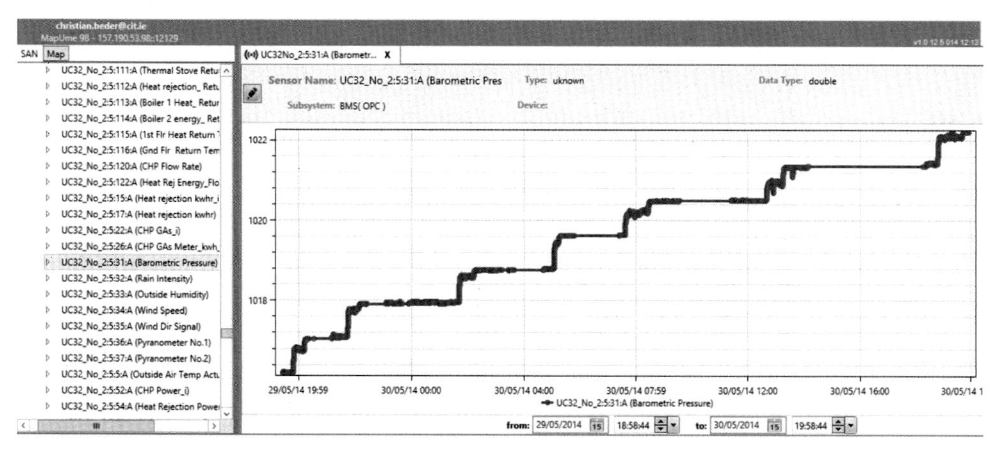

Figure 5.20 ***Data aggregator management GUI.***

resources as well as the data aggregator (similarly to step one of the operation process described previously) for input data related DR. such as real-time energy prices that can be used in a DR scenario.

5.1.4 Visualization Service

A data aggregator management GUI has been implemented on NICORE for the visualization of the various data point in the individual BMS and neighborhood systems (Fig. 5.20). The data aggregator management GUI connects to the NICORE middleware and provides a list of all available data points. The available data points are listed on the left-hand side of Fig. 5.20 and the data point reading history over a period of time is presented in a graph on the right-hand side.

5.2 Energy Services—Residential

This section details services implemented for an apartment complex in the Bishopstown test-bed, Parchment square. The section elaborates on how these services address the COOPERaTE use-cases. More generally the section illustrates the enabling impact an IoT based energy management system can have in residential environments with respect to demand-side management (DSM) and DR.

In the absence of any monitoring and/or control technology an end-to-end system was deployed to support DSM and DR services, Sections 5.2.2 and 5.2.3. The solution deployed precedes and partly aligns to the Intel IoT reference model and utilizes many of the same commercially available assets. The Intel IoT reference model is an end-to-end reference model and family of Intel products that works with third party solutions to provide a foundation for connecting devices and delivering trusted data to the cloud. The reference model consists of an "edge" gateway platform (left of Fig. 5.21) as well as a representative backend "cloud" management system (right of Fig. 5.21). It

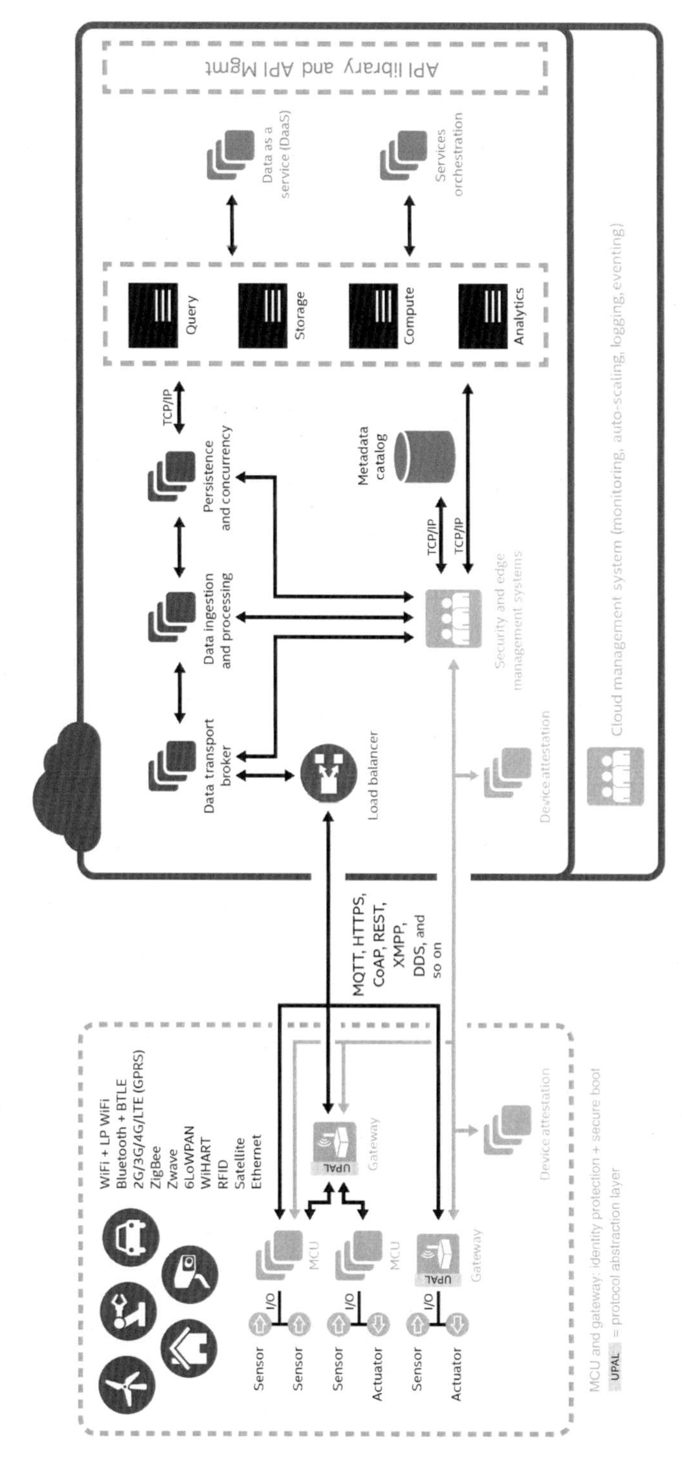

Figure 5.21 ***Intel's IoT reference model*** (Intel, 2015).

is also aligned to the conceptual IIC layers as outlined in Section 2. More information available at:

- http://www.intel.com/content/www/us/en/internet-of-things/iot-platform.html.
- http://www.intel.com/content/www/us/en/internet-of-things/white-papers/iot-platform-reference-architecture-paper.html

The focus within COOPERaTE was primarily the edge tier/layer for example, the gateway as a mechanism for enabling edge based intelligence (as in locally hosted compute and services) and as a link to cloud hosted compute and services. Fig. 5.22 later gives a more specific system overview of the solution deployed. It is this system that support the data tagging service outlined in Section 2.2 and energy services outlined in Section 5.2.3. The approach leverages API's and the NIM service when communicating beyond the boundaries of the deployed system as part of the wider EPN.

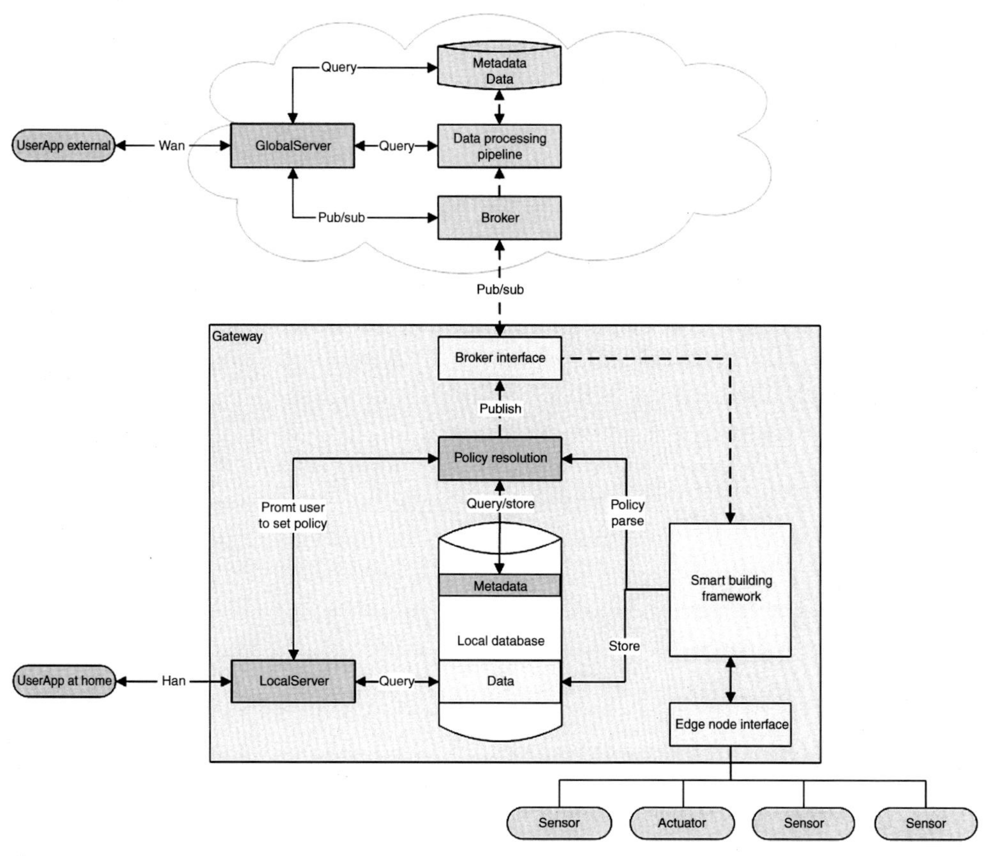

Figure 5.22 ***System overview.***

5.2.1 User Defined Data Access

At the time of deployment and to a large degree at the time of writing, many IoT solution vendors in the built environment, particularly residential, focus on providing full service stacks (sensors to cloud applications) in narrow verticals (security, utilities control, lighting etc.). However, the approach here was driven by the trends toward increasingly ubiquitous multivendor wireless sensing and maturing data privacy legislation, and thus the likely decoupling of sensors and services. This decoupling will likely require gateway devices and data management software that provides a user with greater control over their data and configuration. This transition from tightly coupled narrow vertical solutions to more loosely coupled solutions is challenged by problems 1–5 listed below.

- *Problem 1*: Currently, IoT solutions, in the built environment, typically assume either local or cloud based services and do not support selective setting of access levels for individual data types.
- *Problem 2*: Typically, IoT solutions do not separate user generated data from operational data required to maintain the solution.
- *Problem 3*: Currently IoT solutions typically assume single ownership and centralized control of resources.
- *Problem 4*: Typically, in IoT home solutions there is no support for user-driven dynamic change in service plan.
- *Problem 5*: Currently, IoT solutions often do not use the location of the client device to dynamically choose between a cloud-based or locally hosted version of a service.

A user defined data access service utilizing the architecture and system of Fig. 5.22 was developed to address such issues. How the deployed system and service addresses these issues is beyond the scope of this chapter and is dealt with in detail in the COOPERaTE deliverable (D3.3 Report detailing dynamic cloud/edge workload exchange strategies, 2015). Here it suffices to say that the system/service mitigates these problems and strengthens the adoption potential for IoT offerings. It supports data privacy by supporting locally hosted services (e.g., lighting control). It allows for locally hosted services to ingest required data without impacting data privacy. It provides third party service providers with a means to communicate the value of their services in the context of the end-users data while maintaining data privacy (e.g., community level utility usage). This approach can therefore be used to incentives end-users to share access to their data with third party service providers who can provide value added services (e.g., energy DR) which is essential for EPNs.

5.2.2 Demand Side Management Services

This subsection outlines the energy services enabled by the system Fig. 5.22 and the data access service of Section 2.2. All services are utilized through an end-user mobile application. The same application service can run globally that is, in the cloud or locally that

Figure 5.23 ***User configurable dashboard view.***

is, on the gateway and supports user identified preferences for example, privacy. What follows describes some of the basic supported services.

Fig. 5.23 shows the configurable dashboard of the GUI. The user can choose from the different metrics charted and send that specific metric-widget to their dashboard. This allows the user to view the information most relevant to them.

Fig. 5.24, illustrates the NRT monitoring capability of the solution. Within the test-bed the following metrics are supported as required by the defined COOPERaTE use-cases—Temperature, Humidity, Motion status, Door/window status, Power, Energy, Relay status, and Battery charge.

Actuation is also supported for relevant end-nodes for example, the electrical relay switches. This data is grouped in terms of "zones", "sensors", and "timescales". "Sensors" are aligned to four categories namely—"usage, comfort, security, and other".

Top right of the home bar, the "cloud strike-through" icon informs the user if they are purely private/local with respect to data egest or if some data types are cloud bound. The "gateway" allows them to switch between gateways should the residence be a larger building with multiple gateway zone. The "notifications" icon indicates if there are events of note for example, "bedroom window is open, temp is below setpoint", "your

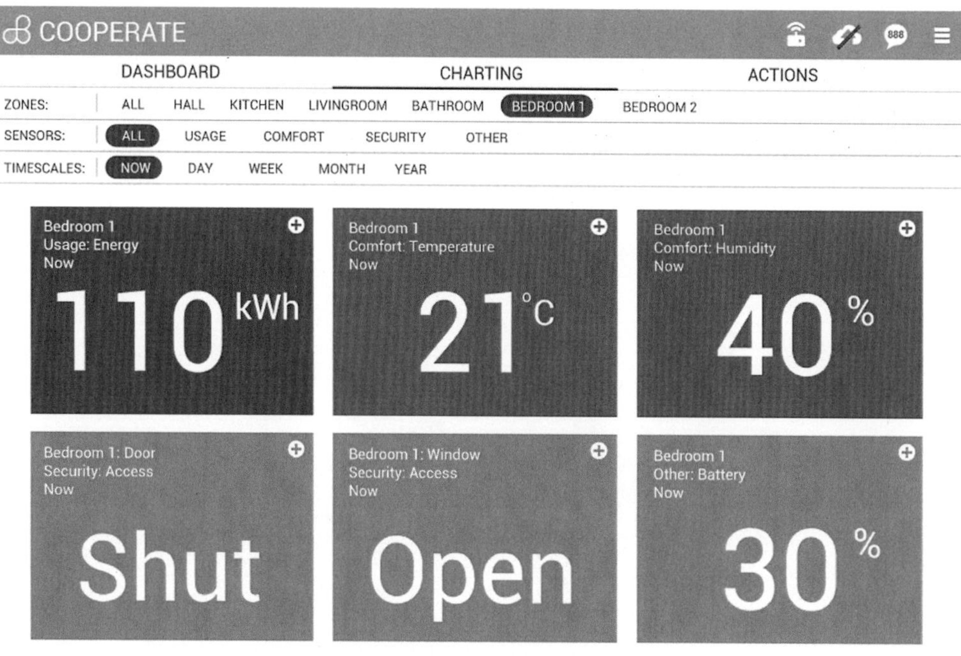

Figure 5.24 ***Charting "now" view, highlighting real-time monitoring.***

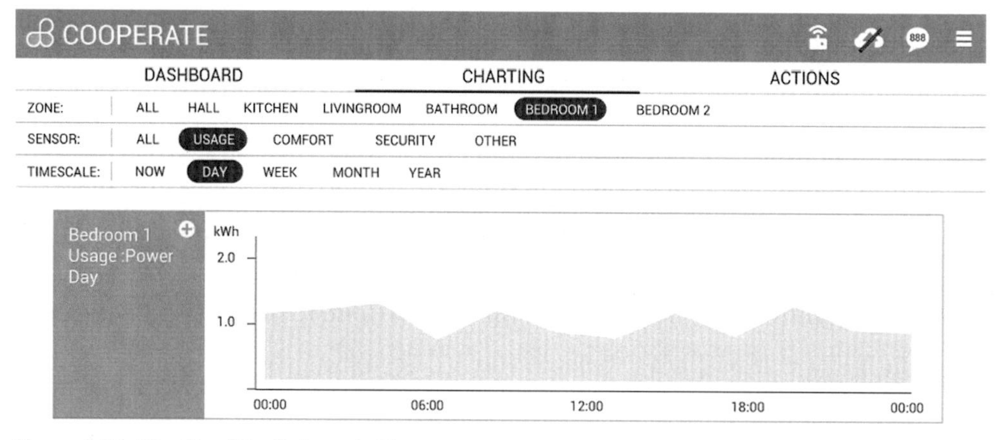

Figure 5.25 ***Charting "Day" view, plotting power.***

building management company needs to gain access to your apartment on Tuesday at 1100", or "there is a pending energy actuation event from your energy aggregator today from 1400–1430".

Fig. 5.25, illustrates the charting capability which allows the user to graph metrics for day, week, month, and year. The approach taken, even when in private mode, is to allow for data ingest which supports services for example, wholesale market pricing for locally run optimization service on the gateway, average neighborhood consumption of users for benchmarking functionality. The only difference between users that are in "private" versus users that are in "shared" mode is the ability of the latter to capitalize on DR, flexibility or other external transaction services. This approach is discussed further in (D3.3 Report detailing dynamic cloud/edge workload exchange strategies, 2015). The overall solution can also utilizes other services for example, existing opensource OpenHAB bindings. The overall functionality of the system and the service described above addresses use-case 1 "real-time monitoring" and use-case 4 "DR" with respect to the residential elements of the test-bed. Use-cases 2 and 3 are partly addressed by the overall system functionality and the optimization service described below, these use-cases are addressed in full, when interfaced to the COOPERTE SoS specifically the NIM/ services.

5.2.3 Optimization/Flexibility Service

The optimization service utilizes an algorithmic service developed outside the COOPERaTE project. The service interface was adapted to accommodate the COOPERaTE test-bed specifics. It decides upon the optimal charge plans for a community of devices to achieve some desired aggregate grid behavior, for example, water heater on, slab heating on. In essence it allows for flexibility services via the NIM to the grid that is, DR services.

Consider the following generic example: four devices (Electric Vehicles EVs, storage heaters, etc.), charging as soon as possible would present the behavior outlined in Fig. 5.26. This naïve charging behavior is detrimental to the grid as it increases the load at peak load times and decreases the voltage levels at critical voltage times. This increases the necessity of further investment in the grid, especially as more and more loads come online. To facilitate an increasing number of loads on the grid without incurring further investment in the grid, one can take a community-centric approach of deciding on the optimal charge plan for each controllable device to achieve an appropriate shape on the aggregate load.

Fig. 5.27 illustrates the aggregate grid behavior one can achieve when the start time and charge rate for all devices in the community are controllable and are informed by a metaheuristic optimization technique. The objective of the optimization technique can incorporate any phenomena that can be mathematically described; for example, eliminating increases in peak load, reducing the overall load levels, eliminating reductions in minimum voltage, maximizing the overall voltage levels, increasing renewable consumption, and so on. The narrow lines in Fig. 5.26 and 5.27 indicate the effect of the devices on the grid load and grid voltage.

For this particular implementation the optimization service was configured to minimize the wholesale energy cost. In such a configuration, the optimization server utilizes Irish whole-island energy wholesale data as one of the inputs to the optimization

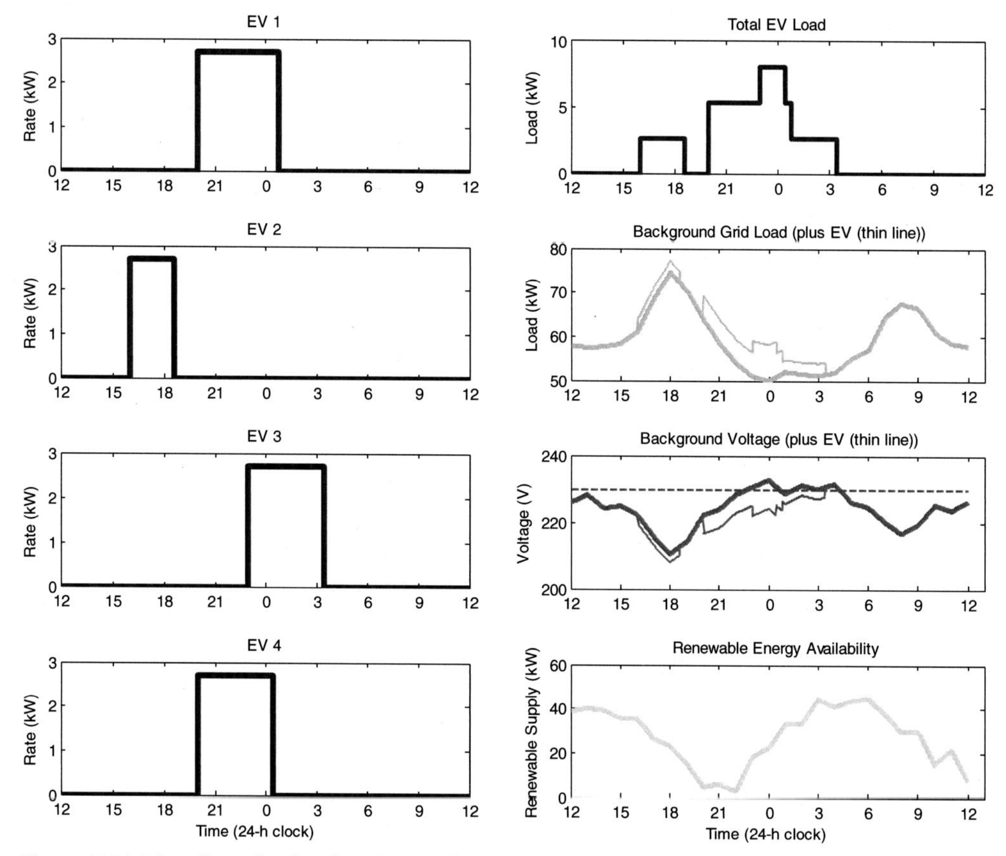

Figure 5.26 ***The effect of naive charging on the entire electricity neighborhood.***

objective function. The objective is to select the charge parameters for each house, which minimize the total cost of charging, subject to all users' time constraints. Hence, by delegating control of their predictably flexible loads to an external party the end-user can experience a cost saving, proportional to the cost benefit which utility stakeholder experiences.

Fig. 5.28 gives a COOPERaTE specific example of how the service can deliver value. The figure graphs the "total power consumption" "water Heater", and "wholesale energy price" for one day in Parchment square. Highlighted by the rectangle boxes one can see it would be cheaper to charge a few hours earlier. This would be automatically moved to cheaper charging periods by the service. Fig. 5.28 is a simplification for illustration and assumes the water heaters are well insulated and the heat energy loss is negligible over 3 h. If loss can't be treated as negligible one can "cost-in" energy loss of the water heaters so that the optimized charge plans would cause more charging in cheaper energy periods while ensuring enough energy remains at the required time, only when doing so would result in a lower net cost of charging.

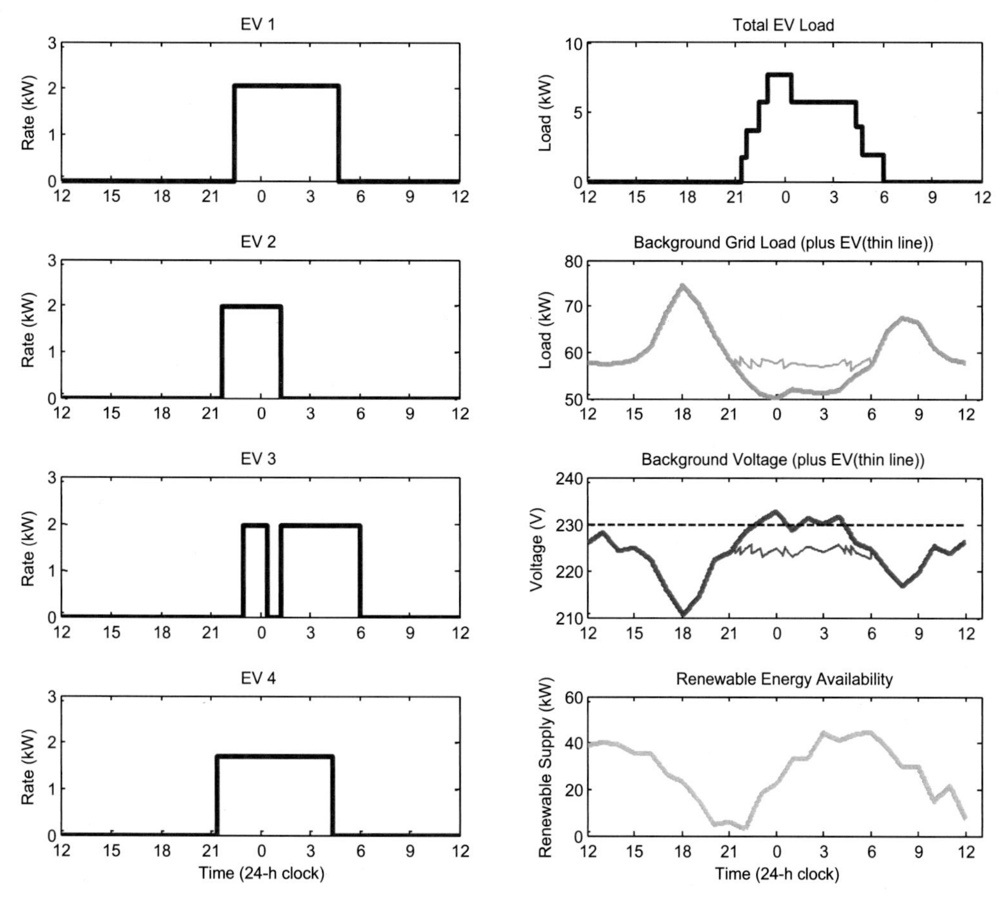

Figure 5.27 ***Effect of load-optimized charging on the grid.***

The service can inform third party aggregators via the NIM about the flexibility in the total energy consumption levels that can be achieved by our energy aggregator service. This can be done by sending complete information about individual controllable device's availability and usage requirements, or by presenting aggregate profiles. These aggregate profiles represent the highest and lowest possible loads that could be achieved at any particular time based on the charge requirements of all participating devices. An energy aggregator can choose a total power level in the exploitable load space Fig. 5.29 that the optimization service will be able to achieve on behalf of the aggregator. Any such flexibility request will require a reoptimization and a dispatch of new charge plans to relevant devices.

5.3 Energy Services—Industrial

This section describes the energy services developed in EMBIX's meter data management platform and implemented specifically within the Bouguyes Challenger industrial test-bed. The EMBIX's platform presents an open API that exposes data to other

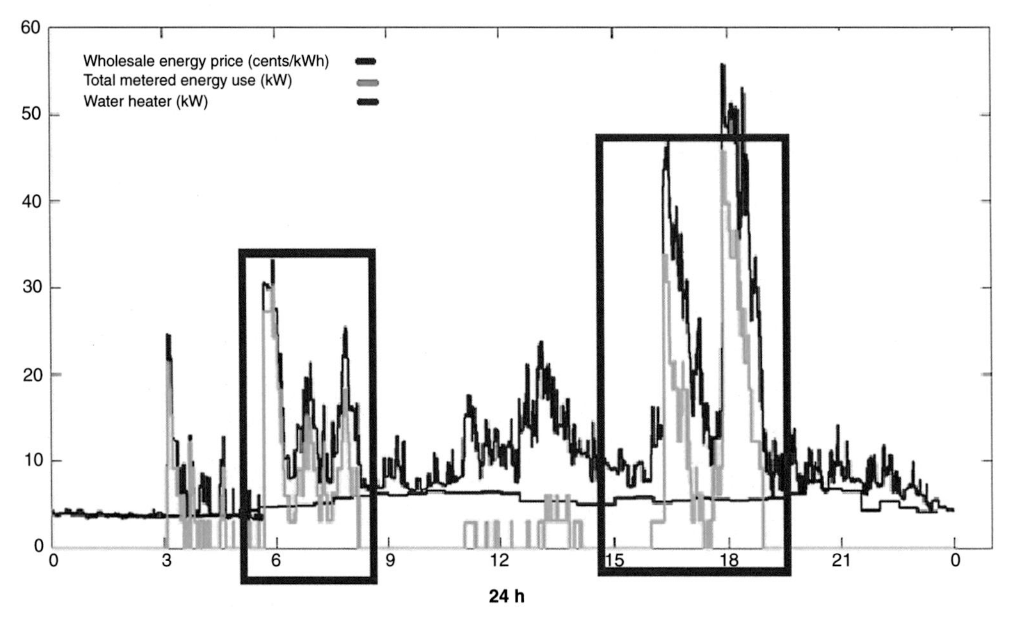

Figure 5.28 ***Day view parchment square.***

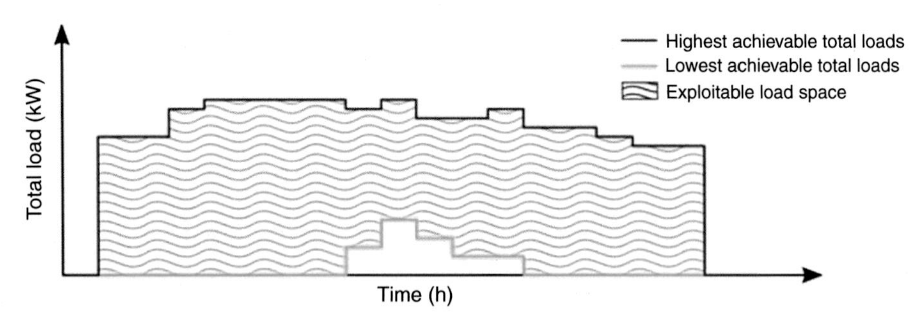

Figure 5.29 ***Opportunity space for demand response (DR).***

services, such as the NIM or any other services developed by a third party, such as forecasting module, analytics module, and so on. *Error! Reference source not found* illustrates an example whereby the system interfaces to a forecasting module (Figs. 5.30 and 5.31).

5.3.1 Visualization Service

To enable any type of analytics it is desirable to have a visualization service available. As such, a tool box (Fig. 5.32) for standard data visualization was developed to allow graphing of stored datasets.

To raise awareness of energy consumption issues and to allow people to act in an informed and "good way", specific dashboards based on the Urban Power open API were also developed specifically for COOPERaTE proof of concept (see details in

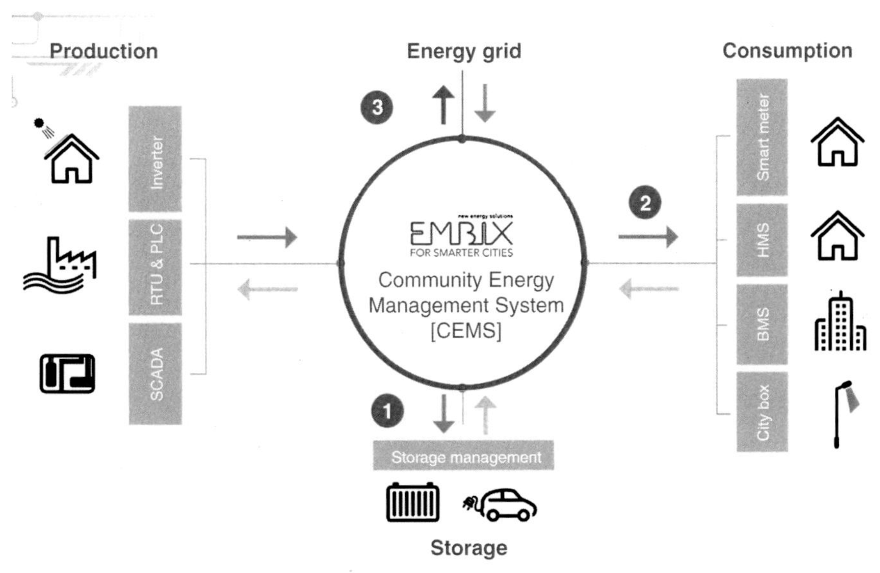

Figure 5.30 ***EMBIX's platform connecting to external forecasting module.***

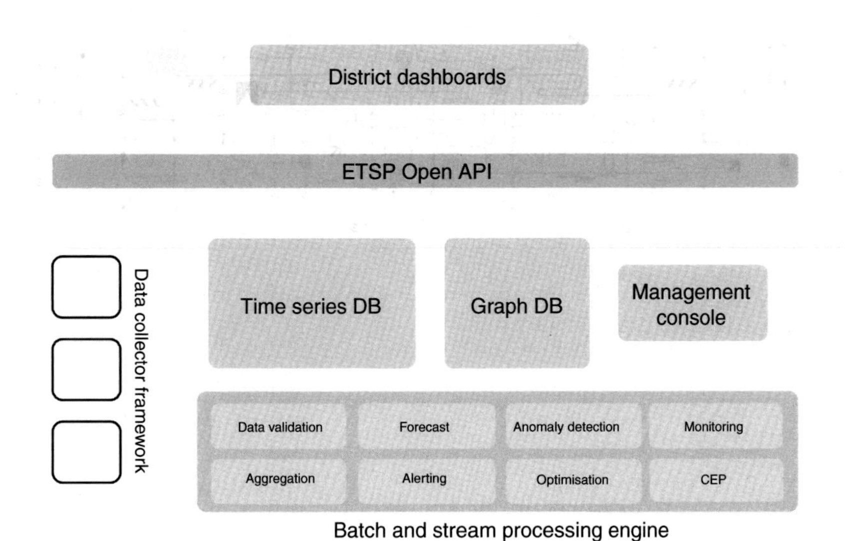

Figure 5.31 ***UrbanPower architecture.***

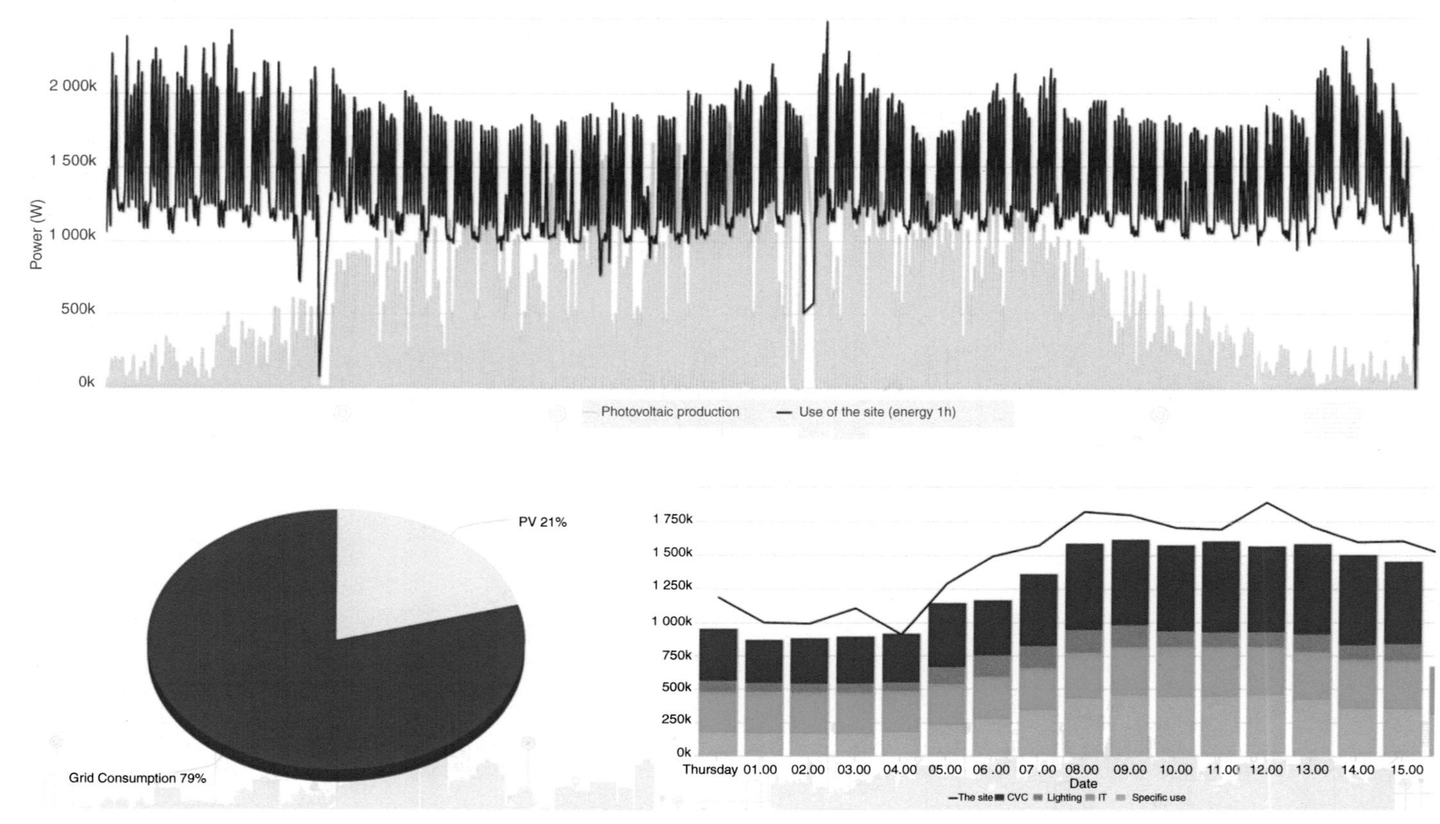

Figure 5.32 *Example of the visualization with the tool visualization box.*

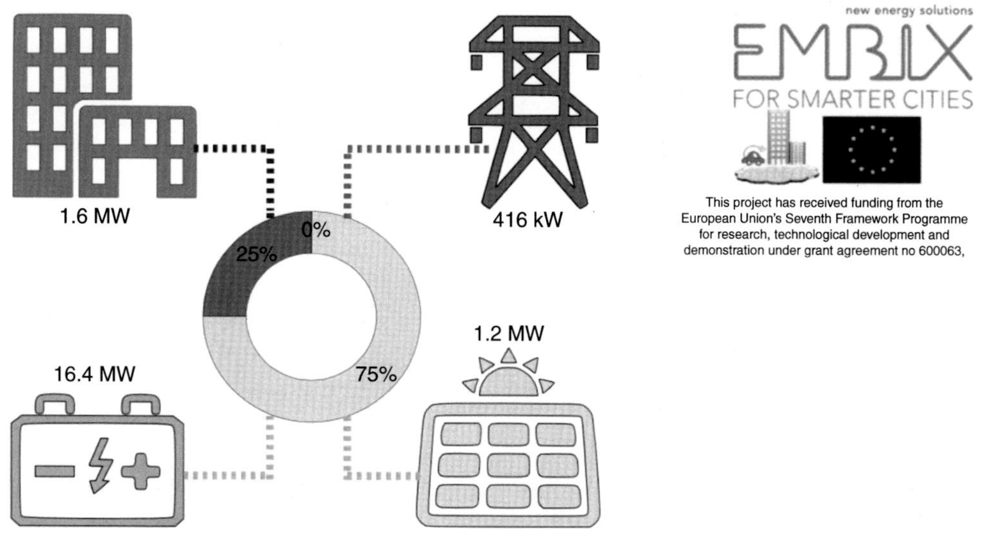

Figure 5.33 ***Challenger Consumption and Production Overview.***

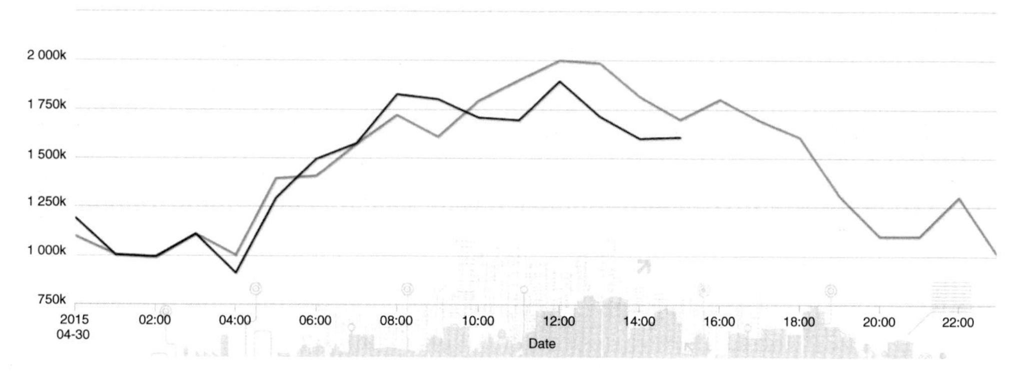

Figure 5.34 ***Consumption forecast versus real consumption.***

Chapter 7). Fig. 5.33 is another example of a visualization service developed within the COOPERaTE project.

5.3.2 Load and Production Forecast

Once extensive historical data is available about energy consumption and production, forecast services can be provided with high accuracy (Fig. 5.34). Forecasts can be made for many usages, such as defining day-ahead flexibility management or the day-ahead market purchase services. For photovoltaic production forecast, different methods have been applied.

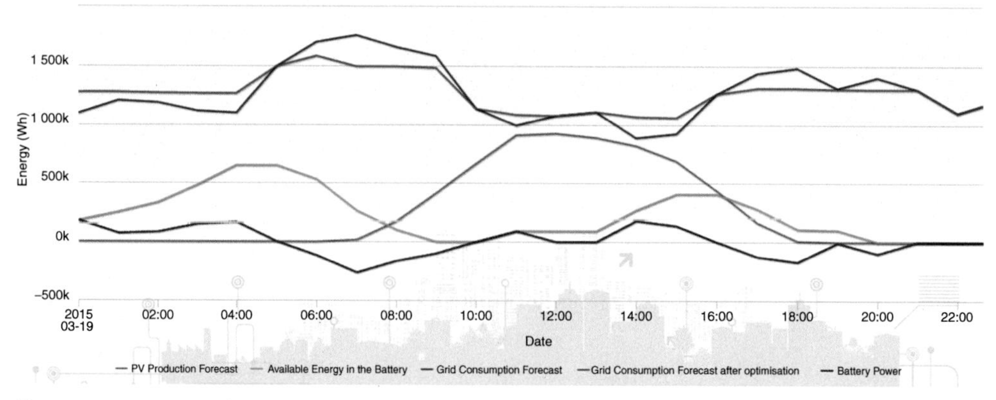

Figure 5.35 ***Battery planning management.***

Through the open EMBIX's platform API, third parties can access historical datasets to analyze and forecast with their own method. The result of the forecasts can then be ingested and visualized within the EMBIX platform.

5.3.3 Flexibility Management Service

A flexibility service allows a campus or neighborhood manager to identify where energy savings can be made and how energy can be traded between different buildings and consumers within the neighborhood or between the neighborhood and the grid. Flexibilities that have been identified include

- Cooling and Heating
- Battery Storage
- Electrical cars

A prototype flexibility service was developed to flatten the grid consumption curve by sending a charge plan every day to the battery system. Fig. 5.35 shows in red the battery power planning and the result on the grid consumption curve (Blue: before optimization and Green: after optimization).

Another flexibility service was tested with the objective to reduce the site electricity bill by considering day-ahead market constraint predictions (Fig. 5.36).

6 CONCLUSIONS

This chapter has presented a case and a prototype for an IoT based SoS approach to implementing an EPN. The current IoT landscape includes a myriad of M2M communication standards, IoT architecture models, reference platforms, and data formats. In order to create an ICT platform for EPNs.

It was proposed by the COOPERaTE project that a single standardized platform was unlikely to be adopted as a solution at distinct scale. Therefore, in order to create

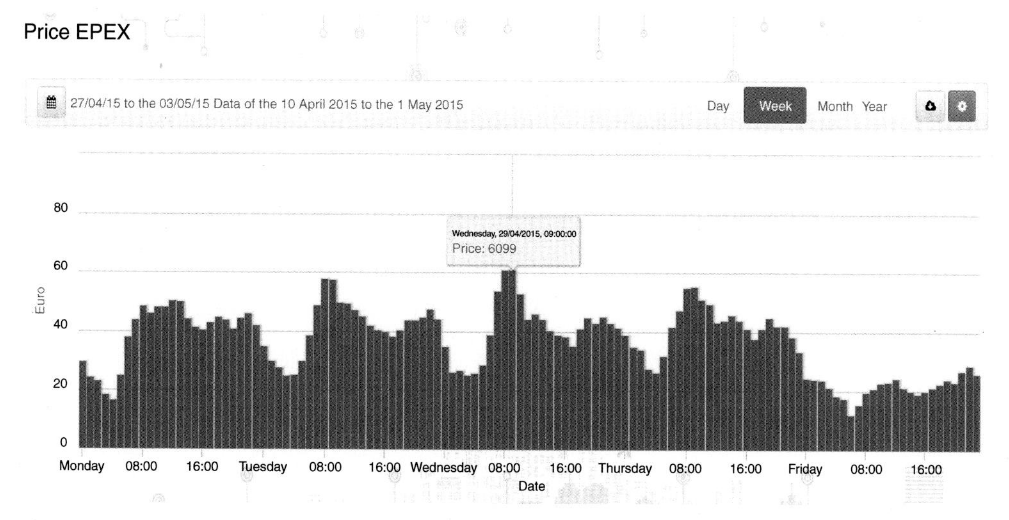

Figure 5.36 ***Electricity Price for Day-Ahead market.***

interoperability across the multiple platforms and communication standards, a data driven semantic operability approach was chosen in the form of the Neighborhood Information Model (NIM) service. The NIM provides interoperability at the data level and can be provided within a single platform, across platforms and also as a service provision by third parties.

A prototype implementation of the NIM was presented followed by a description of example services implemented across three neighborhood management platforms as part of the EU FP7 COOPERaTE project. The efforts of the project demonstrate the feasibility of the chosen approach and how such an approach can create a SoS management architecture and platform to enable EPN.

Some further commentary on the experiences relating to the proposed approach are given in Chapter 8 "Barriers, Challenges, and Recommendations related to development of Energy Positive Neighborhoods and Smart Energy Districts'.

ACKNOWLEDGMENTS

The authors are grateful for the inputs provided by all partners in the Cooperate project, and to the European commission for the funding support under grant no. 600063 (Cooperate).

REFERENCES

Boardman, J., Sauser, B. (2006). System-of-systems—the meaning of. *System-of-systems Engineering, 2006 IEEE/SMC International Conference on*, Los Angeles, CA, USA.

Bonagura, N, Folco G, Kolding M, & Laurini, G. (2012). Analysis of the demand of cloud computing services in Europe and barriers to uptake. Available from http://ec.europa.eu/information_society/newsroom/cf/dae/document.cfm?doc_id=3983

Bradshaw, D, Folco, G, Cattaneo, G, & Kolding, M. (2012). Quantitative Estimates of the Demand for Cloud Computing in Europe and the Likely Barriers to Take-up. Available from http://cordis.europa.eu/fp7/ict/ssai/study-cc_en.html

BSI. (2014). *Smart city concept model- guide to establishing a model for data interoperability*. London, UK: British Standards Institute.

BuildingSmart, IFC4 Add1 Release. (2016a). Available from: http://www.buildingsmart-tech.org/specifications/ifc-releases/ifc4-add1-release

BuildingSmart, ifcXML releases. (2016b). Available from: http://www.buildingsmart-tech.org/specifications/ifcxml-releases

BuildingSmart, ifcOWL. (2016c). Available from: http://www.buildingsmart-tech.org/future/linked-data/ifcowl

Carrez, F., Bauer, M., Boussard, M., Bui, N., Jardak, C., De Loof, J., Magerkurth, C., Meissner, S., Nettsträter, A., Olivereau, A., Thoma, M., Walewski, J.W., Stefa, J., & Salinas, A. (2013). Internet of Things—architecture IoT—A Final architectural reference model for the IoT v3.0. http://www.iot-a.eu/public/public-documents/d1.5/at_download/file

Cisco. (2013). Building the Internet of Things. (pp. 2013–2014), IoT World Forum Reference Model.

CityGML. | OGC. Available: http://www.opengeospatial.org/standards/citygml. [Accessed: 18-Jan-2016].

Compton, M., Barnaghi, P., Bermudez, L., García-Castro, R., Corcho, O., Cox, S., Graybeal, J., Hauswirth, M., Henson, C., Herzog, A., Huang, V., Janowicz, K., Kelsey, W. D., Le Phuoc, D., Lefort, L., Leggieri, M., Neuhaus, H., Nikolov, A., Page, K., Passant, A., Sheth, A., & Taylor, K. (2012). The SSN ontology of the W3C semantic sensor network incubator group. *Web Semantic Science Service Agents World Wide Web*, *17*, 25–32.

Corrado, V., & Ballarini, I. (2012). Report on the Accessible Energy Data. FP7 SEMANCO project public deliverable 3.1. http://semanco-project.eu/index_htm_files/SEMANCO_D3.1_20120921-1.pdf

Fowler, M. (2010). *Domain-specific languages*. Pearson Education.

Gamma, E., Helm, R., Johnson, R., & Vlissides, J. (1994). *Design patterns: Elements of reusable object-oriented software*. Pearson Education.

Greifenberg, T., Look, M., Rumpe, B., & Ellis, K. (2014). Integrating Heterogeneous Building and Periphery Data Models at the District Level: The NIM Approach. In: Proc. 5th Workshop on eeBuilding Data Models (eeBDM) (part of 10th European Conference on Product & Process Modelling (ECPPM) 2014). Vienna, Austria.

Guibene, W., Nolan, K.E., & Kelly, M.Y. (2015). Survey on Clean Slate Cellular-IoT Standard Proposals. Computer and Information Technology; Ubiquitous Computing and Communications; Dependable, Autonomic and Secure Computing; Pervasive Intelligence and Computing (CIT/IUCC/DASC/PICOM), IEEE International Conference on, Liverpool, UK, pp. 1596–1599.

Krcmar, H., Reussner, R, & Rumpe, B (Eds.), (2014). Trusted Cloud Computing. Springer International, Schweiz.

IEC 61970-501:2006. (2016). IEC Webstore. Available from: https://webstore.iec.ch/publication/6215

Industrial Internet Consortium, 2015. Industrial Internet Reference Architecture. Available from: http://www.iiconsortium.org/IIRA.htm

Intel, 2015. Intel IoT Reference Model. Available from: http://download.intel.com/newsroom/kits/iot/insights/2014/gallery/images/INTEL_04_iot-01-1-01.jpg

International Electrochemical Commission, CIM Standards, Geneva, Switzerland. Available from: http://www.iec.ch/smartgrid/standards/.

ITU-T. (2012). *Recommendation Series Y: Global Information Infrastructure, Internet Protocol Aspects and Next Generation Networks*. Geneva, Switzerland: International Telecommunications Union.

Karsai, G. et al. (2009). Design guidelines for domain specific languages. In: Proc. of 9th OOPSLA workshop on domain-specific modeling. Orlando, FL, USA (pp. 7–13).

Krahn, H., Rumpe, B., & Völkel, S. (2010). MontiCore: a framework for compositional development of domain specific languages. *International Journal on Software Tools for Technology Transfer, 12*(5), 353–372.

Look, M. & Greifenberg, T. (2013). D1.2 Report detailing Neighbourhood Information Model Semantics. COOPERATE Control and Optimisation for energy positive Neighbourhoods. http://www.cooperate-fp7.eu/files/cooperate/downloads/COOPERATE_D12.pdf: s.n.

Look, M. & Greifenberg, T. (2013). D1.5 Report on validation site refined Neighbourhood Information Model. COOPERATE Control and Optimisation for energy positive Neighbourhoods. http://www.cooperate-fp7.eu/files/cooperate/downloads/COOPERATE_D15.pdf: s.n.

Look, M. & Greifenberg, T. (2013). D1.7 Report detailing final Neighbourhood Information Model architecture. COOPERATE Control and Optimisation for energy positive Neighbourhoods. http://www.cooperate-fp7.eu/files/cooperate/downloads/COOPERATE_D17.pdf: s.n.

McGibney, A., Beder, C., & Klepal, M. (2012). MapUme Smartphone Localisation as a Service—a cloud based architecture for providing indoor localisation services. In: Proc. International Conference on Indoor Positioning and Indoor Navigation (IPIN), Sydney, Australia.

OMG, (2011). UML 2.4.1 Infrastructure. Available from: http://www.omg.org/spec/UML/2.4.1/Infrastructure/PDF/

Open Interconnect Consortium. (2015). Oic Core Candidate Specification Project B.

OWL 2 Web Ontology Language Document Overview. 2015. Available from: http://www.w3.org/TR/2009/WD-owl2-overview-20090327/

Pesch, D., Ellis, K., Kouramas, K. & Assef, Y. (2013). D1.1 Report on Requirements and Use Cases Specification. COOPERATE Control and Optimisation for energy positive Neighbourhoods, http://www.cooperate-fp7.eu/files/cooperate/downloads/COOPERATE_D11.pdf: s.n.

Postscapes. 2015. IoT Alliances round-up. Available from: http://postscapes.com/internet-of-things-alliances-roundup/.

RDF Schema 1.1. Available from: https://www.w3.org/TR/2014/REC-rdf-schema-20140225/

Rumpe, B. (2011). *Modellierung mit UML: Sprache, Konzepte und Methodik*. Berlin, Germany: Springer Verlag.

Rumpe, B. (2012). *Agile Modellierung mit UML: Codegenerierung, Testfälle, Refactoring*. Berlin, Germany: Springer Verlag.

Schindler, M. (2012). Eine Werkzeuginfrastruktur zur agilen Entwicklung mit der UML/P. Ph.D. thesis. Shaker Verlag.

eeSemantics (Login required). (2016). Available from https://webgate.ec.europa.eu/fpfis/wikis/display/eeSemantics/Home

SPARQL Query Language for RDF. Available from: http://www.w3.org/TR/rdf-sparql-query/#basicpatterns

SWRL: A Semantic Web Rule Language Combining OWL and RuleML. Available from: https://www.w3.org/Submission/SWRL/

Völkel, S. 2011. Kompositionale Entwicklung domänenspezifischer Sprachen. Ph.D. thesis. Shaker Verlag, Aachen, Germany.

Yan, L., Zhang, Y., Yang, L. T., & Ning, H. (2008). *The Internet of Things: From RFID to the Next-Generation Pervasive Networked Systems*. Boca Raton, FL, USA: CRC Books, Auerbach Publ.

Zimmermann, H. 1980. OSI Reference Model—The ISO Model of Architecture for Open Systems Interconnection. IEEE Transactions on Communications COM-28, No. 4.

Business Cases

N. Good, E.A. Martínez Ceseña, P. Mancarella
The University of Manchester, School of Electrical and Electronic Engineering, Manchester, United Kingdom

Contents

1 INTRODUCTION

A Business Case (BC) is defined as the underlying logic and quantitative assessment (through various metrics, see Section 2.5) of any physical or commercial intervention. In the context of an Energy Positive Neighborhood (EPN), examples of a physical intervention may be addition of on-site generation or electricity storage, while commercial intervention may involve accessing and participating in various energy-related markets (e.g., energy wholesale or capacity markets, see Section 4.1). Both such interventions are ultimately related to the previously explored concept of flexibility (see Chapter 2), either to increase district flexibility, or to confer the ability to exploit such flexibility. The intervention under assessment may feature both physical and commercial elements. Indeed this will commonly be the case for EPNs, which may require Information and Communication Technology (ICT) investment to provide the monitoring, communication, computation, and actuation to exploit EPN resources to take part in various energy-related markets.

In liberalized energy systems the identification of an attractive BC should be the primary assessment criteria for any intervention as, without a positive BC (by whichever metric deemed appropriate, see Section 2.5), an entity driven solely by economic considerations has no incentive to invest, and the system will not benefit from access to the flexibility available from the district. That is not to say that other considerations may not factor in the decision. Depending on the status, ownership, and objectives of an entity, social, environmental, or political factors may feature to some extent. For example, a social housing provider may wish to secure lower energy bills for its residents, while an avowedly "green" electricity retailer might procure wholesale electricity above the market rate to satisfy environmental objectives. Further, different entities could consider less immediate economic factors, which are not easily quantified. An example may be a predicted first-mover advantage, as a company foresees some future commercial advantage from developing skills or supply chains which may reap reward in the future. However, without specific information on an entity's preferences, in a liberalized system, the BC should clearly be the default criteria for assessment. Further advantage of adopting this BC view is its generalizable nature, which enables adaptation to any district, and any commercial context. Therefore, understanding the BC for an intervention is, in a liberalized system, analogous to understanding its *viability*. Hence, determination of the BC for an EPN should be considered a crucial part of the EPN process (see Chapter 3).

As shall be described in this chapter, determination of the BC for an EPN requires several steps. First, the energy resources of the EPN must be identified, as well as the markets that an EPN may partake in (see Sections 2.1 and 2.2). This step is far from straightforward as the physical characteristics of EPN resources may be complicated to define and, even though the rules and data for relevant markets are well-defined and deterministic, the amount of information required to properly analyze and understand the market can be substantial. The most significant challenges to assessment of BCs arise from two related characteristics of the energy system and relevant contexts: *connectedness* and *uncertainty*.

Outside the EPN, the energy system can be highly *connected* (or *networked*), both commercially and physically. In the commercial realm, EPNs are connected not only to actors who are directly related to their energy provision, that is, energy retailers, network operators, government (as tax raisers or incentive providers), but also to those who are part of the wider network. The latter could include other retailers, who may buy excess EPN electricity, or bulk fuel suppliers, who may suffer a knock-on effect if an EPN reduces grid electricity import and hence production by large thermal generators. Understanding these commercial "network effects" is a significant and important task, for both the actors concerned (to understand the impact of any intervention by another actor on their BC) and also for those charged with regulatory and political oversight of the energy system (who are concerned with general system efficiency and progress toward political objectives, such as sustainability, affordability, and security objectives). Formulation of practical and complete methods for dealing with these commercial network effects, within the described assessment framework, is hence a particularly useful contribution (see Section 3).

Physically, EPNs are *connected* to other consumers and suppliers of energy through the various (electricity, gas, and heat) networks. Particularly on the electricity network, which must be balanced moment-by-moment (as opposed to gas and heat networks, which have considerable storage in the networks themselves, through the tolerance around network pressure/temperature set-points), physical network effects can be substantial, given network thermal and voltage constraints. Understanding these physical network effects is a requirement of network operators, who have a statutory duty to maintain the reliability of their networks. Although consideration of the effect of interventions on physical networks is outside the scope of this work (given the complexity of network assessment, this would be a substantial extension), this may be a salient aspect for EPNs, given the paucity of options available to deal with distribution network constraints which may occur if there is substantial electrification of heating and transport within the neighborhood.

Besides the network effects that arise from interactions outside the EPN, there are also very salient interactions that occur within the EPN. Given the uncertainty on the commercial relationship between the consumers within the neighborhood and the commercial agent that represents the neighborhood in the market (and the nature of that agent), in this chapter the EPN is defined as both the neighborhood consumers and the relevant commercial agent. A significant aspect of EPN BC assessment is thus the rules of the interaction between the EPN commercial agent, and the EPN consumers. In more common terms, these rules are the contract terms, and they can be significant factor in determining the overall value of the BC (see Section 2.4).

For all extra- and intra-EPN, commercial and physical interactions the *uncertainty* of the nature of those interactions is a crucial factor, which must be considered in the assessment of EPN BCs. In the operational realm uncertainty will lie in market prices, energy service demand and environmental factors (which will determine RES electricity production, as well as thermal constraints for electricity networks, and losses from heat networks). Therefore, the explicit consideration of uncertainty in EPN optimization is a key feature in BC assessment. Over the longer term, uncertainty in regulation and in the behavior of energy system actors is an important aspect to consider in the assessment of BCs. This is due to possible reaction of regulators and other actors to EPN activity (a system feedback), and other extraneous factors. This may include adjustment of regulation to increase alignment with political objectives, and of pricing/trading/investment strategies by actors to maintain/increase profitability.

2 BUSINESS CASE ASSESSMENT PROCESS

Overall, the BC assessment process is analogous to the wider EPN implementation process (see Chapter 3), which uses the SUDA (Sensing, Understanding, Deciding, and Acting) framework. As demonstrated in Fig. 6.1, the first step is identification of the relevant resources of the EPN, while the second step is identification of the markets in which the EPN might participate. These steps can be considered part of the *sensing* phase. The third step is assessment of the regulatory context, and the fourth is assessment of possible commercial arrangements. These steps can be considered part of the *understanding* phase. The fifth step is the quantitative assessment of considered BCs, through production of Cost Benefit Analysis (CBA) metrics, as deemed appropriate by the assessor. The results of this step inform the *deciding* phase, indicating how the EPN should select which BCs to pursue (in the *acting* phase).

2.1 Identify Resources

The *first step* for assessment of the BC for an EPN intervention is to identify the (current and proposed) EPN energy-related resources. Such resources may be characterized as either *static* or *flexible*. *Static* resources are those which are not schedulable or fully dispatchable. Examples may be solar Photovoltaic (PV) plant or wind turbines,[1] whose operation is dependent on the abundance of sun and wind. The BC for *static* resources will derive from the reduced per unit cost of energy produced, and the avoidance of taxes

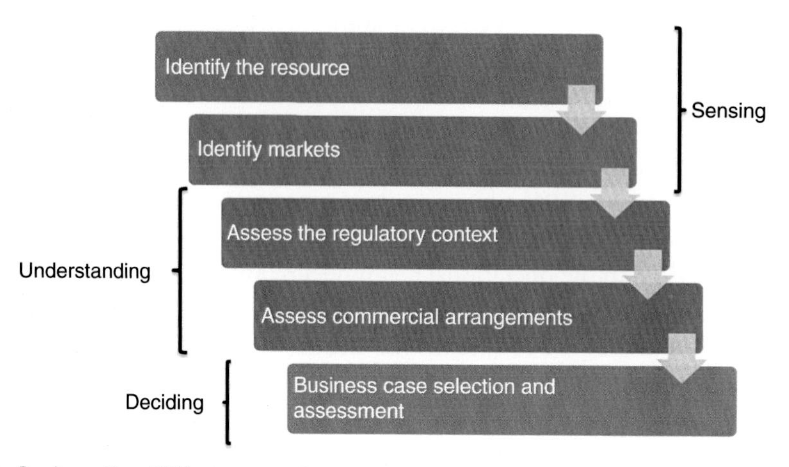

Figure 6.1 ***Business Case* (BC) *assessment process.***

[1]Such resources can be curtailed in extreme cases, at substantial cost to the operator. Given the focus on flexibility, such a situation within an EPN is unlikely, and not considered further.

and charges incorporated in energy import tariffs (see Section 4.1). *Flexible* resources are those which are schedulable and dispatchable. The operation of these resources may be optimized, with respect to price signals from relevant markets, to minimize EPN energy-related costs. Examples may be electrical batteries, or Electrical Heat Pumps (EHP), with associated Thermal Energy Storage (TES), which can use storage to enable flexible operation. Given the previously detailed system benefit of *flexibility* (see Chapter 2) as a better feature for EPN than energy positivity itself, and the potential for profit with low investment (given the requirement for addition of relatively inexpensive enabling technologies, see Chapter 5), BCs based on flexible resources can be considered of primary interest, to both the EPN and to system operators, regulators, and governments.

The focus on flexibility, which in the context of demand-side resources, such as EPNs, may be termed *Demand Response* (*DR*) (Losi, Mancarella, & Vicino, 2015), motivates systematic analysis of categories of DR. As already introduced in Chapter 2, EPN DR can be classified into three categories, which may (as demonstrated in the COOPERaTE case studies, see Section 4) be deployed together:

1. Energy *shifting* in time. Unlike other common EPN energy vectors, the value of electricity can vary significantly over short timescales. Thus shifting import of grid electricity/local electricity generation in time, through use of storage, can produce a positive BC. By explicitly considering a multienergy context (Mancarella, 2014), storage may also be used to shift use of derivative products (such as heat, in the case of TES), in time, to alter the operating schedule of multigeneration devices, such as Combined Heat and Power (CHP) (Kitapbayev, Moriarty, & Mancarella, 2015).
2. *Substitution* of energy vectors. Differences in the cost of energy vectors can motivate shift of the provision of some service or intermediate product (e.g., thermal comfort or heat) from one device/energy vector to another. An example may be shift in provision of heat from an EHP to a gas-fired CHP unit, resulting in a substitution of electricity with gas (Mancarella & Chicco, 2013b; Capuder & Mancarella, 2014).
3. *Trading* of consumer *utility*. If a consumer is willing to put a price on the energy service that they are consuming (explicitly or implicitly, through acceptance of a contract with provision for degraded utility, under specific circumstances), then operation of energy consuming devices may be curtailed, if the cash value of curtailment exceeds the value of foregone utility. An example may be trading of thermal comfort, by accepting a reduced set temperature (Good et al., 2015; Althaher, Mancarella, & Mutale, 2015).

2.1.1 Resource Types

At a practical level, there are multiple energy resources which can be deployed in an EPN to deliver the various categories of DR detailed. These can be summarized as:

1. *Consumers*: The neighborhood may comprise a wide variety of consumers including households, residential, and commercial buildings and small businesses, among others.

ICT will provide customers with information regarding their energy consumption and expenses, and signals from a Neighborhood Energy Manager (NEM, see later) to encourage energy consumption changes (e.g., price signals, incentives, or direct control). The flexibility of customers to modify their energy consumption (through curtailment of energy services) and their response to different signals will vary based on their individual preferences, activities, socioeconomic level, and other factors. Thus, the NEM will require a portfolio of signals to maximize energy flexibility from the EPN while addressing each customer's circumstances.

2. *Distributed Energy Storage (DES)*: Different energy storage technologies may be available within the neighborhood, such as Electrical Energy Storage (EES) and TES. EES from battery banks and other fixed infrastructure will be permanently within the EPN, whereas EES from mobile resources, such as electric vehicles will only be available at certain times (e.g., at night). DES provides the options of storing surplus generation that can be used at times of generation shortages and/or store cheap energy that can be used or sell at times of high prices. The latter under the assumption that some form of time based price signal is available.
3. *Distributed heat (and cooling)*: Heating, cooling and domestic hot water can be provided within the neighborhood with Heat Pumps (HPs) that operate with either electricity or gas. Another popular technology is CHP, which produces both thermal [i.e., heat and/or cooling, if we also include trigeneration (Mancarella, 2006)] and electrical energy in a single process. These technologies provide the option to meet heating requirements either by importing heat directly (if connected to a heat network) or gas for CHP or electricity for the EHP (i.e., substitution of energy vectors). Other options are also available, including (gas fired) absorption heat pumps, engine driven heat pumps, and so forth (Mancarella & Chicco, 2009).
4. *Distributed Generation (DG)*: Small fuel–based or renewable energy–based DG can be installed throughout the neighborhood. Gas–based DG tends to be highly efficient and relatively environment friendly (i.e., low CO_2 emissions), and can be used by the EPN (substituting DG generated electricity for grid electricity) whenever the tradeoff between electricity and gas prices is convenient. Renewable energy–based DG (e.g., solar PV systems installed on rooftops) are not driven by volatile prices but on an uncertain energy resource. The output of these resources has to be forecasted in order to facilitate the use of renewable energies in the management (scheduling and then real-time dispatch) of the EPN (e.g., to ensure that surplus renewable energy can be exported or stored when needed).

The EPN will optimize its performance by coordinating proper actions from the aforementioned actors whenever technically and economically feasible. For example, during periods of high electricity prices, the EPN might decide to meet demand requirements by reducing energy consumption, using cheaper electricity from storage, generating power from DG, reducing EHP electricity consumption by using gas, or any other combinations. As a result, the EPN will have flexibility to meet internal energy requirements whereas following different criteria (e.g., costs and CO_2 minimization and renewable energy use maximization) and provide a wide range of services. How such action is controlled and coordinated is an open point (see Section 2.4). Whatever actor eventually takes this role, it is useful to define and describe the general role of *NEM*, a proposed new role for an entity that manages energy and nonenergy services for a single or for several neighborhoods. While most of the neighborhood control (e.g., energy consumption and generation) will be handled automatically via the COOPERaTE platform and associated services, the NEM takes a supervisory role as a human-in-the-loop for handling unpredictable events and stakeholder management. The NEM may own and operate the platform, just operate it, or just take the role of a market operator within the EPN whereas ownership lies with (and related investments taken by) other actors (e.g., retailers, aggregators, and existing neighborhood actors).

2.2 Identify Markets

Given identification of EPN resources, the *second step* in the assessment of EPN BCs is to identify the markets through participation in which EPN resources may realize value. Throughout this chapter the definition of markets is expanded, to include any mechanism which an EPN is directly or indirectly obligated to participate in [e.g., the tax system, use-of-system (UoS) fee charging regimes]. Although the identity and characteristics of markets vary by system, they can generally be defined as *energy*, *flexibility*, or *capacity* markets (Six et al., 2015). The distinction is useful as each market type has different characteristics. These characteristics are:

1. call frequency, length, and magnitude;
2. call uncertainty and lead-time; and
3. the method of DR call remuneration.

For *energy* markets, DR can be used to arbitrage across time periods and/or energy vectors, to minimize the cost of energy service provision. DR in energy markets will typically be price-based (Siano, 2014), and will thus be frequently exercised. The length and magnitude of DR exercise are dependent on the providers. Therefore, revenue and any loss of comfort can be compared, and managed according to user preferences. For *capacity* market (CM), DR may be used to mitigate generation or transmission/distribution capacity (Martínez-Ceseña & Mancarella, 2014b; Martínez-Ceseña & Mancarella, 2014c; Martínez-Ceseña, Good, & Mancarella, 2015). In CMs DR can be used to reduce peak demand (Aalami, Moghaddam, & Yousefi, 2010). At the distribution level, where there is little competition to provide flexibility, districts may defer or avoid distribution network enhancement (Poudineh & Jamasb, 2014; Martínez-Ceseña et al., 2015). The frequency of capacity-related DR calls can be expected to be rare (Martínez-Ceseña et al., 2015), while the length and magnitude may be large, especially for uncertain, reliability-related capacity services. However, given the high value of reliability, these services will also generate significant revenue. For *flexibility* markets [for services which support the operation of the electricity system by providing balancing/reserve (Vlachos & Biskas, 2013; Schachter & Mancarella, 2015; Ma et al., 2013)], DR may prove a cheaper resource than more traditional alternatives (i.e., large thermal generators), effectively operating as a "real option" in substitution of other system/market resources (Schachter & Mancarella, 2015). These services will fall between those in energy and capacity markets, in terms of frequency of calls, which are usually uncertain and have minimal lead-time. The services are called during contracted periods (those most likely to be imbalanced) and contracts often specify a maximum number of calls per year (Schachter & Mancarella, 2015). Remuneration is generally quite high, as it is related to reliability, though generally less than capacity-related services.

Later the common relevant markets for the three main energy vectors under consideration (i.e., electricity, gas, and heat) will be explored. However, once again, it is important to notice that most value will be extracted from markets related to electricity, since it is there that the real flexibility-driven value opportunities lie (in line with the discussions of Chapter 2), whereby flexibility can be generated by the interaction with other energy vectors indeed (Mancarella, 2014; Capuder & Mancarella, 2014; Mancarella & Chicco, 2013a).

2.2.1 Electricity Power Market

Traditionally, electricity trading has been conducted based on two premises. First, electricity has to be produced and consumed in real time as large scale electricity storage is deemed economically unfeasible. Second, most electricity is generated far from zones of consumption and thus the power grid must transport large volumes of energy through long distances from the generation sites to the zones of consumption (transmission level) and distribute the energy throughout the zones (distribution level) (the EPN will be connected at this level).

In accordance, in liberalized environments the power market has been designed for the trading of energy in advance (i.e., ensuring that enough energy is available when needed), availability of additional energy for contingency purposes (i.e., ensuring that any generation–consumption mismatch can be corrected), and ancillary services (providing energy-related services, such as voltage, capacity, and frequency support, and ensuring that all parts of the grid function as reliably, efficiently, and economically as possible[2]). Additionally current markets have been designed to internalize costs of transmission and distribution (i.e., the price of electricity at the distribution level is higher than at the

[2]It is important to note that not all ancillary services are market tender and some actors, such as generators have the responsibility to provide certain services. For the case of the UK consult (National Grid, 2013).

transmission level)[3] and "protect" consumers from price volatility (i.e., consumers perceive fixed price rates rather than real-time prices). This market philosophy is optimal under traditional practices where all energy flows in one direction (i.e., from generation to consumption) and the demand-side is not meant to actively participate in the management of the system. Both assumptions are not compatible with the EPN concept and, more in general, within the context of extracting benefits from DR (Losi et al., 2015).

In practice, the electricity market comprises several markets for the trade of electrical energy and related service at different time periods. It would be impractical to describe all existing market structures, as the specific characteristics of the markets vary in each country. Nevertheless, a reasonable general description of key markets can be provided as presented in the following.

- *Wholesale market*: Traditionally, most energy is traded in advance in the wholesale market. Energy can be traded from several years in advance until gate closure (gate closure is 1 h before delivery in the United Kingdom). This market is typically pool based, meaning that the market electricity price is set as a function of the prices and power offered by generation and retailers and it might consider network constraints at the transmission level (i.e., nodal prices). Alternatively, energy can be traded without regard for network constraints via unrestricted bilateral contracts as in the United Kingdom (Elexon, 2012).
- *Balancing market*: The balancing market typically operates between gate closure and time of delivery and is managed by the system operator [the Transmission System Operator (TSO) takes this function in the United Kingdom]. In this market, generators offer to increase or decrease their output and retailers and demand-side aggregator offer to increase or decrease the consumption of their consumers if requested by the system operator for balancing the system close to real time (e.g., 30 min before actual delivery).
- *Imbalance settlement*: Actors participating in the markets may not generate or consume the exact amount of power that they traded. These actors might be held responsible for introducing imbalances to the market (balancing responsible actors) and be penalized accordingly. Balancing penalties are calculated ex post based on the position of the balancing responsible actors and the costs for balancing the market (these costs are taken from the balancing market). Settlement processes vary but often prices are designed such that actors are penalized for imbalances, providing an incentive to balance their position before delivery.

[3]Uses of system charges are applied for the use of the transmission network and the different levels of the distribution network, namely high, medium, and low voltage. Thus, at any given time, a consumer connected to the low voltage distribution network will tend to pay more for energy than customers in the medium distribution network, and so on.

- *Ancillary services market*: This market is again typically managed by the system operator, which is the single buyer, and it is used to trade ancillary services beyond the obligation of actors (National Grid, 2013). That is, actors that have fulfilled their ancillary service obligations with the system operator may trade additional services in the market.
- *Retail market*: This market exists whenever different retailers can supply customers in the same zone (e.g., single customers or aggregators). In this case, customers have the option to contract the supplier that best meet their needs (e.g., lower costs or renewable energies support[4]). As retailers can make agreements directly with consumers, it is in principle possible to have different retailers operating within the EPN.

In addition to electricity markets, such as the ones listed previously, a conceptual *constraint management market* is considered in this work. This market would be managed by the Distribution Network Operator [DNO, thereby giving the DNO system management responsibilities in the outlook of becoming a Distribution System Operator (DSO) (Saint-Pierre & Mancarella, 2016)] for the trading of energy services to meet constraints, increase reliability, and relax reinforcement requirements at the distribution level, among other services. The introduction of this market is necessary to enable BCs in which the EPN sells services to the DNO.

2.2.2 Gas Market

Gas, like electricity, is traded as a commodity in liberalized market environments. Like electricity it must be transmitted from the producer to the consumer through transmission and distribution networks. Also like electricity, security of service must be maintained by respecting certain constraints on the operation of the network. However, unlike the electricity sector, large-scale storage is economically viable and the provision of balancing and ancillary services in the gas sector is not a critical issue in the short term.

Gas network pressure constraints, analogous to electricity network voltage constraints, are relatively relaxed. In fact, the ability to vary the network pressure (known as "linepack") can be considered an inherent storage capability. As a result, there is no need for balancing or ancillary services close to real time to ensure the security of the system [while an electricity system must be balanced on a timescale of seconds, gas networks are balanced on timescales ranging from 1 h to 1 month, with typical daily time scales (Keyaerts, 2012)].

Similarly to the electricity market, the gas market comprises wholesale, balancing, ancillary, and retail markets. Gas is traded in advance in a wholesale market to ensure it is available when needed. Unlike the electricity market however there is no "gate closure" at which gas system users (known as shippers) must finalize their position. Instead

[4] When bilateral trading is possible, some retailers can make a compromise to secure a percentage of energy from renewable sources.

shippers (who must balance their position or face penalties) continue to trade up until delivery to ensure a balanced position, competing with the gas system operator who is seeking to maintain system security. In the balancing market, shippers and the gas system operator can procure gas balancing services, in the form of buy/sell options. Often these services are procured bilaterally over the long-medium term, though there are some instances of shorter market based procurement (KEMA, 2009). The ancillary market is mainly used to trade a type of ancillary service from large industrial customers who accept interruptible contracts. These customers are curtailed at times of high stress if all balancing options have been exhausted. The retail market allows customers to contract the retailer in their area that best meets their need.

2.2.3 Heat Market

Unlike electricity and gas, heat (in the form of hot water and sometimes steam) is not suited to transport over large networks due to the large losses experienced. Although heat losses can be mitigated to enable city scale networks and heat sources several kilometers from their demand, most heat networks are much smaller. Currently, there are more than 5000 district heating systems in Europe supplying more than 9% of total European heat demand (DHC+ Technology Platform, n.d.), though penetration varies greatly by country. Heat networks may, however, supply several EPNs, if they are located close enough together, enabling exploitation of trade-offs between heat and other energy vectors (e.g., gas and electricity). The localized nature of heat networks means that there are no common codes for organization of heat markets (where markets exist; many heat networks operate as monopoly heat providers).

2.2.4 Emissions and Efficiency Markets

Increasing concerns about environmental threats and energy demand growth have directed attention to the usage of energy, particularly to the associated efficiency and CO_2 footprint. Accordingly, several countries have developed emissions and efficiency markets, or other mechanisms, to incentivize energy efficiency improvements and/or CO_2 reductions. Emissions reductions and efficiency improvement mechanisms are closely related and may be mutually exclusive to avoid overlapping as energy efficiency improvements also result in CO_2 reductions. An example of this is the United Kingdom's carbon reduction commitment energy efficiency scheme, which only targets efficiency improvements from actors that are not already covered by climate change agreements and the EU emissions trading system (DEFRA, 2010).

The use of markets for the trade of emission reductions or efficiency improvement (in the form of import energy reductions) certificates is meant to facilitate achievement of the underlying objectives (e.g., binding targets at the national or European level) at the lowest cost. That is, actors that can reduce their CO_2 footprint and energy consumption at low costs are incentivized to obtain certificates that would be bought by actors who would otherwise incur significant costs to meet their mandatory targets. The CO_2/energy reduction targets, as well as the allocation of certificates are based on baselines set for each actor.

The baselines are key factors for the success of the aforementioned markets, as they are crucial in the determination of the magnitude of the emissions/energy savings achieved by each actor, and thus the allocation of certificates. The baseline can be set as static or dynamic. The estimation of the former is relatively simple as it can be based mainly on historical data, but the accuracy of such baseline can be low. Conversely, dynamic baselines are deemed to produce the best results but their estimation is complex, as the baseline has to be calculated periodically in consideration on factors that would affect underlying emissions/consumption (e.g., changes in environmental conditions, available technology, and consumer behavior, among others). In this regard, the information captured by the EPN platform should favor the use of dynamic baselines.

In Europe CO_2 is traded as a commodity by power stations and large industrial users in the EU emissions trading system. However, to reduce regulatory burden, installations with emissions less than 25,000 tons CO_2 per year and (if combustion activities are taking place) a rated thermal input of 35 MW as well as hospitals are omitted from the scheme, given the implementation of equivalent measures (EU, 2003). Energy efficiency certificates can also be traded as commodities in some countries (e.g., Italy and France) by large actors, such as Energy Service Companies (ESCos) and gas distributors (Sorrell et al., 2009).

Even though there are no emissions or efficiency markets for small applications, there are bespoke support schemes and obligations for small actors (e.g., at the household level), such as feed-in tariffs (OFGEM, 2014). These schemes incentivize or oblige small actors to invest in energy efficiency improvement mechanisms (e.g., insulation) or renewable generation systems (e.g., domestic PV systems). The use of these schemes, instead of exposing the actors to the markets, implies that small consumers are deemed unable or unwilling to handle price volatility from the markets (this is not the case for an EPN).

2.3 Assess the Regulatory Context

In order to appreciate the value of any BC that derives from participation in a market, it is crucial to acknowledge that this value is highly sensitive to market rules and regulations, which dictate how DR resources can participate. More broadly, regulation which determines the cost-reflectiveness of market prices (alternatively, the degree to which costs are socialized across some group) can affect the potential to profit from deploying DR (Haring, Kirschen, & Andersson, 2015). Therefore, the *third step* in the assessment of EPN BCs is the assessment of the impact of market structures and the regulatory context on the viability and value of the BC. Salient aspects include: (1) any rules governing the ability of consumers to partake in energy-related markets; (2) if and how externalities (such as greenhouse gas emissions) are priced and charged; (3) how energy network access is charged for or, equivalently, how operation, maintenance, and expansion of networks are funded; and (4) how reliability of supply is maintained.

Regulations necessarily exist to designate the framework under which regulated entities can operate. Additionally, they can be used by relevant governments as tools to achieve policy objectives. As discussed in Chapter 2, the regulatory framework in which the EPN operates can have a significant impact on the ability to achieve the energy positivity/economic value.

The nature of the regulatory frameworks may vary (possibly significantly) by system. In order to study the potential effect that different regulatory frameworks can have on the various BCs an EPN might pursue, it is useful to characterize the frameworks in terms of the energy system aspects which regulation can impact. Here arrangements will be summarized for four aspects for which regulation can have a significant impact: (1) market participation of EPNs, (2) CO_2 reduction, (3) network fees, and (4) reliability of supply. Considering that the nature of any aspect of regulation may vary widely, these aspects of the regulatory frameworks are characterized based on a conservative–liberal spectrum.

An aspect of regulation is deemed conservative when it is highly prescriptive, limiting uncertainty and risk. This type of regulation favors "traditional" arrangements while restricting innovative developments, which may otherwise increase value. Liberal regulations on the other hand are typically not prescriptive, allowing parties to exchange various commodities and services without central price control. This tends to increase uncertainty and risk for parties, but also increases value in general, and particularly for parties who are appropriately engaged and have the flexibility to trade commodities while managing risks (e.g., EPNs).

Clearly, exploring all points on a conservative-liberal spectrum for all aspects of regulation is impractical. Accordingly, in order to explore the effect of the regulatory context on the explored BCs in a practical way, only two regulatory scenarios will be explored, namely traditional and liberalized. The traditional scenario is largely intended to be broadly equivalent to current arrangements, while the liberalized scenario is intended to explore the effect of pushing liberalization as far as possible, and may be considered representative of possible regulatory arrangements in the near future. For each scenario, the effect of regulation on EPN operation will be explored in terms of local generation, renewable generation, grid efficiency, and grid security.

2.3.1 EPN Market Participation

2.3.1.1 Traditional

In a traditional regulation scenario, electricity and gas consumers must buy energy from retailers at flat or simple two-time band time-of-use prices. Wholesale market risk lies with the retailer, who includes risk premiums in their retail tariffs. If the EPN includes DG, net exports from each Grid Connection Point (GCP) are remunerated at a regulated, fixed rate. Ancillary service markets may be available to EPNs through their retailer or an aggregator who may exercise direct control over resources. Ancillary services are limited to DG (rather than storage or flexible demand) due to the complexity of baselining storage and demand.

Traditional regulation with respect to EPN market participation may encourage local energy generation, especially for risk averse owners (as returns are somewhat certain). However, traditional regulation may make more marginal DG unviable by restricting direct access to the market, especially for more risk-affine operators. By isolating DG from market signals it is likely that the traditional regulatory scenario may restrict CO_2 reductions, as price signals may indicate times of high/low grid carbon intensity. Similarly, insulation from energy market signals is likely to reduce grid efficiency. Allowing some access of EPN resources to ancillary service markets may partially support the grid security objective, though the barriers to participation for storage and flexible demand restrict the potential contribution.

2.3.1.2 Liberalized

In a liberalized regulation scenario an active demand-side party, such as an EPN, can access energy and ancillary service markets directly. They will also have freedom to contract with parties, such as aggregators or ESCos to participate in such markets on their behalf.

Such liberalization may encourage local energy generation by increasing the economic value achievable by DG operators, although this may also increase transaction costs (as the number of transactions increases, especially if an intermediary is engaged). DG profitability may particularly be boosted by removing risk premiums charged under the traditional scenario. However, this may result in more risk being shouldered by the DG operator. Overall CO_2 emissions should decrease, as visibility of market prices (which will generally correlate with grid CO_2 emission) may motivate EPN DG, in conjunction with EPN storage and flexible demand, to reduce emissions (e.g., increasing CHP operation at times of high grid CO_2 emission and reducing CHP operation at times of low grid CO_2 emission). Grid efficiency is expected to increase as the liquidity of energy markets improve, while security should also improve, as more resources (storage and flexible demand, as well as DG) can contribute to system ancillary services.

2.3.2 CO_2 Reduction and Low Carbon Energy Integration

2.3.2.1 Traditional

In a traditional regulation scenario, CO_2 reduction and low carbon energy integration is incentivized at the EPN level by subsidies for low carbon electricity and, possibly, heat generation (feed-in tariffs, for instance). A carbon market is applicable for large-scale electricity generation.

At the local level, the traditional CO_2 reduction regulation is likely to increase local energy generation, by increasing the attractiveness of local (low carbon) generation (such as CHP). However, such an incentive may also encourage increased electricity consumption from low carbon heating (such as EHP) regardless of the actual demand (potentially resulting in undesirable heat-spilling).

A carbon market at the bulk level will serve to increase local generation by adding costs to grid electricity. With regards to CO_2 reductions, such regulation will generally facilitate CO_2 emissions reduction. However, by incentivizing "low carbon" generation at all times, it may, in fact, encourage increasing (local) CO_2 emissions by incentivizing CHP generation when grid electricity is actually lower carbon (e.g., due to high wind or solar penetration). Also, such blunt incentives may reduce grid efficiency by distorting market signals. Grid security should not be significantly affected.

2.3.2.2 Liberalized

In a liberalized regulation scenario CO_2 reduction and low carbon energy generation is incentivized through market mechanisms. The carbon market is extended down to the local, EPN, level. This may reduce local energy generation, as the incentive to operate local generation will vary according to the grid/local CO_2 intensity spread. Accordingly, CO_2 reductions are expected to become more effective, as local and grid emissions would be evaluated on the same basis. Grid efficiency should improve as distorting blunt subsidies are removed. Grid security should not be significantly affected.

2.3.3 Distribution Network Fees

2.3.3.1 Traditional

In the traditional regulation scenario, Distribution Use of System (DUoS) fees are predominantly charged on the energy component. Charges may be time-banded but are static through the year. Such an arrangement will generally encourage local generation, as the widespread application of DUoS fees will provide a consistent incentive to operate local generation to reduce DUoS charges. Whether such regulation increases or decreases CO_2 emission will depend on the carbon intensity of local generation. Given the general correlation between network stress (i.e., when spare network capacity is scarce) and generation scarcity, such a blunt DUoS fee regime is unlikely to promote network efficiency. The failure to dynamically signal network stress may also have a negative effect on network security, as parties on the distribution network would not be encouraged to react at times of high network stress.

2.3.3.2 Liberalized

In the liberalized regulation scenario, high resolution (typically half-hourly) DUoS fees reflect distribution network stress more accurately. Such fees may also become dynamic, signaling system stress to users. This may remove some incentive for local generation, as networks are typically stressed for only limited periods of the year, limiting the periods when DUoS fees will be charged and, thus, limiting the opportunities to benefit from avoiding them. Assuming that local generation generally has lower carbon intensity than grid generation, this may also increase CO_2 emissions. Grid efficiency and security should be improved as distribution network stress (which generally coincides with generation scarcity) is accurately signaled to users.

2.3.4 Reliability of Supply

2.3.4.1 Traditional

In a traditional regulation scenario, supply reliability at the local level is the responsibility of DNOs. DNOs as regulated entities will generally aim at meeting preagreed reliability targets in terms of average energy not supplied, the average customer curtailment index, customer interruptions, and customer minutes lost among other reliability metrics (the specific metrics may vary in different countries). For this purpose, DNOs tend to spend large amounts of capital in redundant infrastructure (and even network automation and reconfiguration systems) to guarantee that electricity supply can be maintained to most customers even after contingencies (e.g., lightning, failure of a distribution line or transformer, and so forth) occur in the system. In other words, as contingencies reduce the capacity of the network (potentially making the network unable to supply all customers), investments in spare emergency capacity are made so that, even after losing capacity due to a contingency, the network can supply most customers. This practice assumes that all reliability support comes from the network itself, without consideration of any support from the demand side that, for instance, EPNs could provide.

2.3.4.2 Liberalized

In a liberalized scenario, DNOs may make agreements with active customers, such as EPNs to provide network support during emergency conditions after contingencies occur and the system has not been fully restored. This fundamentally changes the rationale by which the distribution networks are planned as, instead of investing beforehand in spare distribution network capacity, the DNOs can request additional capacity from EPNs (in exchange for a payment). The EPNs would then release network capacity by (1) reducing their electricity imports, (2) isolating themselves from the network if the EPNs can also work as a microgrid, and (3) even exporting electricity to nearby customers outside the neighborhood further releasing network capacity (Syrri, Martínez-Ceseña, & Mancarella, 2015). In principle, network reliability support from EPNs can facilitate more economically effective distribution networks as the DNOs could choose the cheapest alternative to increase network reliability between investing in spare network capacity and paying active customers for network support.

2.4 Assess Possible Commercial Arrangements

As described, regulation often dictates how consumers can partake in energy-related markets. However, as energy-related markets, at all levels, from wholesale to retail levels, are liberalized there may be multiple options open to an EPN as to how they exploit their flexibility to provide DR. They may prefer to contract with an ESCo, aggregator, or similar, who may exploit flexibility within the terms of the contract (including level of service obligations, e.g., for thermal comfort), forgoing some profit in order to transfer risk to the commercial partner. Alternatively, an EPN may wish to accede to the relevant markets themselves, assuming the attendant price risk completely, and keeping all profit. Or, the EPN commercial arrangement will fall somewhere between these two extremes. Determination of the commercial arrangements for exploiting EPN flexibility is hence the *fourth step*, in assessment of EPN BCs.

Following convention, the EPN commercial partner can generically be referred to as the "aggregator" (see Losi et al., 2015, for a detailed exposition on this actor). Given the separation of roles from actors, it is possible for various actors to take the role of aggregators. This would have a significant impact on the BC of EPNs. If the role of an aggregator providing services to an EPN is to be fulfilled by an already existing actor, that actor would be one which already has a strong relationship with the EPN. That is, either the DNO (by virtue of its physical connection and likely role[5]) or the retailer (who already has a commercial relationship with the EPN) would take the role of aggregator. As the DNO, as a monopoly provider of access to the distribution network, is necessarily a regulated entity in liberalized power systems, it is unlikely to fulfill the role of aggregator.[6] The DNO would have to forego its current role to become an aggregator. Retailers on the other hand do not have impediments to take both a retailer and an aggregator role and indeed may be well placed to exploit the synergies available from providing energy from the centralized power system to the consumer while also providing flexibility services from the consumer to the power system.

It is important to note that the role of aggregator could also be fulfilled by new actors that are currently not active in the power system. This could include any actor with a strong relationship with EPNs (e.g., local government bodies, property developers, equipment manufacturers, or facilities management companies) though there would be synergistic advantages for those actors which can exploit technical knowledge (e.g., equipment manufacturers) or existing ICT systems (e.g., facilities management companies). Equally, the aggregator role could be fulfilled by independent actors unrelated to the EPN. Such actors may focus on aggregation alone with responsibility stopping at

[5]The DNO might be in charge of the installation and operation of EPN enabling infrastructure, such as smart meters.

[6]Regulated actors are forbidden to compete with nonregulated actors.

the power system side of the meter, or they may exploit synergies that are enabled by additionally taking the role of an energy services provider. Such an actor may be termed an "aggrESCo" (concatenation of aggregator and ESCo) and could assume (through a contractual agreement) the responsibility for provision of certain energy services (e.g., heating), possibly through the assumption of an outsourced NEM role. Indeed if the NEM is willing to bear the costs (e.g., license and transaction costs) it could fulfill the aggregator role itself. Clearly there are many options for an aggregator that is not joined with some other energy system actor.

It is also important to note that an EPN may be achieved without central coordination of interaction with the power system (i.e., electricity purchasing and selling of flexibility). The NEM role may be reduced or eliminated if parties within the EPN retain/establish their own commercial relationships with retailers/aggregators. If there is demand for intra-EPN energy/flexibility markets the NEM may be the operator for such markets. Alternatively, if no intra-EPN markets are to exist, the NEM role may become peripheral with the concept of energy positivity being achieved through the direct interaction of parties within the EPN with various retailers and aggregators in the wider power system.

For any arrangement where the EPN (through the NEM) is not acting in markets directly, a salient question is: how are the benefits (i.e., profits) of EPN BCs to be shared between the EPN and aggregator. Aspects that should be covered should include investment (i.e., who pays for enabling infrastructure), maintenance (i.e., who pays for maintenance of BC-related infrastructure), operation (i.e., who controls relevant plant), and profit share and risk (how should profits be shared, and how should any risk be shared).

2.5 Quantitative Assessment

The *fifth and final step* of the BC assessment process is the quantitative BC assessment. Bringing together information of the EPN resources, the relevant markets, the regulatory environment and the EPN commercial arrangements, and using the appropriate physical models of the EPN and appropriate optimization techniques (see Chapter 4), the expected value of analyzed BCs, assessed through various metrics, can be calculated.

A CBA consists of comparing costs and benefit flows associated with a given investment project from the perspective of a given actor (the selected actor is not necessarily the actor that makes the investment), with the objective of identifying whether or not the benefits from the underlying investment offset the corresponding cost. Thus, the CBA provides an indicator of whether or not a given investment project is beneficial or not for a given actor. In the context of the COOPERaTE project, several CBAs were performed from the perspective of different actors (e.g., consumers, DNOs, retailers, and so forth) under different market frameworks and considering different attributes for the EPNs (e.g., infrastructure and associated services) with the objective of identifying the conditions that engender or deter attractive BCs for EPNs.

The CBA can be performed based on a variety of well-known and widely used criteria based on the time-value of money, such as Net Present Value (NPV), Internal Rate of Return (IRR), payback time, among other criteria based on other premises (e.g., carbon footprint). A brief description of the time-value of money principle, the aforementioned criteria, and multicriteria analysis is provided as follows.

2.5.1 Time-Value of Money

The time value of money principle argues that money has a greater value in the present than in the future. That is, a rational investor (and most people) prefers to receive money in the present than to be promised the same amount of money in the future. Similarly, a rational investor would rather pay money in the future than pay the same amount immediately.

One reason for this perception of time-value of money is that capital secured in the present can be saved or invested to produce revenues in the future. Another reason is that capital devaluates with time due to inflation. Moreover, if a given amount of money is promised in the future, there is always some risk that the right amount cannot be delivered (e.g., less or no money might be received). Accordingly, in order for rational investors to sacrifice money (invest) in the present; the expected returns must include a premium that compensates for the time-value of money effect. The premium required to compensate for the time-value of money effect is usually expressed as a factor, namely discount rate (d).

Based on the time-value of money consideration, it is possible to quantify series of cash flows (i.e., costs and/or benefits) that occur in different periods (e.g., all cash flows within the same year) by referencing them to the same year (usually the present) based on the discount rate. This procedure is called discounting.

The time-value of money and related discounting procedure may be better explained with an example. Assume that the time-value of money is considered to be 10% (i.e., $d = 10\%$, thus a 10% premium per year is required from future money). That is, an investor would be indifferent between receiving €100 today, €110 in a year [i.e., €100 × (100% + 10%], €121 in 2 years [i.e., €100/(100% + 10%)^2], and so forth. Accordingly, an investment that offers to provide €110 in a year plus €121 in 2 years has a discounted value referenced to the present (or present value) of €200 [i.e., €110/(100% + 10%) + 121/(100% + 10%)^2]. In general terms the present value of a cash flow can be expressed as follows:

$$\text{Present value} = \sum_{t}^{T} \frac{\text{Cash flow}_t}{(1+d)^t} \tag{6.1}$$

where t represents the different time periods (in this case, years) during which the cash flows occur and T represents the last time period under consideration (usually the end of the operational lifetime of the investment project).

2.5.1.1 Net Present Value

The NPV criterion consists of comparing the costs and benefits associated with an investment project (during the expected operational lifetime of the project) in terms of their present value. The NPV can be expressed as follows:

$$\text{NPV} = \sum_{t}^{T} \frac{\text{Benefits}_t - \text{Costs}_t}{(1+d)^t} \tag{6.2}$$

This criterion produces a straightforward assessment of the project in the sense that the investment project is deemed beneficial as long as the NPV is positive (the project is considered inconvenient otherwise). This occurs because a positive NPV implies that the expected benefits not only exceed costs but also provide a premium that surpasses the discount rate. That is, the actor is better off investing in the project (or incurring the costs associated to the project) than placing the same amount of money elsewhere (e.g., the bank) even when receiving a premium equivalent to the discount rate for the money (10% in the example).

Even though the NPV criterion is clear, it does not provide much information about the investment. For example, it does not specify the amount of years that the actor has to wait before the benefit from the project offsets the costs (e.g., payback time), nor the real premium that the project is providing.[7] Additional information can be obtained from other metrics (e.g., payback time and IRR).

2.5.1.2 Payback Time

The payback time criterion is the period (normally years, but it can be expressed in weeks, months or other time periods) that is required for the project to become profitable (i.e., the minimum time needed to render the NPV equal to zero). Thus, the payback period can be calculated as the minimum value of the time period T that satisfies the following equation:

$$\sum_{t}^{T} \frac{\text{Benefits}_t - \text{Costs}_t}{(1+d)^t} \geq 0 \tag{6.3}$$

This tends to be an important metric as most actors are not only interested in highly profitable investment projects, but on profitable projects that payback in the short term.

2.5.1.3 Internal Rate of Return

The IRR is an indication of the exact premium rate that an investment project is offering in exchange for all costs incurred. The IRR can be estimated as the discount rate (d)

[7]A positive NPV indicates that the premium is higher than the discount rate (10% in the example), but it does not indicate the exact value of the premium.

that renders the present value of both benefits and costs the same. That is, the discount rate that satisfies the following equation:

$$\sum_{t}^{T} \frac{\text{Benefits}_t}{(1+d)^t} = \sum_{t}^{T} \frac{\text{Costs}_t}{(1+d)^t} \quad (6.4)$$

Conversely to the NPV criterion that centers on absolute profits, the IRR criterion shows the efficiency of an investment with regard to the premium rate that it produces compared to the costs. That is, whereas the NPV criterion favors investments with high profits regardless of the costs, the IRR criterion favors projects that produce more benefits in comparison to costs.

3 NETWORKED INTERACTIONS

A further step that can be of interest to other actors within the energy system can be analysis of the "networked interactions" relevant to any EPN BC. Energy systems are in fact highly networked, and increasingly so as demand-side actors, such as EPNs, actively participate in markets, and create further links between various energy vector markets (e.g., for electricity and gas). Such effects can significantly affect the BCs of non-EPN energy system actors, and hence their quantification is of interest.

To describe the network effects of an EPN intervention in the energy system it is first necessary to understand the actors external to the EPN. Given the uncertainty of the relationship between the aggregator, retailer (who may take the role of aggregator), and the EPN, the EPN is bundled together with the retailer (electricity and gas) and aggregator, to form a "Retailer–Aggregator–EPN" (RAEPN) in this chapter. Energy system actors that are external to the RAEPN actor are defined as follows.

3.1 Energy System Actors

Based on the scope of this work, and the characteristics of the electricity, gas, heat, and emissions/efficiency markets, the external actors that will have a greater influence on the business of EPNs are:

- *TSO*: The TSO is a regulated entity, that is responsible for the management and sometimes the development of the transmission grid and for the operation of most ancillary service markets (the TSO might own or only operate the network). TSOs provide physical access to the electricity market to different players (e.g., generators) according to nondiscriminatory and transparent rules and might take the role of market operators. In addition, TSOs ensure the security of supply, safe operation and maintenance of the system, and generation–consumption balance via services provided by different actors in accordance to the grid code or contracted in the balancing and ancillary markets (National Grid, 2013). The costs of buying balancing services are usually transferred to the parties who are unbalanced; whereas the costs of real-time balancing are typically transferred to consumers [the wholesale electricity price is increased by a Transmission Use of System (TUoS) charge].

- *DNO*: The DNO (electricity or gas) is a regulated entity, who is responsible for the transport of the energy on the distribution networks between transportation/transmission networks and the consumers. Similar to TSOs, DNOs provide physical access to the distribution network to consumers according to nondiscriminatory and transparent rules. The DNOs are also in charge of the safe and economic operation of the network and for investments in new infrastructure. Distribution costs are externalized to consumers in the form of DUoS charges and connection charges. Although, depending on the characteristics of the consumers and regulation, the charges may be paid directly by consumers or by retailers (i.e., retailers represent small consumers).
- *Producers*: Bulk generation comprises traditional large generators connected to the transmission system, such as steam turbines, wind parks, or big hydro plants. These generators can participate directly on the wholesale, balancing and ancillary markets. This type of generation tends to have an advantage in the wholesale market due to their economies of scale. However, their fitness to participate on the balancing and ancillary markets depends on the generation technology (e.g., generation technologies with a slow and expensive response, such as nuclear are not adequate to provide balancing or ancillary services).
- *Gas supplier*: The gas supplier is the only actor in the gas sector that is addressed in detail in this book. The main function of the gas supplier is to trade gas with consumers (e.g., the EPN). For this purpose the supplier has to either extract gas from the network or contract another actor that has the rights to extract gas from the network.
- *Gas shipper*: Gas shippers are responsible for transporting gas through gas networks. They have to sign a contract with the gas system operator and are balancing responsible in the sense that they must ensure that the amount of gas they inject into or extract from the system balances with their contract.
- *Government*: Governments are responsible for collecting taxes and obligations on energy consumption. Generally, a Value-Added-Tax (VAT) is applied, while various Environmental and Social Obligations (ESO) may also be charged. Governments may also provide Low Carbon Incentives (LCI) to motivate investment in and operation of low carbon technologies.

3.2 Value Mapping Methodology

Given the actors defined in the previous section, for the considered context and the markets relevant in that context, a value mapping methodology should be used to graphically represent the connections in the energy system. This mapping can then be used to inform analysis of the network impact of an EPN intervention. It can also be useful in

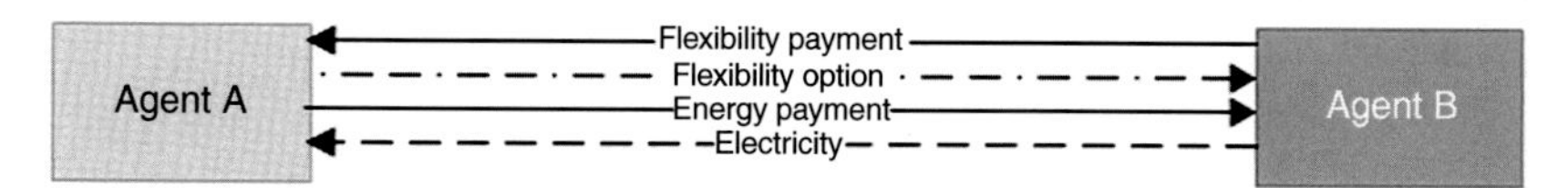

Figure 6.2 ***Generic value exchange mapping for flexibility BCs.***

the identification of markets, and could be undertaken in tandem with step 2 of the BC assessment process (see Section 2.2).

The foundation for the mapping methodology is the e^3 value methodology. The e^3 value methodology is an interdisciplinary methodology for the evaluation of BCs in highly networked environments, which maps all relevant interactions with the aim of facilitating an understanding of the effects on all actors and to assess the BC profitability with respect to those actors (Pudjianto et al., 2010). As hinted previously, a key aspect of the e^3 methodology is mapping of commodity and cash flows as exchanges; every commodity flow must have an associated remuneration.

The e^3 methodology describes BCs to a high level of detail utilizing complex sets of ports, agents, and triggering actions (stimuli). However, the level of detail may be considered excessive for strategic, practical assessment as the one presented in this book. Therefore, a simplified version of the value mappings is utilized here.

As in the e^3 methodology, actors and their relevant roles are specified and the interactions between the actors are represented by the relevant exchanges of commodities and cash (Fig. 6.2). It is vital to note that while the e^3 methodology focused on exchanges related to the power system, here exchanges of all energy-related commodities are mapped, which is an innovative development.

This mapping framework was developed for multicommodity exchanges both outside and within the neighborhood.

3.2.1 Energy System Actors/Roles

The actors in the energy system may take several roles simultaneously in related markets external to the EPN, which have to be clearly defined when creating the mappings. An example of this is the Electricity TSO (ETSO), which can normally fulfill two different roles (Fig. 6.3).

For the sake of simplicity, a pattern code is adopted for the representation of roles taken by relevant actors. As shown in Fig. 6.4, roles are broken down into four groups.

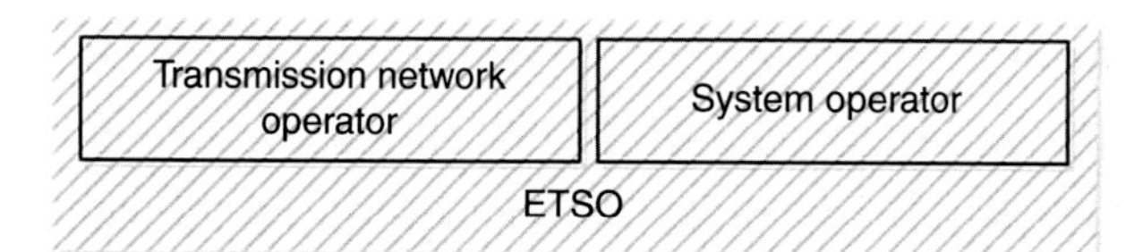

Figure 6.3 ***The ETSO actor and its roles.***

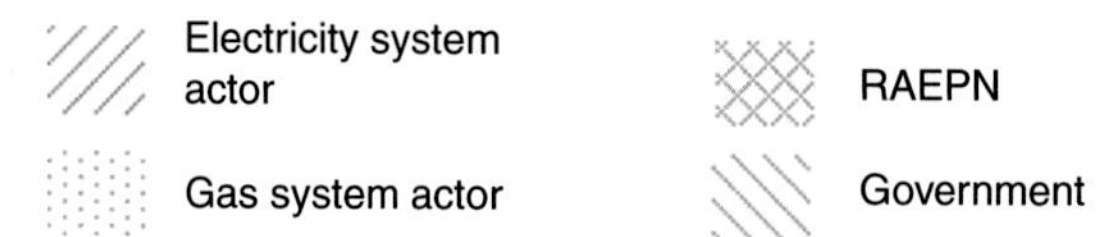

Figure 6.4 ***Representation of actors in the energy system mapping.***

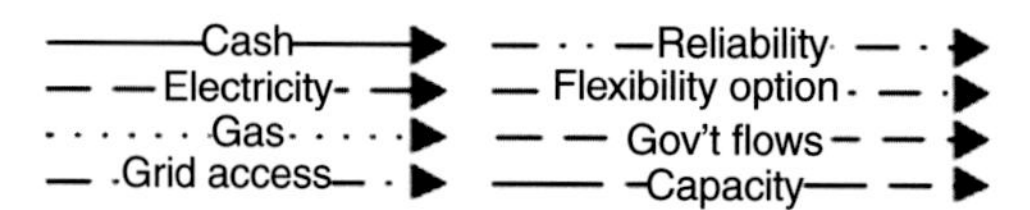

Figure 6.5 ***Representation of flow types in the external EPN mapping.***

3.2.2 Energy System Flows

The various flows between the energy system actors are represented with arrows of different patterns as shown in Fig. 6.5.

In this list electricity and gas are tangible flows as the commodities are directly measurable from existing markets and the attendant cash flow can be simply calculated by applying a per unit cost. The remaining commodity flows are intangible. The flow could be guaranteeing access to a network to allow physical delivery of another commodity (i.e., grid access), guaranteeing the ability of that network to operate so that that commodity could be delivered (e.g., reliability) or be an option to enact the exchange of a commodity (flexibility option). Government flows are related to collection of taxes or other obligations, while capacity flows are related to payments for provision of generation capacity.

4 CASE STUDIES

The described BC assessment process is illustrated by application to assessment of various interventions on two real neighborhoods, one in France and one in Ireland. In this section the various relevant energy prices for each context are identified and presented, before operational (annual) cash flows, for the RAEPN actor and other energy system actors, are presented, for various BCs. Subsequently, methods for splitting BC benefits between the EPN and the retailer–aggregator are presented, before an economic assessment for each neighborhood is quantified.

4.1 Price Modeling

In this section, energy price components (which derive from the relevant markets and charging regimes) and relevant energy-related incentives shall be described for each considered context. These price components are as experienced by the RAEPN and, accordingly, retailer cash flows (e.g., metering and staff costs, and profit margins, among others) are not considered. These retail-related components, which will be passed on to the EPN, are dependent on the exact commercial relationship between the Retailer–Aggregator (RA) and EPN. Some indication of how such costs can be passed to the EPN will be explored in Section 4.3.

The analysis of the energy price and incentives presented in this section is based only on variable (€/kWh) price components, as fixed/annual costs (e.g., connection charges and standing charges) are unlikely to change through pursuance of any given BC. Subsequent to the description of the relevant components, the price profiles for the utilized model days shall be shown.

4.1.1 Ireland

The Irish campus is connected to both the electricity and gas grid, and, hence, the price components of both energy vectors should be considered in addition to relevant operating reserve (OR) prices. Other incentives, such as LCI in Ireland that are applicable to wind generation within EPNs are not considered in this work. The rationale for this is that, as wind generation is not flexible, these particular LCI do not affect the marginal impacts of a BC.

4.1.1.1 Electricity

As shown in Table 6.1, there are several (variable) components of the electricity price faced by a RAEPN. Broadly speaking, these components relate to consumption at (1) the "commercial" level, which is independent of the physical location of consumption; (2) the GCP level, which is the point of connection to the public grid at which UoS fees are charged, and (3) the premises level, at which consumption fees should be charged (Good et al., 2016).

At the commercial level, wholesale electricity prices and capacity charges are applicable. As opposed to more liberalized systems, there is no imbalance electricity component, as retailers are not balancing responsible (i.e., they are not required to declare their consumption before delivery). Hence, all consumption risk is transferred to the ETSO. At the GCP level, Electricity Distribution UoS (EDUoS) and Electricity Transmission UoS (ETUoS), and imperfections charge are applicable. At the premises level, VAT is applicable. More details on the given price components is provided in Table 6.1.

Table 6.1 Irish variable electricity price components

Level	Component
Commercial	Wholesale electricity
	Capacity
GCP	ETUoS fees
	EDUoS fees
	Imperfections charge
Premises	VAT

Table 6.2 Irish variable gas price components

Level	Component
Commercial	Wholesale gas
GCP	GTUoS fees
	GDUoS fees
Premises	VAT
	ESO

4.1.1.2 Gas

As with electricity, there are several (variable) components of the gas price faced by the RAEPN at the commercial, GCP, and premises levels (Table 6.2) (Good et al., 2016). At the commercial level, wholesale gas prices are valid. At the GCP level, Gas Distribution and Transmission Use of System (GDUoS and GTUoS) fees are relevant. At the premises level, VAT and a gas carbon tax, a type of ESO, is applicable.

4.1.1.3 Operating Reserve

When there is a drop in system frequency, due to load or production variation, the droop control some electricity generation machines immediately react based on their own inertia and capacity. If the power disturbance is too large and more power is needed, reserve power from generators (ramping up generation) or demand (turning down demand) is required. This is called incorporating OR and Replacement Reserve (RR). Reserve can be called Primary OR (POR), Secondary OR (SOR), or Tertiary OR (TOR) depending on the relevant response time, with RR operating at slower response times (Fig. 6.6). The Irish definitions are reported according to the Irish grid code (EirGrid, 2013) in subsequent sections. All reserve products are required 24 h a day.

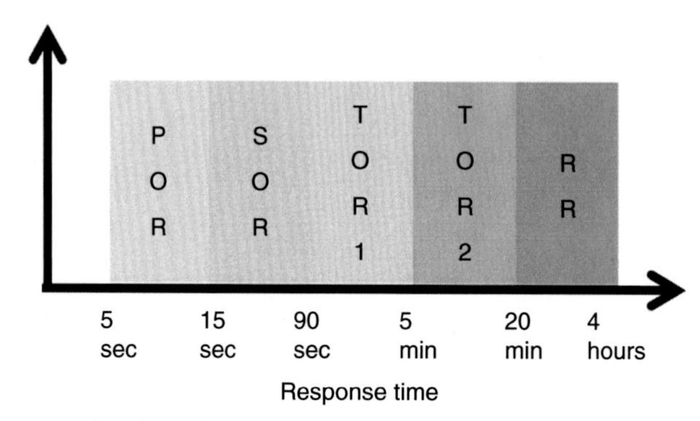

Figure 6.6 ***Reserve operating time.***

4.1.1.4 Price Profiles

As shall be described in Section 4.2, the various BCs shall be assessed by considering 7 model days, namely summer weekday and weekend, shoulder[8] weekday and weekend, winter weekday and weekend, and "peak" day. Utilizing aforementioned data, the RAEPN prices/price profiles for the various model days and disaggregated by price component can be detailed. Fig. 6.7 shows the utilized prices by season and component for electricity, while Fig. 6.8 shows the equivalent for gas. Also included in Fig. 6.7 is the availability price for POR and TOR2. For the components which vary day-to-day (wholesale energy costs and capacity fees), the most typical daily profiles for the season are selected.

4.1.2 France

Conversely to the Irish case, the French campus is only connected to the electricity grid. The price components of electricity are detailed in subsequent section. In addition, a new CM in which the resources of an EPN might participate is also described. As in the Irish context, although there are LCI in France, the only technology on the test bed which is eligible is solar PV. Thus, once again the LCI are neglected in these studies, as solar PV generation is not flexible and the incentives do not have any impacts on relevant BCs for EPNs.

4.1.2.1 Electricity

As discussed in Section 4.1.1.1, there are various components of the RAEPN electricity price applicable at the "commercial," GCP, and premises levels. At the commercial level, wholesale and imbalance electricity prices are applicable. At the GCP level, UoS fees are charged for use of the public grid. ESO and VAT are charged at the premises level. The relevant electricity price components, together with their level assignation are presented in Table 6.3.

4.1.2.2 Capacity Market

From year 2016/17 a CM will operate in France. This market will be open to generation and demand-side participants. Capacity owners must be available during peak hours (determined 1 day in advance by the grid operator), and are expected to occur on 10–25 days, 0700–1500 and 1800–2000 (Rte, 2015). Capacity prices will vary in the region €0–60 per kilowatt if security of supply is not threatened (there will be an administered price if security is threatened). For this work a price of €30 per kilowatt is assumed.

Operation of the CM will prompt an obligation on electricity consumers, according to consumption over peak hours. Given the lack of historical data on the level of this obligation, this obligation will not be considered in this work.

[8] Representing spring/autumn.

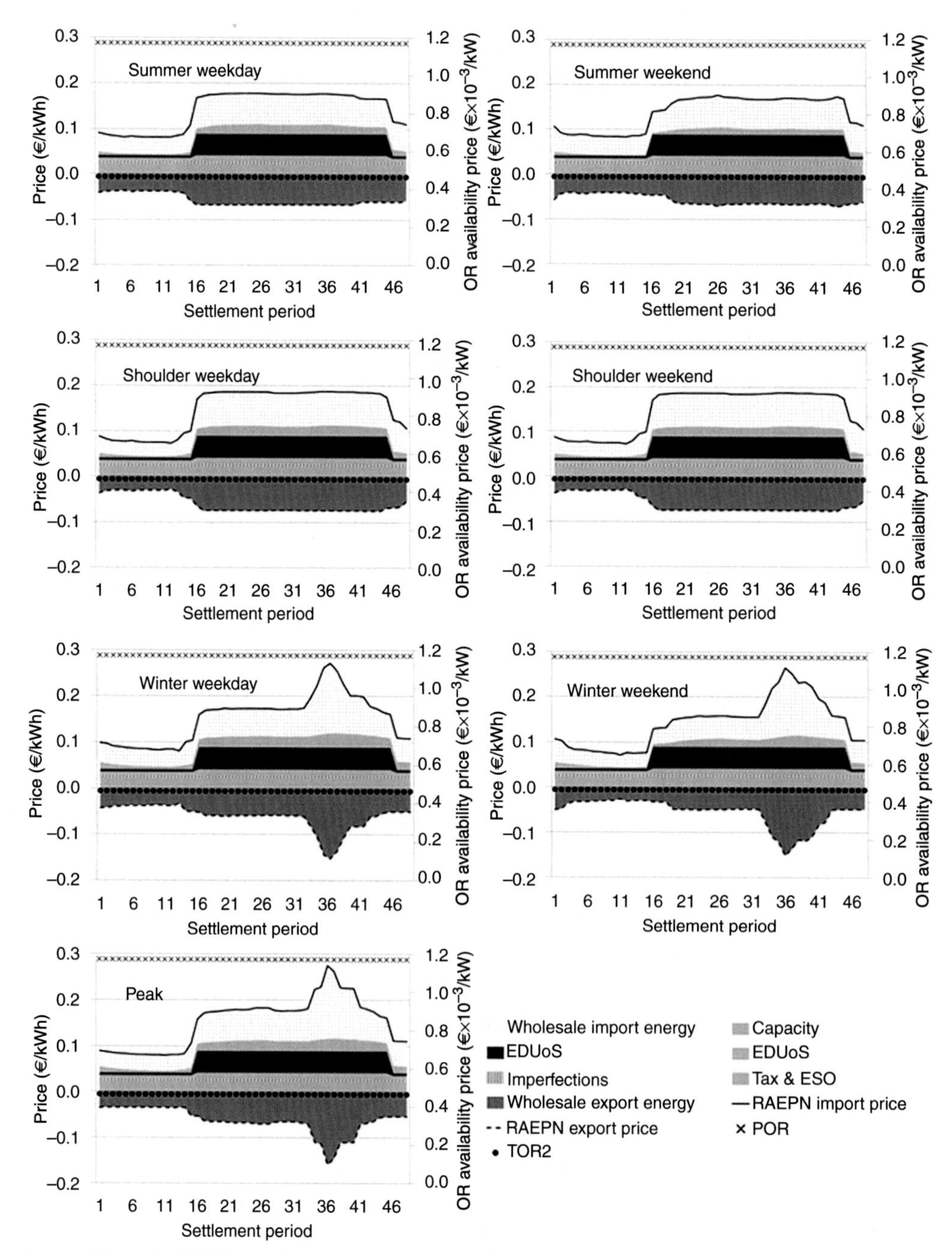

Figure 6.7 ***Irish RAEPN electricity prices, by season, by component.***

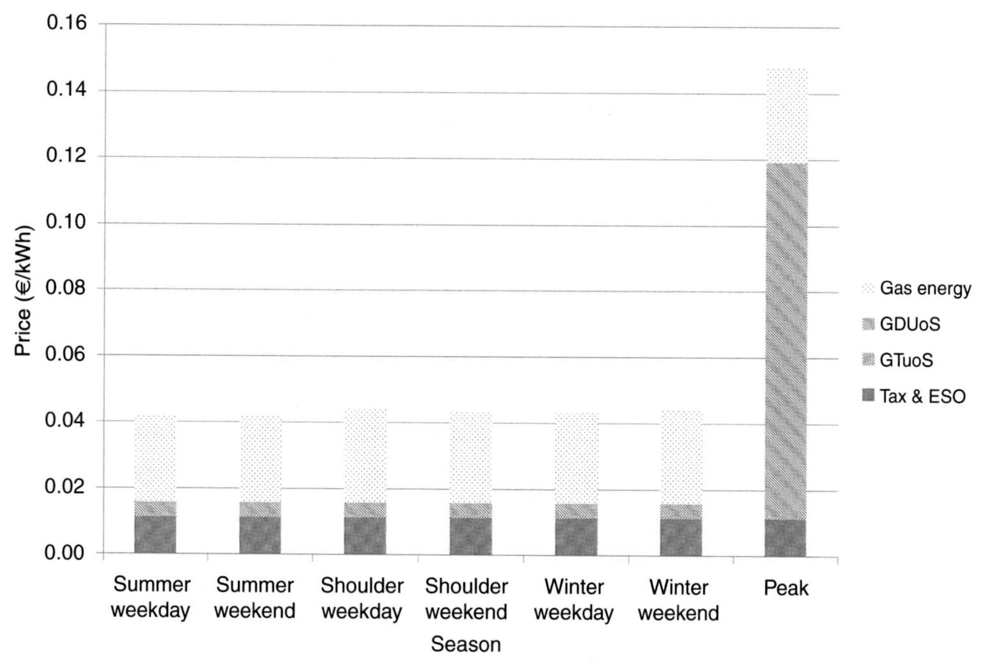

Figure 6.8 ***Irish RAEPN gas prices, by season by component.***

Table 6.3 French variable electricity price components

Level	Component
Commercial	Wholesale electricity
	Imbalance electricity
GCP	Grid UoS fees
	BRE fees
Premises	ESO
	VAT

4.1.2.3 Price Profiles

As described in Section 4.1.1.4, price profiles disaggregated by price component should be detailed for 7 representative days. Again, for the components which vary day-to-day (in this case, wholesale and imbalance energy costs), the most typical daily profiles for the season are selected. Note that there is no UoS component in the graphs of Fig. 6.9 as, for the tariff being considered, UoS fees are charged based on the maximum power and annual consumed energy of the user. The CM availability price shown in Fig. 6.9 is set based on a capacity price of €30 per kilowatt (the midpoint of the expected range, see Section 4.1.2.2), and assumes that a capacity call is equally likely over the 15 days of

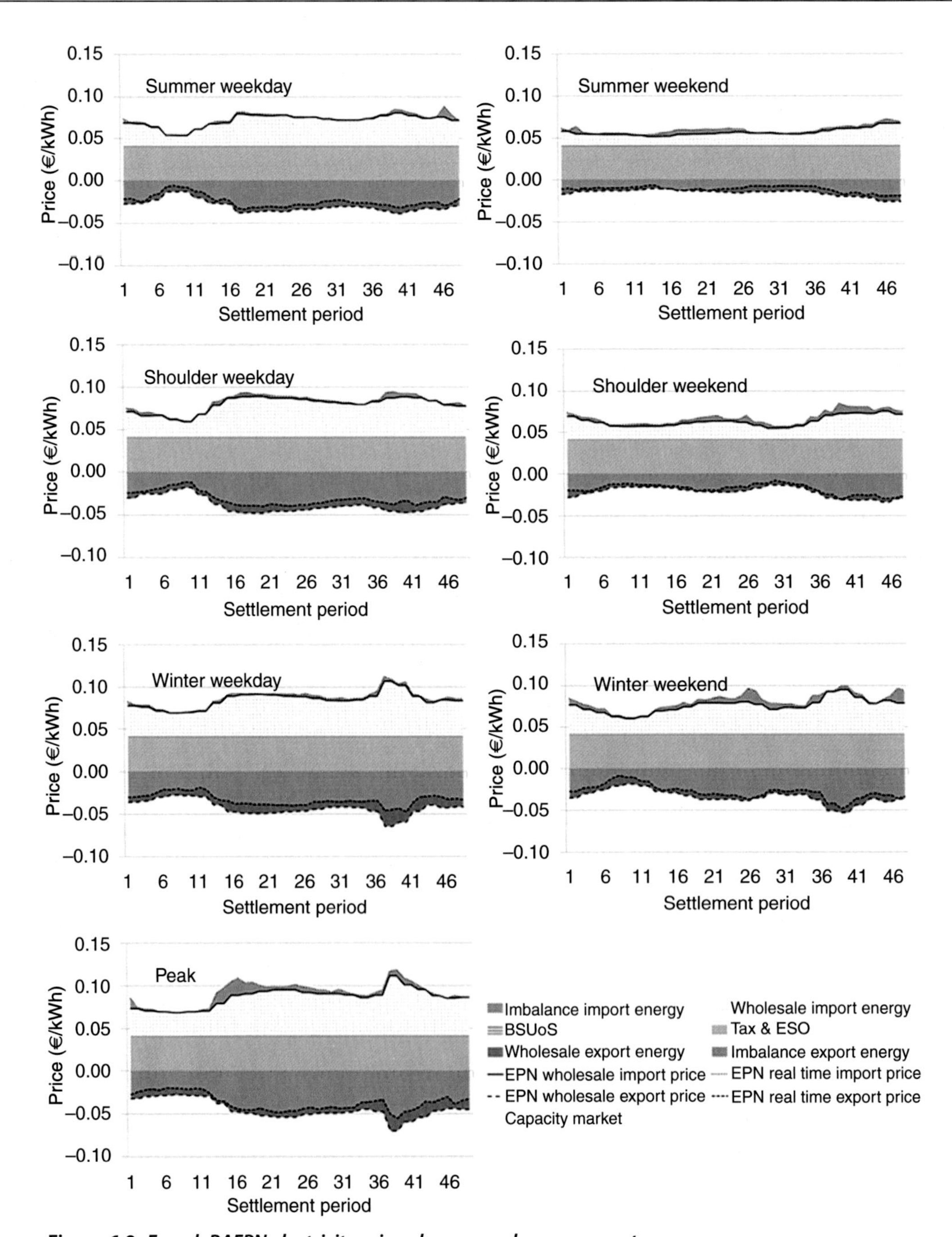

Figure 6.9 ***French RAEPN electricity prices, by season, by component.***

the modeled peak period, and the 10 h per day (0700–1500 and 1800–2000) in which capacity may be called.

4.2 Operational Cash Flows

The estimation of relevant operational cash flows is crucial for the assessment of the value of various EPN BCs. In this work, such cash flows were estimated by performing an optimization (with respect to the appropriate price signals) on a model of the test sites. The utilized test site models included description of all flexible (electrical and thermal) devices (e.g., EHP, CHP, gas boilers), electricity generation (e.g., solar PV and wind turbines) and relevant electrical and thermal storage. The utilized models were informed by electricity and thermal demand profiles from the developed physical models described in Chapter 4. The utilized optimization model is stochastic, capturing the uncertainty of demand.

The exact (commercial) definition of an EPN should not be fixed, as there are multiple commercial arrangements through which energy positivity may be achieved. For example, the individual (building) constituents of an EPN may contract with an existing energy retailer to perform the NEM function, who may also then adopt the role of aggregator, exploiting flexible resources within the EPN to partake in relevant markets. In this case, the relationship of the EPN constituents to their retailer may not change considerably from present, with the retailer–aggregator–NEM continuing to bill the EPN constituents on a per unit of energy consumed basis. Accordingly, benefits from exploited flexibility could be passed back to the users either through a reduction in their tariffs, or through annual payments. At the other extreme, an independent NEM may become a balancing responsible party to directly partake in the relevant energy markets. In this way the EPN would take on the roles of retailer and aggregator itself, and face all relevant market prices, UoS fees, taxes, and so forth.

It is due to this wide spectrum of possible commercial arrangements for the EPN that, in this work, the RAEPN actor is defined, which encompasses the roles of energy retailer, aggregator and EPN. BCs are assessed on the basis that optimization (with respect to the relevant exogenous price signals) is conducted by this abstract actor. This actor is exposed to, and perceives, the various relevant price signals, as detailed in Figs. 6.10 and 6.11, for the Irish and French contexts, respectively. The exact nature of the relationships between the RAEPN constituents shall not be explicitly addressed, although indicative scenarios of how benefits may be shared between a retailer–aggregator and the EPN shall be explored (see Section 4.3).

Later, RAEPN cash flows are presented for the various considered cases (as detailed in Table 6.4). In each figure the cash flows presented are the *change* in cash flow, for the case compared to the appropriate base case, which is specificed in each figure. This enables the incremental net impact of the case, compared to the appropriate base case, to

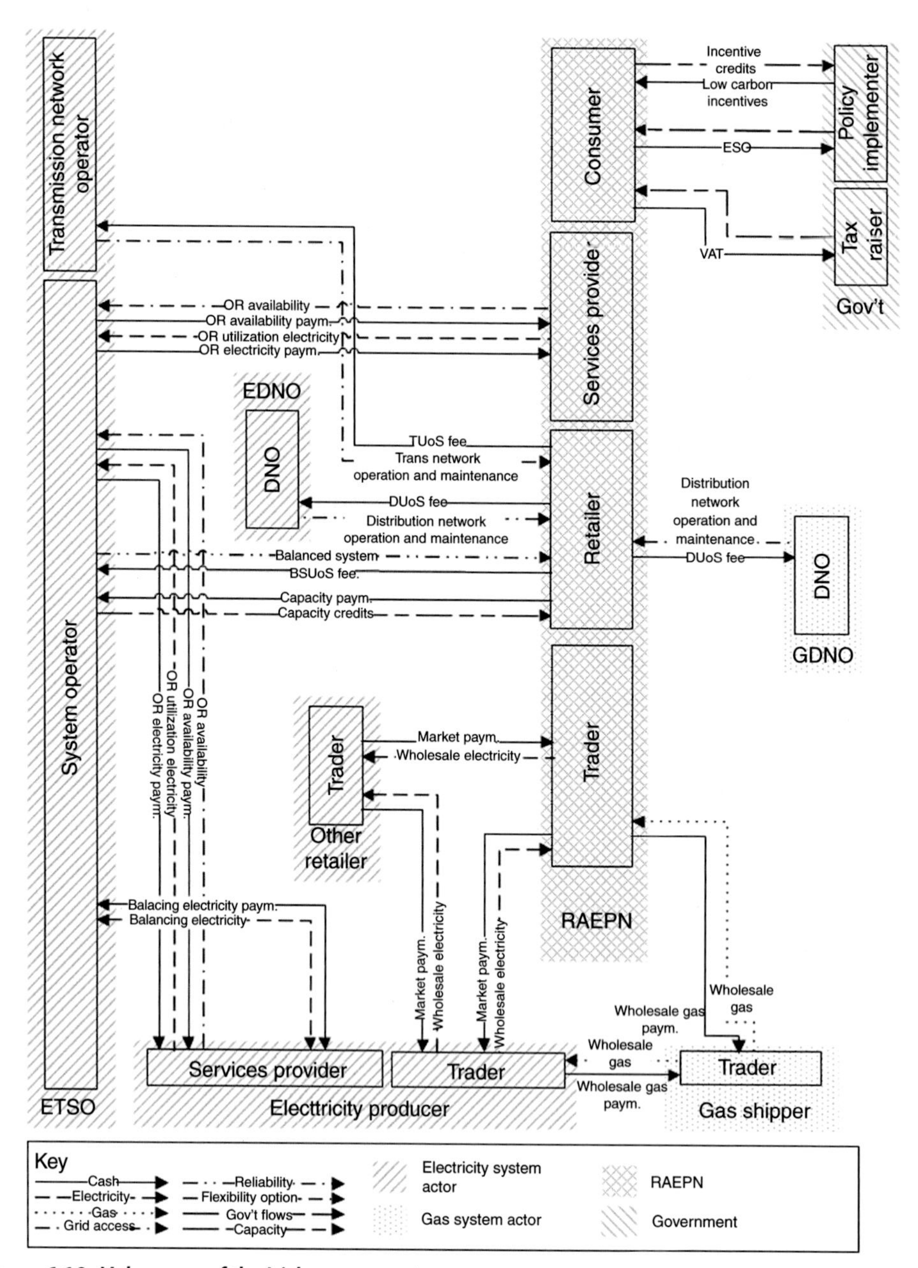

Figure 6.10 ***Value map of the Irish energy system.***

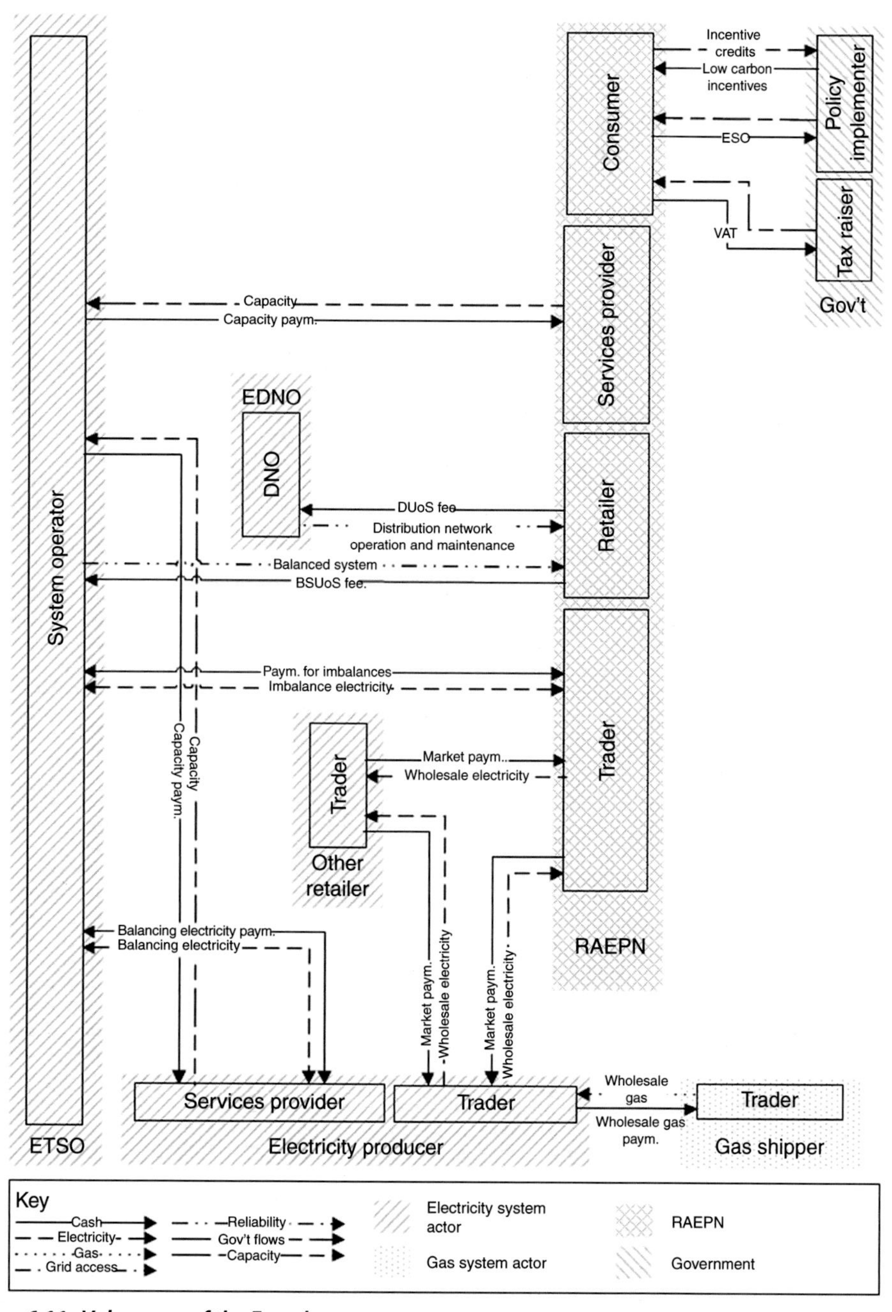

Figure 6.11 ***Value map of the French energy system.***

Table 6.4 Summary of considered cases

Group	Case name	Acronym	Description
Base	Load following	LF	Heat generators follow load, no use of storage
	Retail price optimization	RO	Minimization of costs with respect to flat retail prices
Central	Optimized purchase on wholesale market	OPWM	Optimization of participation in wholesale electricity markets
	Minimization of imbalance penalties	MIP	Minimization of imbalance penalties in addition to optimization of wholesale market participation
	Operating reserve	OR	Maximization of revenue from provision of operating reserve (TOR2, see Section 4.1.1.3)
	Capacity market	CM	Maximization of CM revenue (see Section 4.1.2.2)
	Distribution network constraint management	DNCM	Optimization of EPN given the ability (due to EPN DR capability) to violate local network capacity constraints
	Capacity payments	CP	Minimization of payment of CP
	Distribution use of system fees	DUoS	Minimization of payment of DUoS fees
	All price signals	All	Optimization on all relevant price signals
Aggregation	Intra-EPN coordination	IEC	Optimization on all relevant price signals with coordination between EPN constituents to optimize trading positions
	Intra-EPN trading	IET	Optimization on all relevant price signals with implementation of a private wire arrangement within the EPN, enabling unimpeded trading of electricity inside the EPN
Sensitivity tests	Increase price volatility	IPV	Optimization on all relevant price signals with 100% increase in WE price variability
	Increase thermal storage	ITS	Optimization on all relevant price signals with 100% increase in thermal storage capacity
	Increase battery storage	IBS	Optimization on all relevant price signals with 100% increase in electrical battery capacity
Regulatory context studies	CHP LCI	LCI	Optimization on all relevant price signals with a LCI applied to CHP electricity production
	Local carbon market	LCM	Optimization on all relevant price signals with a price on local carbon emissions
	Reliability	Rel	Optimization on all relevant price signals with requirement to island the EPN for 8 h at peak times

be discerned. Subsequently, for brevity, the acronyms given in Table 6.4 will be used to describe the case. The cases are separated into groups.

- "Base" cases represent conditions that are possible without the EPN concept;
- "Central" cases are possible under current conditions, although some changes to market rules (e.g., minimum unit sizes), or new markets [e.g., for Distribution Network Constraint Management (DNCM)] may be required;
- "Aggregation" cases show the effect of various types of aggregation;
- "Sensitivity tests" study the sensitivity of results to changes in the amount and value of flexibility;
- "Regulatory context studies" explore the effect of changes to the regulatory context.

In the remainder of this section:

- The change in operational cash flows for the Retail price Optimization (RO) case [compared to the Load Following (LF) base], is presented. This case is presented as, while implementation of the EPN concept is not necessary to optimize with respect to retail prices (as all consumers have access to their retail prices), it may be that, in practice, a naïve neighborhood may focus on energy rather than cost minimization (neglecting the benefits of self-consumption of locally generated energy, which avoids UoS fees and taxes).
- The changes in operational cash flows for BCs relating to the central BCs, compared to the RO case, are also presented. Also shown are the cash flows for the "All" case. The effects that the EPN could have on other actors within the energy system are also shown.
- The effect of aggregation is studied, through considering the change in revenue of the Intra-EPN Coordination (IEC) and Intra-EPN Trading (IET) cases, compared to the "All" case, for the Irish test bed.
- The sensitivity of the IEC case, to changes in the main determinants of BC value: (electricity wholesale) price signal volatility (which dictates the profitability of demand shifting), and storage size (which dictates the capacity to shift demand), is studied, by showing the change in revenue of this case, compared to the "All" case.
- Finally, in this section, the effect of the regulatory context (with respect to the particularly salient regulation aspects identified in Section 2.3) is explored to demonstrate how operational cash flows of various actors can be affected by conservative and liberal regulatory tools. This is done by showing the change in revenue for the appropriate cases compared to the IET case.

4.2.1 Optimization on Retail Prices

Figs. 6.12 and 6.13 show the change in annual cash flows for the Irish and French cases, when the EPN is optimized with respect to retail prices. The changes are relative to a simple (heat) load following policy. In each optimization, a capacity limit (based on the maximum capacity required in the base LF case) is implemented, implying the local network is operated close to firm capacity (enabling illustration of the DNCM BC, see Section 4.2.2).

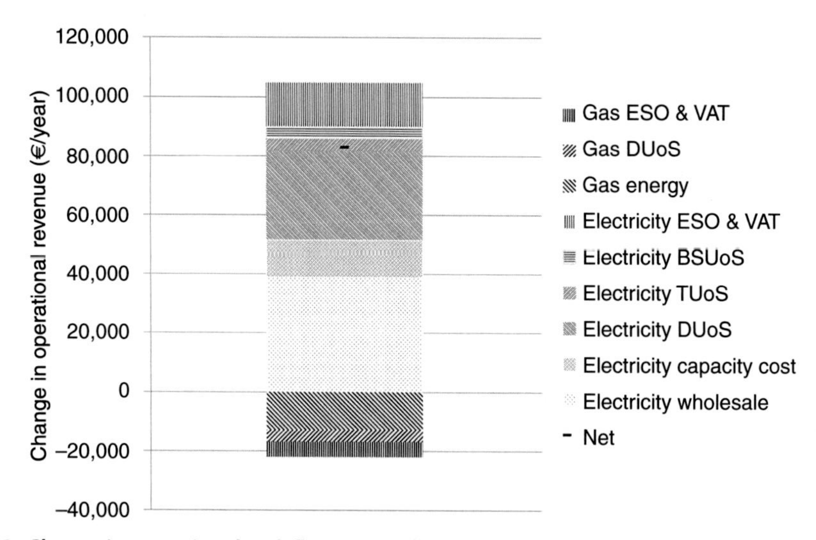

Figure 6.12 ***Change in operational cash flows—retail price optimization relative to the load following base—Irish case.***

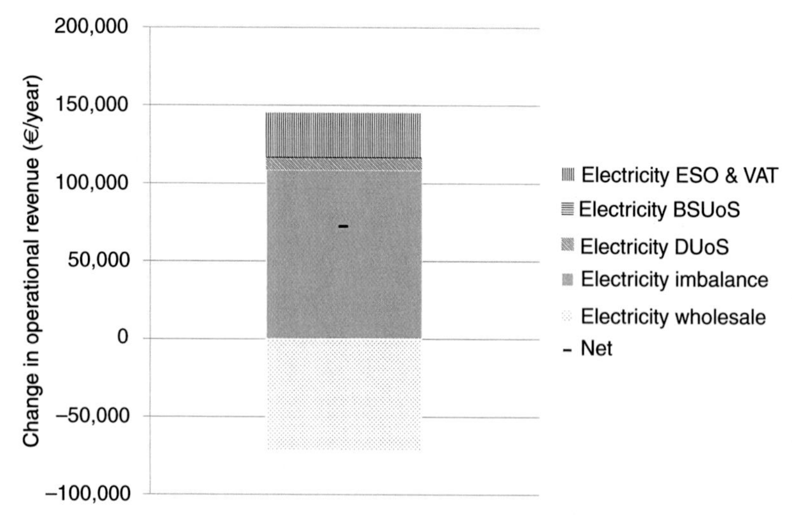

Figure 6.13 ***Change in operational cash flows—retail price optimization relative to the load following base—French case.***

4.2.1.1 Ireland

In the Irish case, the potential savings from optimization on retail prices are €82,800 per year (Fig. 6.12). This corresponds to a saving of 16.3% on the estimated retail bill. Savings can be attributed to the shifting of electricity consumption and generation (and using EPN battery storage) to maximize EPN self-consumption of generated electricity (thus minimizing import of electricity, with the associated UoS and tax costs). Electricity wholesale and capacity costs are reduced as the incentive to maximize self-consumption has a disproportionate impact at times of high energy demand (as there are more electricity imports in the base), which are correlated with times of high wholesale energy/capacity price.

4.2.1.2 France

In the French case (Fig. 6.13), the potential savings from optimization on retail prices, compared to a load following policy could result in savings of €72,500 per year. This corresponds to a saving of 6.8% on the estimated retail bill. As in the Irish case, the benefit is derived from the shifting of electricity consumption and generation (and using EPN battery storage) to maximize EPN self-consumption of generated electricity. Also, tax and UoS costs are reduced as electricity consumption is generally reduced as the more efficient EHP is favored over electric resistive heating. There are also significant savings as the incentive to self-consume reduces variability of electricity imports across scenarios (as energy demand and PV generation scenarios are somewhat correlated), and hence steer EPN electricity import/export similarly. This change in imports/exports results in more use of the wholesale market, less use of imbalance electricity, and reduced cost overall.

4.2.2 Central Business Cases

4.2.2.1 Ireland

For the Irish case, five central BCs were considered. In the Optimized Purchase on the Wholesale Market (OPWM), OR, Capacity Payments (CP), and DUoS BCs, the value (in terms of operational cash flows) for the RAEPN actor is created by responding to the relevant price signals. For the DNCM BC, value is created by allowing the EPN to exceed the security based capacity constraints considered in the previous cases (set according to peak import/export levels in the base case; implicitly assuming that the local network is operating at its limit[9]). Import/export limits are set to 610 kW for Irish case, and 2877 kW for the French case. The rationale for this service is that the EPN can access the emergency capacity of the network (typically only used during contingencies) in exchange for actively managing network constraints (e.g., releasing emergency capacity should a contingency occur). All BCs are presented compared to a "retail price" optimization, where the RAEPN optimizes with respect to retail price signals. A final case, in which all price signals are considered (in addition to the capacity limit removal), is also shown.

[9]The value of the modeled DNCM case can be considered a maximum, as the value will be less if the local network has some spare capacity.

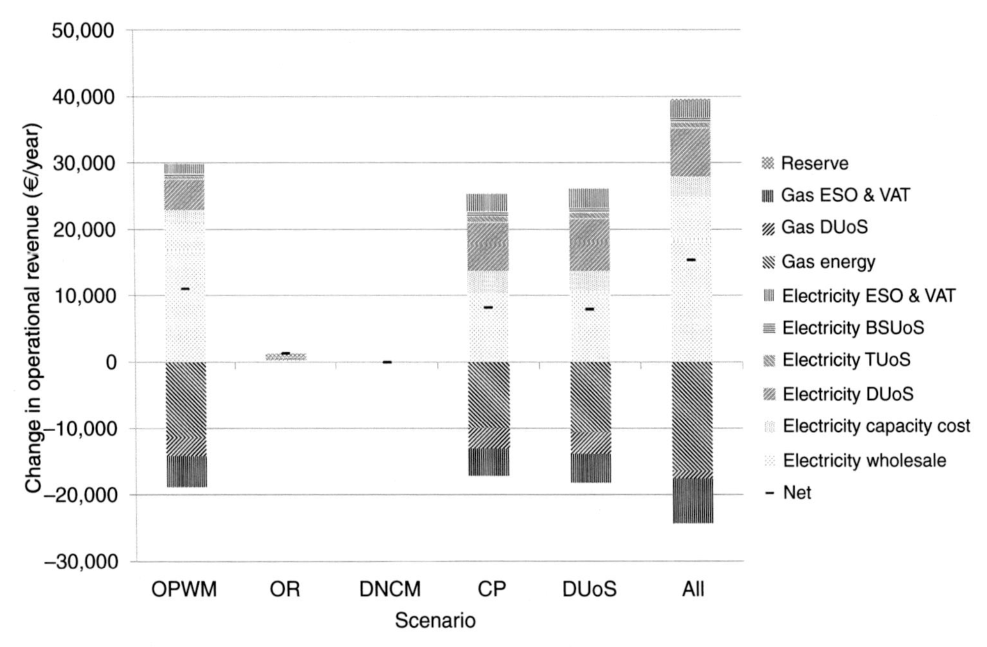

Figure 6.14 ***Change in operational cash flows—various BCs relative to the retail price optimization base—Irish case.***

Note that the Market Index Price (MIP) BC cannot be demonstrated here as, in Ireland, demand is not required to assume balance responsibility, and therefore there are no imbalance penalties to manage.

For the various BCs, RAEPN cash flows, broken down by price component, together with the relevant net cash flows, are shown in Fig. 6.14. As shown in Fig. 6.14, the OPWM BC results in an increase of operational revenue of €11,000 per year. This is due to two factors. First, the shift in electricity import away from times of high price, to lower price periods (the "peaky" nature if winter Irish wholesale electricity prices makes time-shifts of even a hour or two profitable during winter periods). Second, there is an increase in CHP operation when economic (at times of high electricity price), which also results in reduced electricity tax and UoS costs, but also means increased expenditure on gas, and associated UoS and taxes.

The OR and DNCM BCs do not produce significant changes to operational revenues. The OR BC produces revenue of €1300 per year. The value of this BC is fundamentally limited by the low value of the TOR2 product (see Section 4.1.1.3), though the requirement that the reserve capacity be consistent across the modeled day also restricts value. If the POR product (which requires response between 5 and 15 s) is considered instead, the value rises to €2200 per year. The change in revenue in the DNCM BC is negligible as the incentive in the retail price optimization (which forms the base case) to maximize self-consumption already serves to reduce peak consumption, making the local network capacity limit, which is relaxed in the DNCM BC, nonbinding. Thus there is nothing to be gained from the relaxation. There may be benefits from pursuing the DNCM case given visibility of the various price signals in the considered base case. This is because such signals will encourage consumption to synchronize at low price times. However, sensitivity tests in this direction indicate that the benefit in such a case is still negligible for several reasons. First, the signals and amount of flexibility are relatively small, which limits the attractiveness of synchronization. Second, the considered base peak demand (which informs the constraint which is relaxed) is the peak across all considered demand scenarios. Thus, with weak signals and low flexibility, the constraint relaxation only results in consumption beyond the base limit for a small number of periods, in a small number of scenarios. This means that the change in behavior is near negligible.

The CP BC produces an increase in operational revenue of €8200 per year. In this BC, revenue is achieved by reducing electricity consumption (on which capacity charges are levied) by increasing CHP electricity generation, and shifting consumption to times of lower capacity charge. It is important to note that the limited variability of the capacity charge through the day limits the potential for optimization (see Section 4.1.1.4).

The DUoS BC produces an increase in operational revenue of €8000 per year. This is due to the shift of electricity consumption away from the 8 a.m. to 10 p.m. period, when EDUoS are highest (see Section 4.1.1.4). This results in reduced expenditure on electricity-related costs and increased expenditure on gas-related costs, as this encourages increased CHP operation.

Consideration of the price signals and constraint relaxations related to the five BCs together results in an increase in operational revenue (on the retail price optimization) of €15,400 per year. Most of these benefits derive from the electricity wholesale market price signals, with some additional benefit derived from the CP and DUoS price signals, which largely complement and accentuate the signals from the electricity wholesale market. Comparing this to estimated retail energy expenditure (in the RO case) of €507,600 per year, it can be deduced that consideration of all price signals could produce an estimated 3% saving on the retail bill.

As shown in Fig. 6.10, changes in revenue flows for the RAEPN affect other actors in the energy system due to its consumption of energy, and use of energy transportation infrastructure. As an example, assume that the marginal electricity producer is a gas plant (with 50% efficiency), and that provision of OR by the EPN displaces OR provided by the electricity producer (at a discount rate of 5%). Accordingly, the impact on the relevant actors from optimizing the EPN on all price signals, compared to a retail price optimization, is shown in Fig. 6.15. The electricity producer loses revenue as the RAEPN buys less electricity, although some of this loss might be offset by reduced

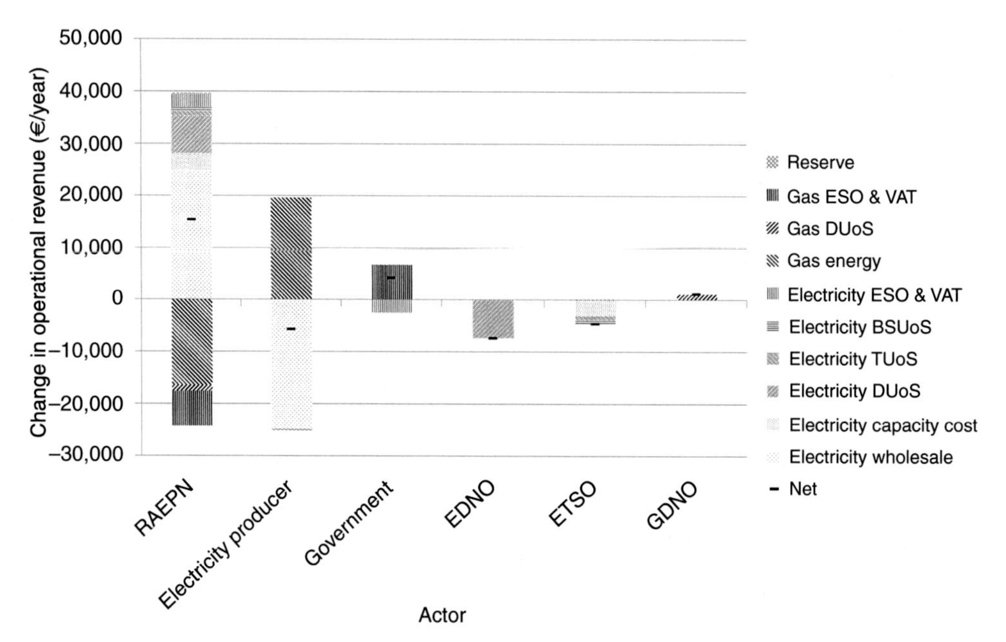

Figure 6.15 ***Change in operational cash flows—all price signals case relative to the retail price optimization base, by actor—Irish case.*** The gas shipper actor was excluded as the costs of the gas shipper in procuring gas from gas producers or importers is not public information, and thus the gas purchase cash flow cannot be modeled.

fuel (gas) costs. The ETSO and EDNO face slight reductions in operational revenue,[10] while the GDNO benefits from increased use of the gas network. The government sees increases in gas VAT/ESO larger than the reduction in electricity VAT.

4.2.2.2 France

For the French case, again, five central BCs were considered. In the OPWM, MIP, CM, and DUoS BCs operational revenue is optimized by responding to relevant price signals. As described in Section 4.2.2.1, the DNCM BC relates to value enabled by relaxation of local network capacity constraints. Again, BC cash flows are presented compared to a "retail price" optimization.

Fig. 6.16 shows the cash flows associated with the various price components and the considered BCs for the RAEPN actor. As shown, the OPWM BC is worth €14,100 per year. This is through reduced imbalance expenditure, with wholesale electricity expenditure actually increasing. This is due to movement away from times of high wholesale

[10]Which will not result in reduction in final revenue, as allowance for grid operators is set by the government, or their agent. A reduction in operational revenue will mean that more revenue will have to be recouped in future periods.

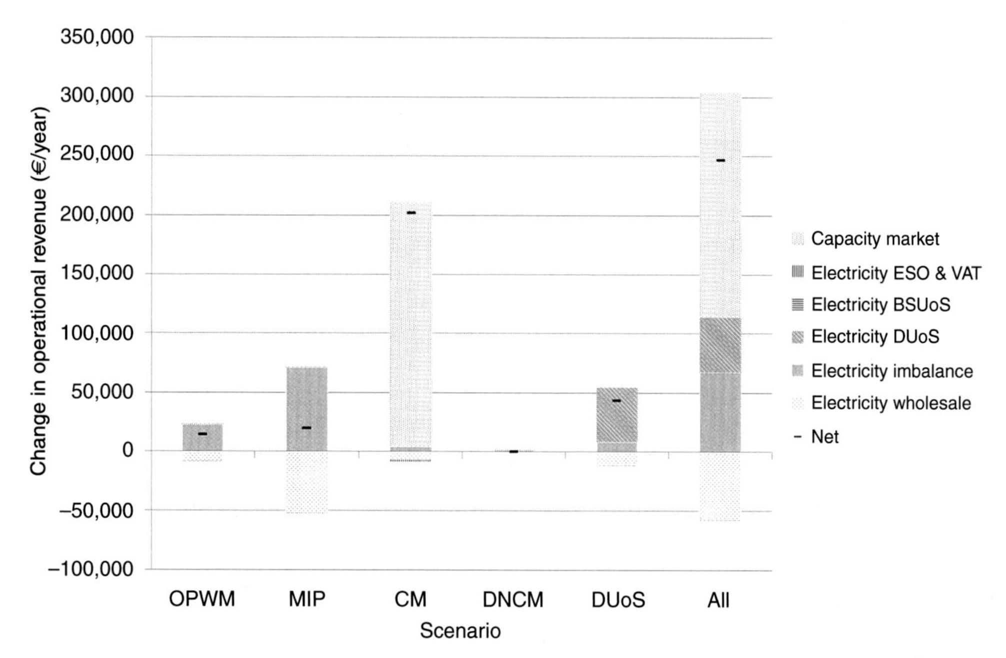

Figure 6.16 ***Change in operational cash flows—various BCs relative to the retail price optimization base—French case.***

price which also reduces exposure to high imbalance prices (higher expected imbalance prices being correlated with high wholesale electricity price, see Fig. 6.9) for the retailer in its market optimization.

Introducing imbalance price signals to the OPWM price signals increases revenue to €19,800 per year. The extra revenue is due to exposure to the expected imbalance prices, particularly at weekends and in the peak season, see Fig. 6.9.

Introduction of capacity price signals from the CM has a more significant impact; producing a net revenue impact of €202,200 per year (assuming a capacity price of €30 per kilowatt, see Section 4.1.2.2). Sixty-four percent of this revenue is attributable to the 4400 kW of back-up diesel generators available at the French test site, with 36% from the 2500 kW of capacity available from flexible electricity consumption (EHP and electric boilers). However, it should be noted that, in the optimization, no constraint on the reduction of consumer utility has been adopted. Though, given that notification of a capacity call occurs day-ahead, any adverse effect on consumer utility could be avoided through building pre-heating and charging of thermal storage. Additionally, there is no consideration of extra maintenance and testing costs that might be required to supply capacity from back-up diesel generators.

As in the Irish case, the effect of the DNCM BC is negligible, as the retail price signals already motivate a reduction in peak consumption, away from the assumed local network capacity. There is revenue gain in the DUoS BC (€43,500 per year), as the peak demand is reduced from 2706 to 2026 kW, which reduces DUoS cost.

Consideration of all price signals, and the constraint relaxation of the DNCM case, results in revenue of €246,800 per year, largely dominated by revenue from the CM. Considering an estimated retail energy expenditure of €1,060,900 per year means that consideration of all price signals could produce an estimated 23% saving. However, it should be emphasized that, as addressed previously, this is assuming that there is no additional constraint related to guaranteeing thermal comfort within buildings. Also, there is no assumption on possible testing costs, for participation of back-up diesel generators. Finally, it should be noted that an assumed capacity price of €30 per kilowatt was used. For guidance, if, alternatively, a capacity price of €15 per kilowatt is used, and only the diesel generators are considered eligible for providing capacity, revenue from all BCs will be €124,600 per year. This would produce an estimated 12% saving on the retail bill.

As shown in Fig. 6.17 the increases in operational revenue experienced by the RAEPN come largely at the expense of the electricity producer, especially through its abstraction of CP. The EDNO can also expect to lose revenue as the EPN reduces its peak import (on which, in the modeled case, DUoS is charged).

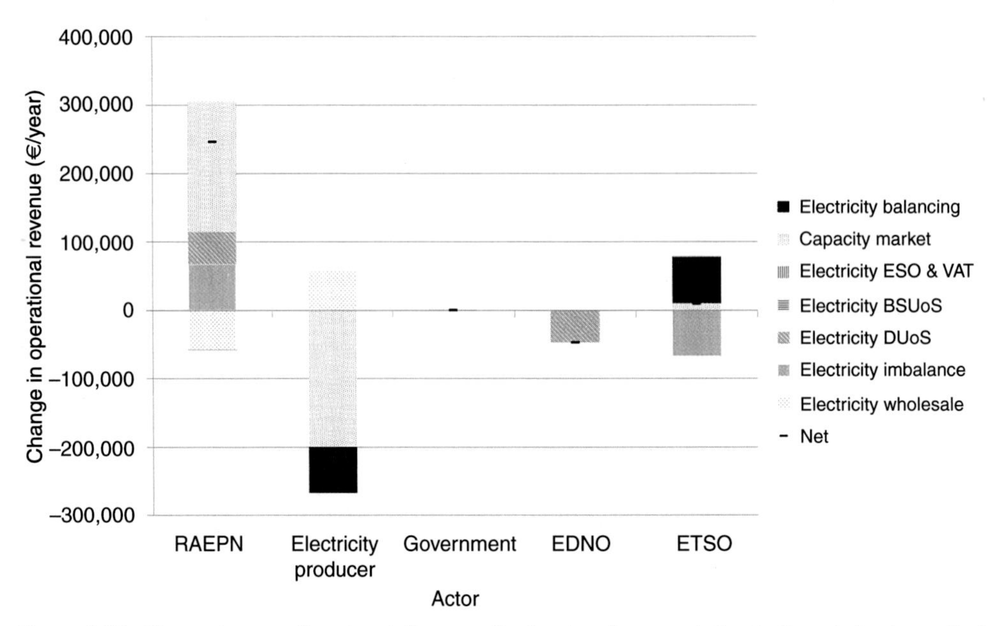

Figure 6.17 ***Change in operational cash flows—all price signals case relative to the retail price optimization base, by actor—French case.***

4.2.3 Aggregation Effect

An EPN may be virtually aggregated (i.e., the actions of the EPN constituents are coordinated) or physically aggregated (i.e., the EPN constituents are local to one another, and share private electricity and/or gas networks) (Good et al., 2016).

While the French test bed is already virtually aggregated (as the site is owned by one party) and physically aggregated (having its own electricity network), it is interesting to consider the impact of IEC (i.e., virtual aggregation) and IET (i.e., physical aggregation) in a different context. Accordingly, IEC and IET are assessed on the Irish test bed, which has several owners and, although all buildings are local to one another, does not share a private-wire network.

Fig. 6.18 shows the additional operational revenues from aggregation, on all price signals base. It can be seen that IEC results in additional revenue of €3800 per year. Mostly, this revenue comes from reduced capacity costs, as electricity generation and consumption is coordinated across buildings to reduce the overall EPN electricity imports. Some revenue from reduced capacity cost is offset by increased gas-related cost, as CHP operation is increased.

IET results in additional revenue of €15,100 per year. Significant revenue is obtained from reduction in electricity UoS costs, as the private-wire arrangement means that it

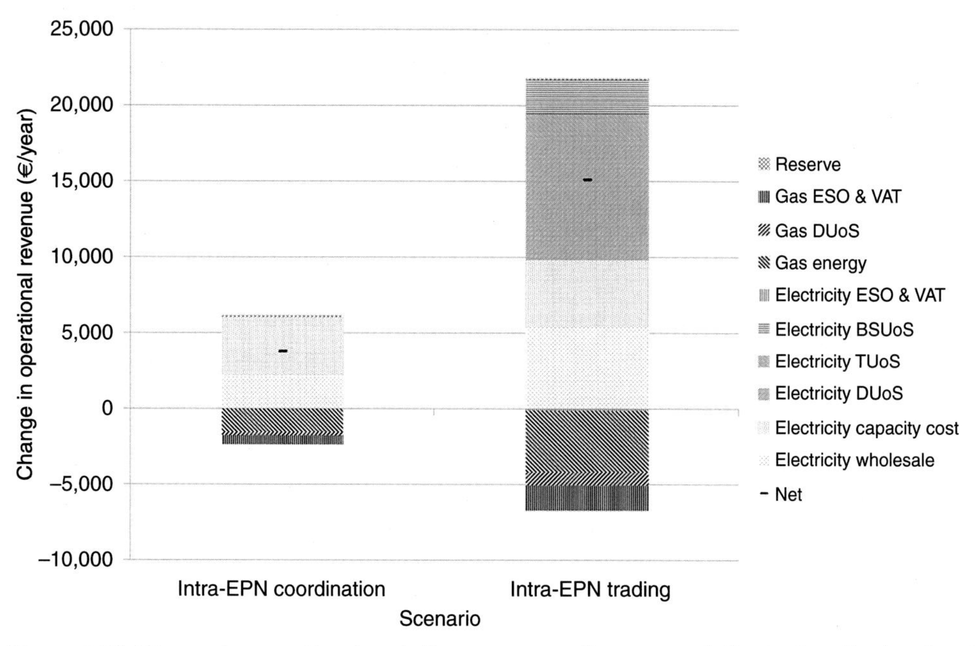

Figure 6.18 ***Change in operational cash flows—aggregation cases relative to the all price signals base—Irish case.***

is the net EPN consumption on which UoS fees are charged, rather than being charged separately on individual buildings.

4.2.4 Business Case Sensitivity Tests

Based on the results presented previously, all BCs under consideration are pursued by using EPN flexibility to move demand away from high price periods (and particularly to avoid electricity import, with the attendant UoS and tax costs), to lower price periods, or to lower price energy vectors. In this light, it can be concluded that the BCs may be sensitive to increase in the amount of energy storage (which enables shifting between periods), and the variability of interperiod price differentials (which reward shifting between periods). To explore the sensitivity of operational cash flows to these factors, tests on price volatility and storage size were carried out for the two test beds, as detailed later. Such tests may be particularly interesting for EPNs as price volatility may likely increase[11] and if investment in increased storage is profitable.

4.2.4.1 Ireland

To investigate the impact of price volatility in the Irish test bed, the variability of the utilized electricity wholesale price profile for each season was increased by 100%. As an illustration, Fig. 6.19 shows the base and new electricity wholesale market price profile for a typical winter weekday.

To investigate the impact of increasing energy storage, two tests were carried out. First, the effect of increasing thermal storage (allowing CHP to shift operation further to high electricity price times) by 100% was considered. Second, battery storage is increased by 100%. All sensitivity tests were assessed on an IEC base, as it is sensible to assume that an EPN would adopt this level of aggregation, given the benefits of coordination (Fig. 6.18) and lack of associated extra costs (as would be present in the IET case).

As shown in Fig. 6.20, an increase in electricity wholesale market price volatility, as modeled, produces an increase in RAEPN revenue of €3600 per year. This is due to an increase in electricity wholesale market revenue, as the reward to shifting grid electricity demand to low price periods increases. As evinced by the decrease in electricity UoS and tax costs, this increase in electricity wholesale market revenue occurs despite increased grid electricity consumption. There is also a reduction in gas-related costs, as heat generation is switched from CHP to gas boilers, as the lower electricity prices (in low price periods) reduce the value of CHP operation at those times.

Also shown in Fig. 6.20, a 100% increase in thermal storage capacity (from a total of 2500 L of hot water storage, and 1500 L of Phase Change Material (PCM) heat storage, to 5000 and 3000 L, respectively), produces an increase in RAEPN revenue of €2500

[11] As penetration of variable, nondispatchable generation, such as wind, increases, and as new large loads, such as electric vehicles and heat pumps, increase.

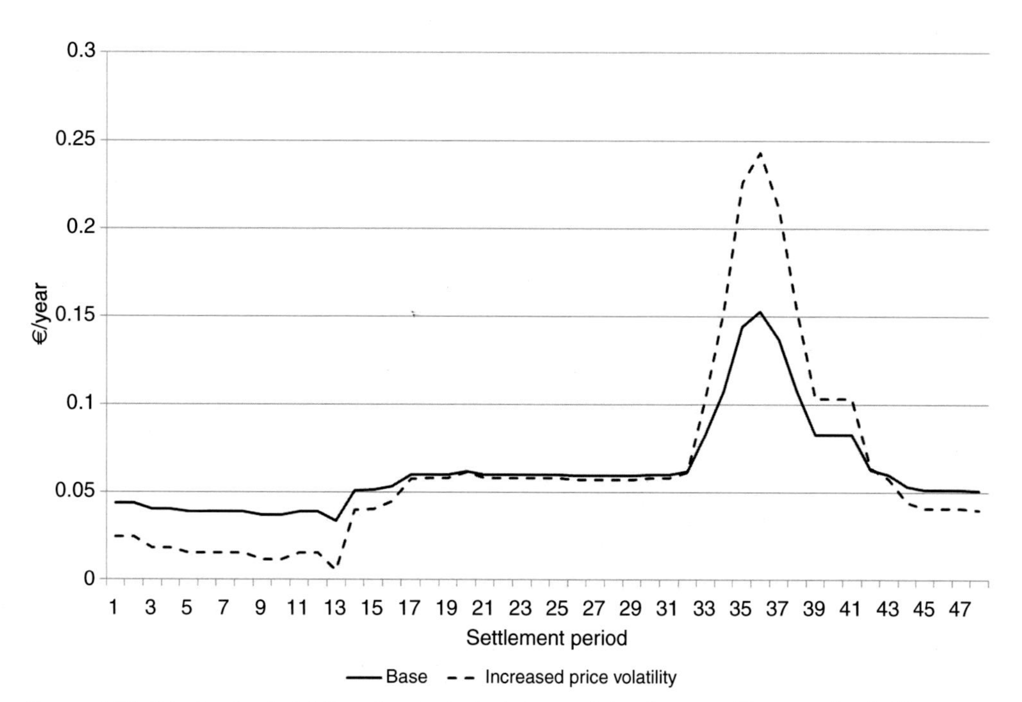

Figure 6.19 ***Wholesale electricity price, typical winter weekday, base, and increased volatility case—Irish case.***

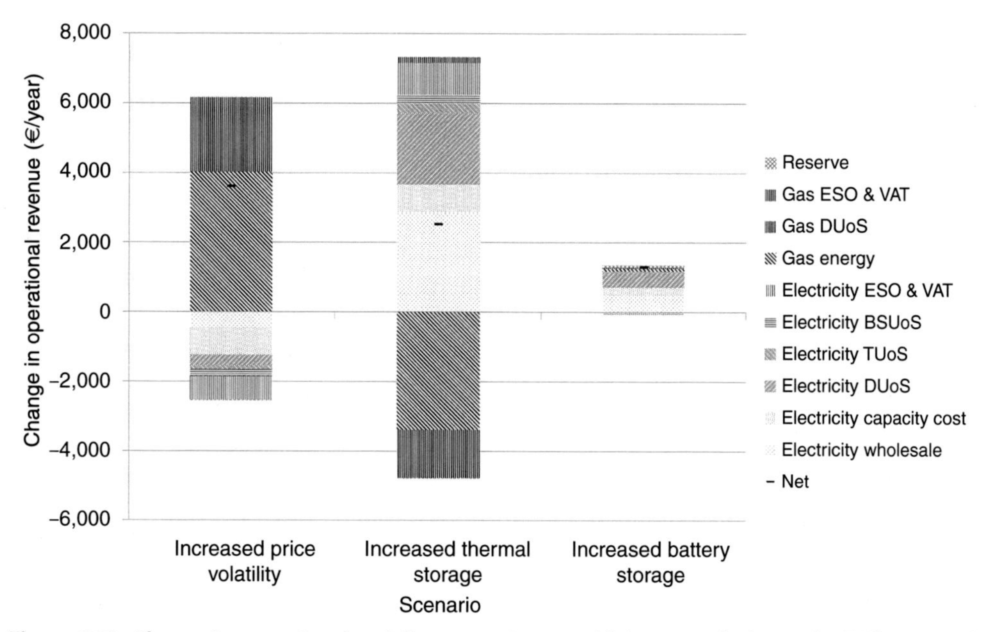

Figure 6.20 ***Change in operational cash flows—various sensitivity tests relative to the IEC base—Irish case.***

per year. This revenue is a result of increased CHP operation at high WE price times, avoiding some UoS and tax costs, as well as reducing electricity wholesale market costs. This is, of course, tempered by an increase in gas energy and tax, and to a lesser extent GDUoS costs.

A 100% increase in electrical battery size (from 35 kWh capacity, 85 kW charge/discharge, to 70 kWh capacity, 170 kW charge/discharge) results in lesser benefits of €1300 per year. This is due to shift from periods of high electricity prices (in electricity wholesale market and UoS components), and a slight increase in reserve availability.

4.2.4.2 France

As with the Irish test bed, three sensitivity tests were carried out for the French test bed. Again, an increased price volatility case, for which the variability of the electricity wholesale market price profile was increased by 100%, was investigated. Fig. 6.21 shows the base and new electricity wholesale market price profile for a typical winter weekday.

Also investigated were an increased thermal storage case, where the available thermal storage was increased from 250,000 to 500,000 L, and an increased electrical battery storage case, where battery capacity was increased from 660 kWh (180/720 kW charging/discharging power) to 1320 kWh (360/1440 kW charging/discharging power).

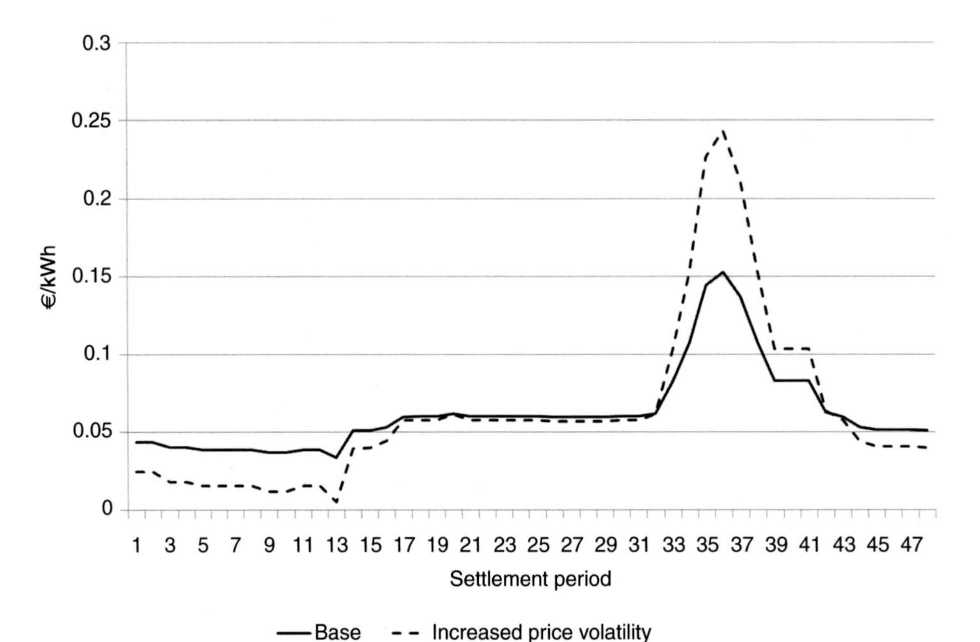

Figure 6.21 ***Wholesale electricity price, typical winter weekday, base, and increased volatility case—French case.***

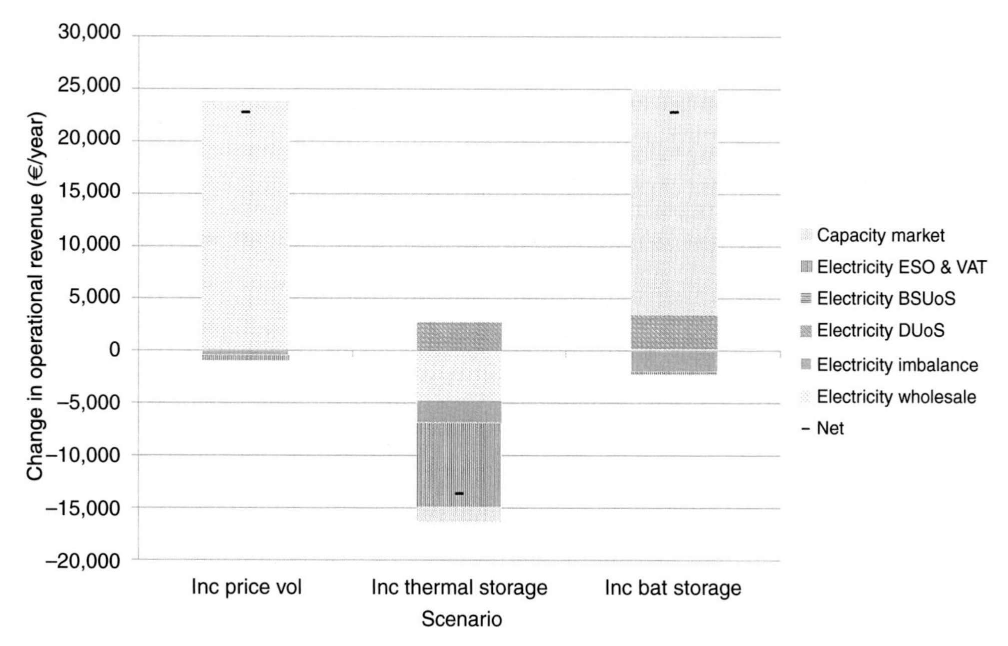

Figure 6.22 ***Change in operational cash flows—various sensitivity tests relative to the IEC base—French case.***

As shown in Fig. 6.22, the increase in electricity wholesale market price volatility produces an increase in revenue of €22,800 per year. This is produced by the increased profitability of shifting electricity consumption away from high price times, toward low price times. The increase in thermal storage results in a reduction in RAEPN revenue of €13,600 per year. This is due to the increased thermal losses associated with the increase of thermal storage. Although this is somewhat tempered by the increased shift of electricity consumption to low price times, the effect is not enough to produce a net positive outcome.

The increase in battery storage produces an increase in RAEPN revenue of €22,800 per year. This is largely achieved through an increase in CP and reduction in DUoS costs, as the additional battery capacity is used to provide capacity and flatten the electricity consumption profile.

4.2.5 Effect of the Regulatory Context

As discussed in Section 2.3, regulation can have a significant impact on an EPN BC. Regulation may improve BCs, by prescribing incentives or market mechanisms (e.g., for CO_2 emissions) to value externalities and direct the value appropriately. Such regulation may provide a revenue stream for an EPN. It may also impede BCs, if regulation stops price signals from a market from reaching the EPN.

In Section 2.3 four aspects for which regulation can have a significant impact were identified: (1) market participation of EPNs, (2) CO_2 reduction, (3) network fees, and (4) reliability of supply. For the first point, the effect of allowing EPNs to partake in energy markets is highlighted by the potential revenue available (to the combined RAEPN actor) from price signal visibility, see Figs. 6.14 and 6.16. For the CO_2 reduction and reliability of supply aspects, examples can be drawn from the employed test beds, to demonstrate the effect of different types of regulation.

4.2.5.1 CO_2 Reduction

The effects of more traditional and more liberalized regulatory approaches on CO_2 reduction can be investigated on the Irish test bed, given the presence of carbon emitting plant (CHP and gas boilers) within the EPN. A more traditional approach, employed, for example, in the United Kingdom, is to incentivize low-carbon electricity generation through feed-in-tariffs, or LCI, on local, relatively low-carbon CHP plant. The effect on all relevant actors attributed to implementing a CHP LCI (set according to the UK equivalent incentive, at 18.61 c/kWh$_{\text{CHP electricity}}$) is shown in Fig. 6.23. An alternative, more liberalized, approach to encouraging CO_2 reduction is to set up a (cap and trade) carbon market, open down to the level of EPNs and similar parties. In this case, the EPN could be assumed to receive allowances equal to its local CO_2 emission in the base energy optimization case, with profit available from selling credits. This effect on the

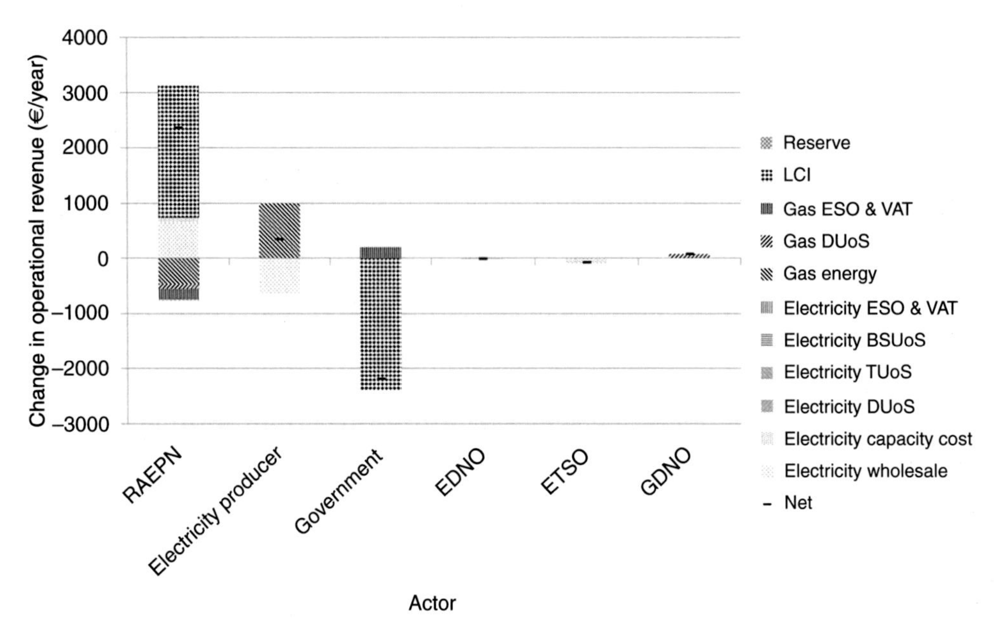

Figure 6.23 ***Change in operational cash flows—CHP LCI case relative to the IEC base, by actor—Irish case.***

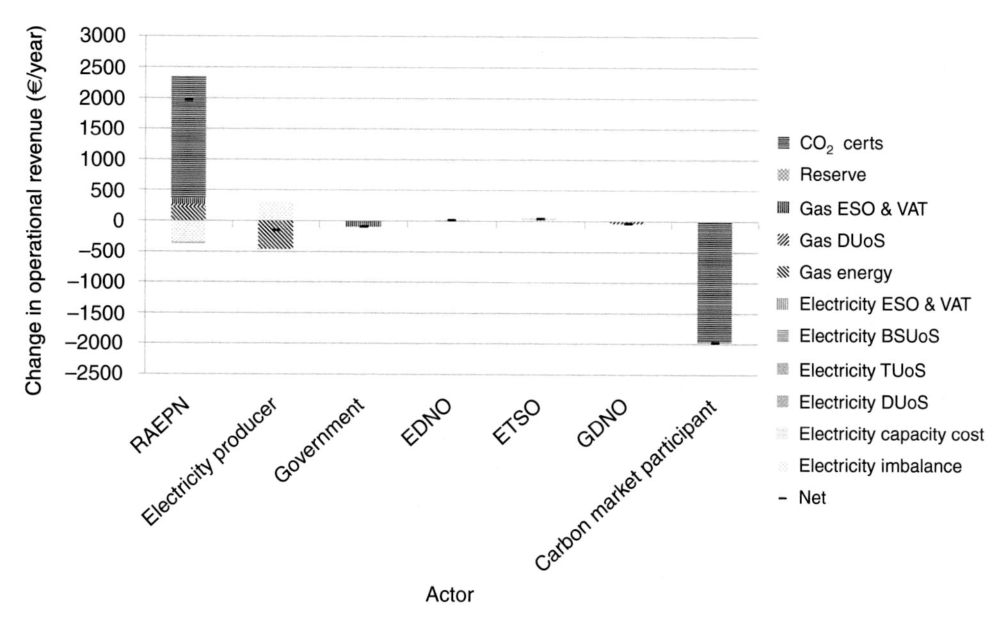

Figure 6.24 ***Change in operational cash flows—LCM case relative to the IEC base, by actor—Irish case.***

relevant actors of this case [with the price of €4.42 per kgCO_2, the average rate on the European Union Emissions Trading Scheme (EUETS) in 2013 (Committee on Climate Change, 2014)] is shown in Fig. 6.24.

As shown in Fig. 6.23, imposition of a CHP LCI has a positive impact on the operational cash flows of the RAEPN, increasing revenue by €2400 per year on the IEC base. Besides revenue from the LCI itself, there is additional wholesale electricity revenue. The fact that there is little gain in electricity UoS and tax elements indicates that this rise in wholesale electricity revenue comes from increases in electricity exports. This rise in electricity revenue is, however, tempered by an increase in gas-related costs, as heat production is shifted from the gas boiler to the (less thermally efficient) CHP. Interestingly, the electricity producer perceives benefits from the imposed CHP LCI under the given assumptions, as the increase in CHP electricity production occurs at times of low electricity price. This means that, under the utilized assumptions, the savings in avoided gas consumption are greater than the loss of electricity sales. Of course, generation plants (especially flexible plant, such as a gas-fired generator) are unlikely to operate at a loss, which occurs in this study due to the simplified assumptions (i.e., the omission of a full system unit commitment). Nevertheless, it is clear that CHP LCI will, for a smart EPN, encourage CHP electricity production at times of low electricity price. As can be expected, the greatest reduction in operational revenue is for the government actor, for whom this policy costs €2400 per year, predominantly in LCI payments. For this expenditure the government, in effect, procures a CO_2 emission reduction (considering both local and grid emissions[12]) of 4800 kg/year (a local increase of 2800 kg, and a grid reduction of 7600 kg).

Fig. 6.24 shows the effect of imposing a carbon market on the EPN. The market is assumed to be a cap-and-trade market, with the EPN receiving permits to cover their CO_2 emission (from CHP and gas boilers) in the energy optimization base case (as described in Section 4.2.1). At the modeled CO_2 price, participation in such a market is worth €2000 per year to the RAEPN. There is some reduction in operational expenditure for the electricity producer (€200 per year), as the carbon price reduces RAEPN CHP operation, and hence reduces RAEPN electricity export. Imposition of a local carbon market (LCM) actually results in an increase in grid CO_2 emission (of 5500 $kgCO_2$/year), with a reduction in EPN emissions (2000 $kgCO_2$/year[13]). This highlights the need for integration of any local CO_2 market with any similar scheme for electricity generators.

4.2.5.2 Reliability of Supply

To ensure reliability of supply, in the traditional approach, the EDNO will make (potentially costly) investments in extra capacity for the distribution network. This emergency capacity is not available during normal operations, as it is meant to ensure that electricity supply can continue in the event of a line outage. A more "liberalized" approach may be for the EDNO to pay a suitably capable microgrid to be able to "island" (self-supply). That is, the microgrid could use the available emergency capacity as long as it releases this capacity (by operating as an island) during emergency conditions (Syrri et al., 2015). For the purposes of this work, it is assumed that the EPN is capable of operating as an islanded microgrid, and doing so enables the EDNO to guarantee reliability of supply to the required standard. In the following analysis on operational cash flows it is assumed that the EPN microgrid is required to have enough reserve capacity (from available generation technologies and storage) to operate as an island during the peak 8 h of the peak season once every 3 years, in accordance with (Martínez-Ceseña & Mancarella, 2014a). If the EPN does not have sufficient generating capacity to island itself, it is assumed that back-up diesel generation will be installed to provide the necessary capacity.

For the Irish test bed, Fig. 6.25 shows the change in annual cash flows for the islanding case, on the IEC base case. This shows that the RAEPN will be €190 per year[14]

[12]Grid emissions are calculated given 2013 Irish grid electricity production, by unit type and unit specific CO_2 emission rates, as described in (EirGrid, n.d.).

[13]Though, being a cap-and-trade market, the net effect across all local CO_2 market participants is assumed to be zero.

[14]On average, as the service is expected to be called once every 3 years.

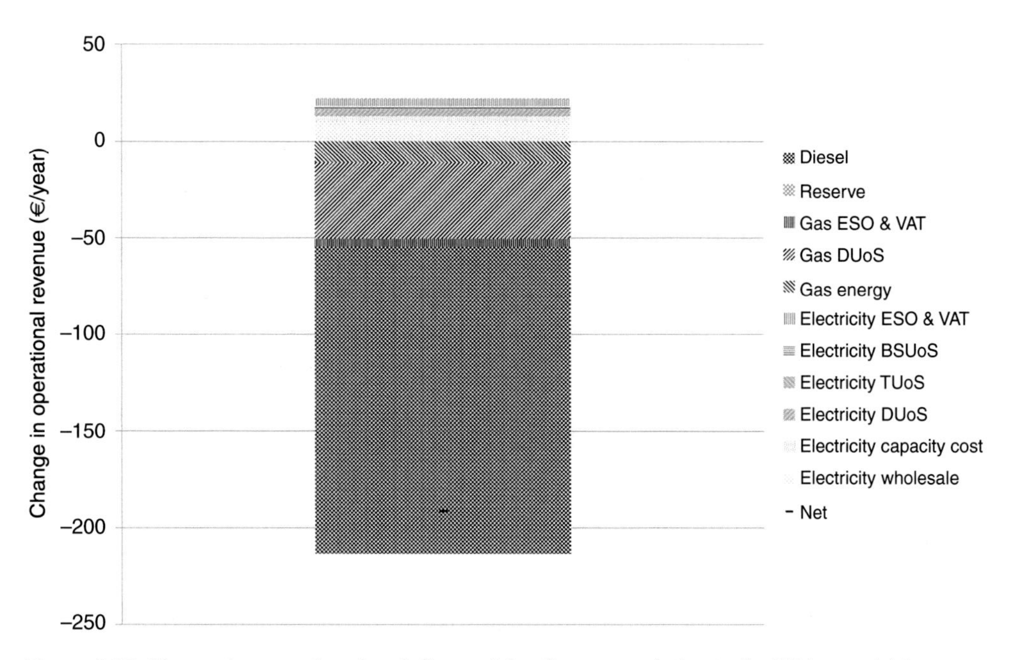

Figure 6.25 ***Change in operational cash flows—islanding case relative to the IEC base—Irish case.***

worse off. Thus the EDNO should compensate the RAEPN, for provision of the islanding service, for provision of reliability, at a rate which covers this €190 per year loss, in addition to the investment cost for installation of the required 750 kW of diesel electricity generating capacity.

For the French test bed, the picture is substantially different (Fig. 6.26). In this case there is substantial operational revenue loss (€89,300 per year), predominantly due to the reduction in capacity available to partake in the CM, as capacity is set aside to provide the capability to island the microgrid.

4.3 Splitting of Benefits Between the Retailer–Aggregator and the EPN

In Section 4.2, results were presented for an abstract RAEPN actor. The presented changes in operational cash flows must then be shared between the constituent retailer–aggregator[15] and EPN actors. The method by which extra benefits (or costs) should be shared, and the proportion that should be assigned to each party is an open question. In reality the method will be agreed through negotiation between the parties. To give some indication of how benefits could be shared, two illustrative methods are presented

[15]For the purposes of this analysis the retailer and aggregator role are fulfilled by one actor, though, in practice, they may be two.

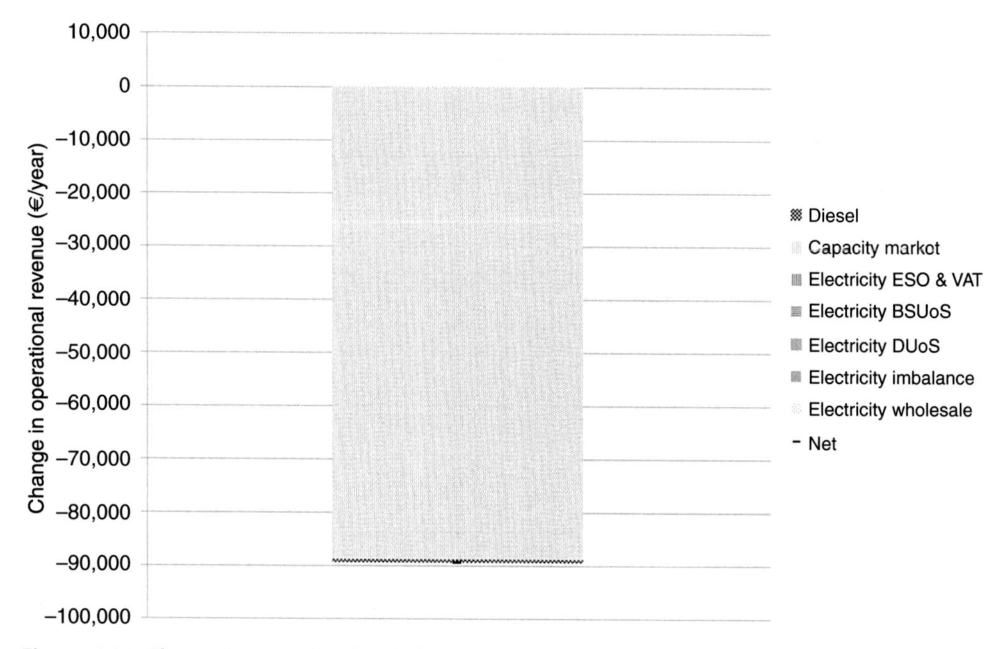

Figure 6.26 ***Change in operational cash flows—islanding case relative to the IEC base—French case.***

in this section. These methods can be considered extremes, as, in reality, any method may fall somewhere between the two. First a "retailer–aggregator led" method is presented. In this case the RA invests in necessary enabling infrastructure and controls the relevant flexible plant, with the EPN remaining a passive user. Second, an "EPN led" method is presented. In this case the EPN is much more active, investing in enabling infrastructure and controlling flexible equipment.

4.3.1 Retailer–Aggregator Led

A central assumption in this method is that the retailer–aggregator will not accept this method if its profit margin from energy retailing declines as a result of any BC. Thus, RA benefits are separated in two parts, namely energy retail related and BC related. This method ensures that all energy retail-related benefits go to the RA. The (flat) retail tariff (which will change according to the case) will ensure this by increasing in response to reductions in demand. Concurrently, it will also decrease in order to allocate the EPN its share of the BC benefits. In this way the RA precludes disruption to their main business (energy retail), while accessing benefits associated with new, flexibility-related BCs.

4.3.1.1 Ireland

The electricity tariffs for various Irish BCs and three benefits splits are shown in Fig. 6.27. The benefits splits considered here are 20% to the EPN, 50% to the EPN, and

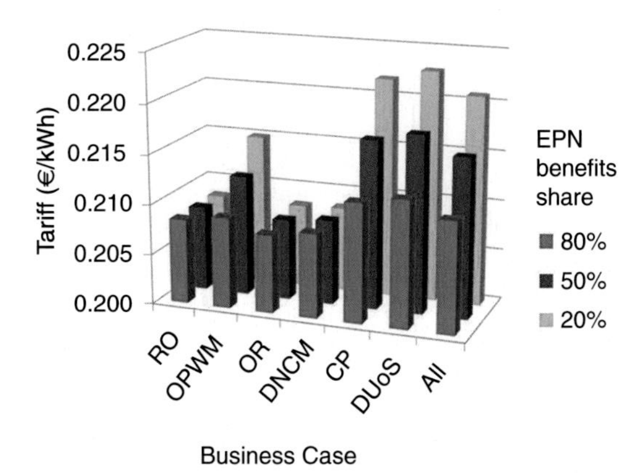

Figure 6.27 ***Electricity tariffs for the three benefits splits—various BCs—Irish case.***

80% to the EPN, with the remainder of the benefits going to the RA. It can be seen that most of the tariffs increase compared to the base case (the RO case). Fig. 6.28 indicates that this is due to the retailer profit margin, which increases in value as the demand reduces, in order to ensure the retailer always gets the same amount of energy-retail profit as for the RO case (note that the retailer profit margin does not change with the benefits splits—it only changes with the BCs, as it is a change in demand that forces it up or down) Fig. 6.27 also show that the tariff increases as the RA share of the benefits increases. This is because a large part of the benefits from a BC are due to a reduction in the overall electricity demand. For the RA to get their share of these benefits, they have to increase the tariffs given to the EPN. Although the tariffs increase from the base case, each BC is profitable for the EPN (Fig. 6.29). The RA profits are the exact opposite of the EPN profits (i.e., when the EPN gets the amount indicated by the green columns

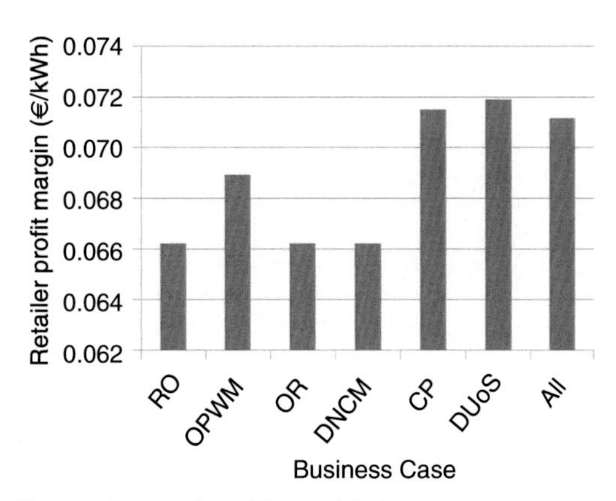

Figure 6.28 ***Retailer profit margin—various BCs—Irish case.***

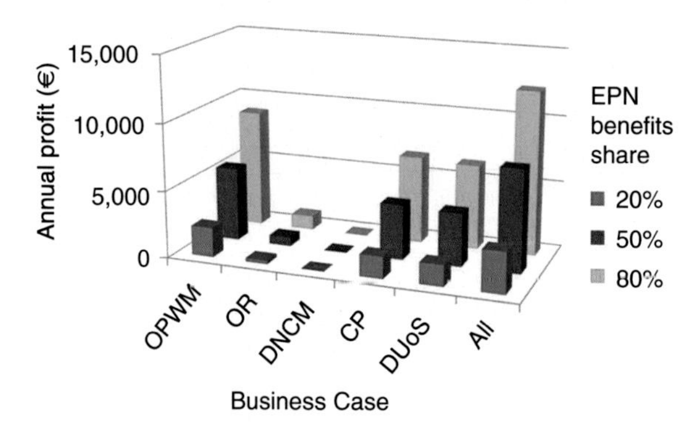

Figure 6.29 ***EPN net cash flow for the three benefits splits—various BCs relative to the retail price optimization case—Irish case.***

which correspond to 80% of the benefits, the RA gets the blue columns that correspond to 20% of the profits and so on).

4.3.1.2 France

Comparing Figs. 6.30 and 6.31 it can be seen that the tariffs for the French case are notably less affected by the changing retailer profit margin, which change in response to changes in demand. This is because (1) the French cases have less of a change in the overall demand than the Irish cases and (2) most of the profits come from optimization on the various signals rather than from a reduced overall demand. For the French case, the CM case stands out significantly from the rest. As indicated in Fig. 6.32, the CM may result in significant revenue for the EPN, even with a low (20%) profit share for the EPN.

4.3.2 EPN Led

For the EPN led method, the EPN receives direct access to market price signals rather than the conventional time-invariant tariff. The EPN pays a fixed annual fee (same as the

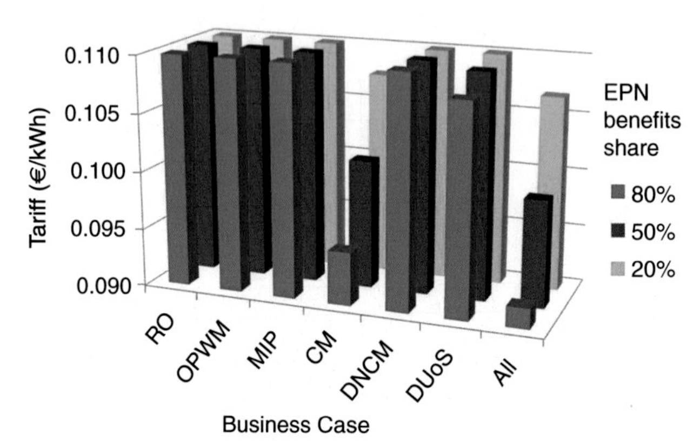

Figure 6.30 ***Electricity tariffs for three benefits splits—various BCs—French case.***

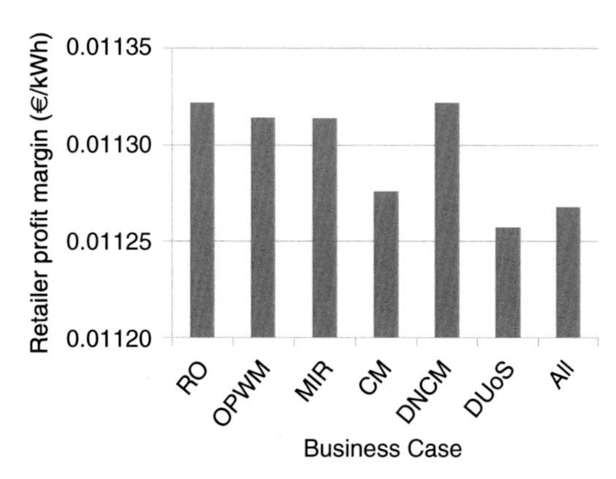

Figure 6.31 ***Retailer profit margin—various BCs—French case.***

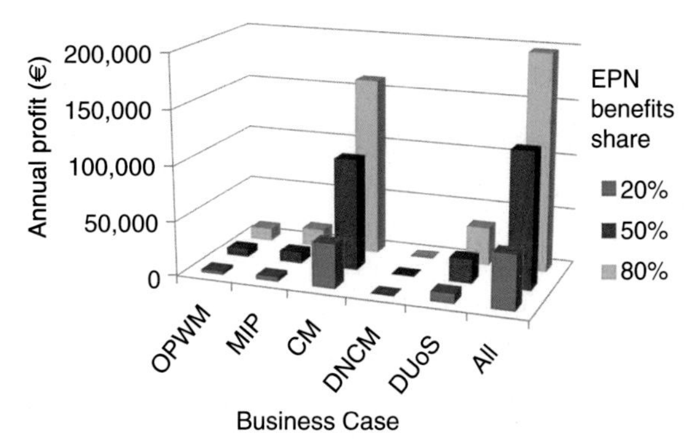

Figure 6.32 ***EPN net cash flow for the three benefits splits—various BCs relative to the retail price optimization case—French case.***

profit margin for the RA led method, but now the fee is in € and not in €/kWh) to the RA for providing direct access to the market price signals, and for fulfilling the necessary statutory functions (with respect to balancing responsibility and so forth). The risk for this method is with the EPN, as it is no longer shielded from the time-varying price signals on the market through a fixed tariff. The RA always gets its annual fee, hence the RA is always sure to recover the retailer costs, leaving the RA in a minimal risk position. The investment costs are taken by the EPN, which also control the flexible equipment. All benefits obtained from a BC go to the EPN.

4.3.2.1 Ireland

Fig. 6.33 shows the cash flow for the EPN for the various energy vectors (relative to the retail price optimization case). The net cash flows are the benefits/costs that the EPN obtains for each BC and it can be seen that these benefits occur as a result of reduced expenditure on electricity imports and increased income from electricity exports. The

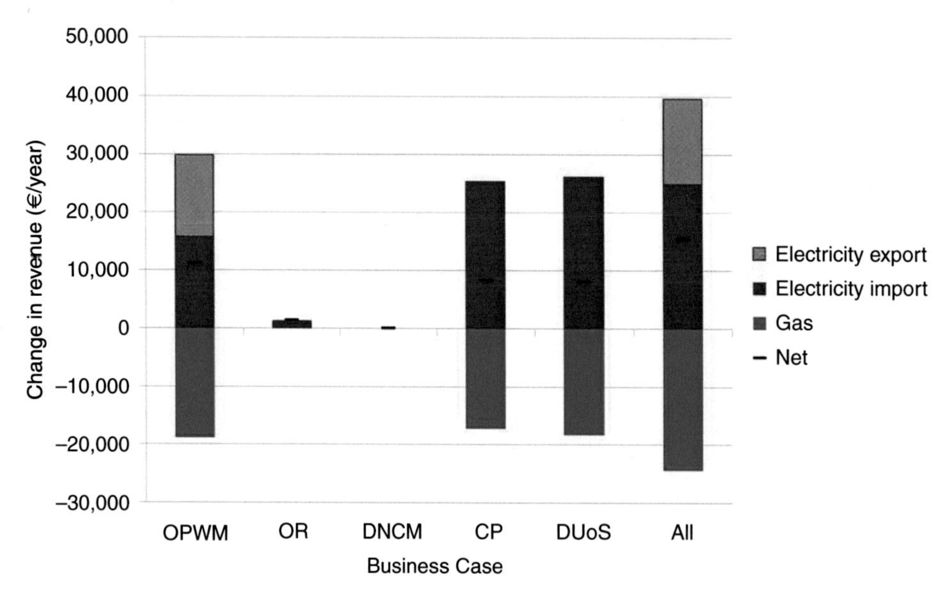

Figure 6.33 ***EPN revenue—various BCs relative to the retail price optimization case—Irish case.***

expenditure on gas imports are increased (due to the shift from gas boiler to CHP for heating), but the overall net cash flow is positive, indicating that the different BCs are beneficial for the EPN from an operational cash flow perspective. The fixed retailer fee is €74,100 per year (determined as equal to the retailer's margin on electricity and gas sales in the base case) for the Irish case. As discussed, the fixed retailer fee does not change with respect to the base case, hence it does not show in Fig. 6.33.

To demonstrate the flow of the retailer fee, from EPN to RA, the absolute cash flows for the EPN are shown in Fig. 6.34.

4.3.2.2 France

Fig. 6.35 shows the cash flow for the EPN for the various energy vectors (relative to the retail price optimization case). The net cash flows are the benefits/costs that the EPN obtains for each BC. It can be seen that all BCs other than the CM case are beneficial to the EPN from an operational cash flow perspective and that these benefits occur as a result of reduced expenditure on electricity imports and increased income from electricity exports. The CM case results in increased expenditure on electricity imports and it is therefore not beneficial to the EPN. The fixed retailer fee is €189,200 per year in the French case. As discussed, the fixed retailer fee does not change with respect to the base case; hence it does not show in Fig. 6.35. Fig. 6.36 shows the absolute cash flows for the EPN.

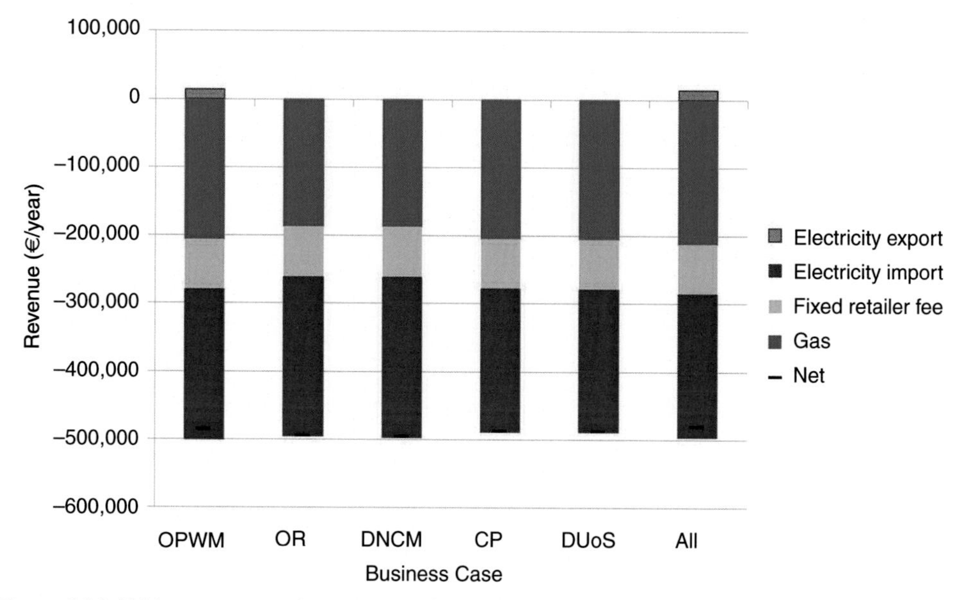

Figure 6.34 ***EPN revenue—various BCs—Irish case.***

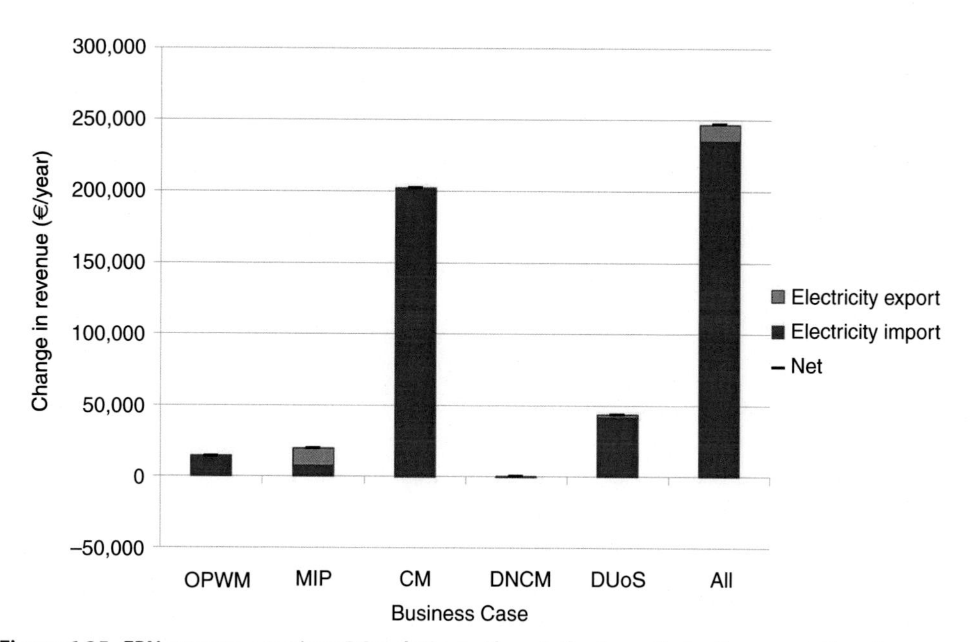

Figure 6.35 ***EPN revenue—various BCs relative to the retail price optimization case—French case.***

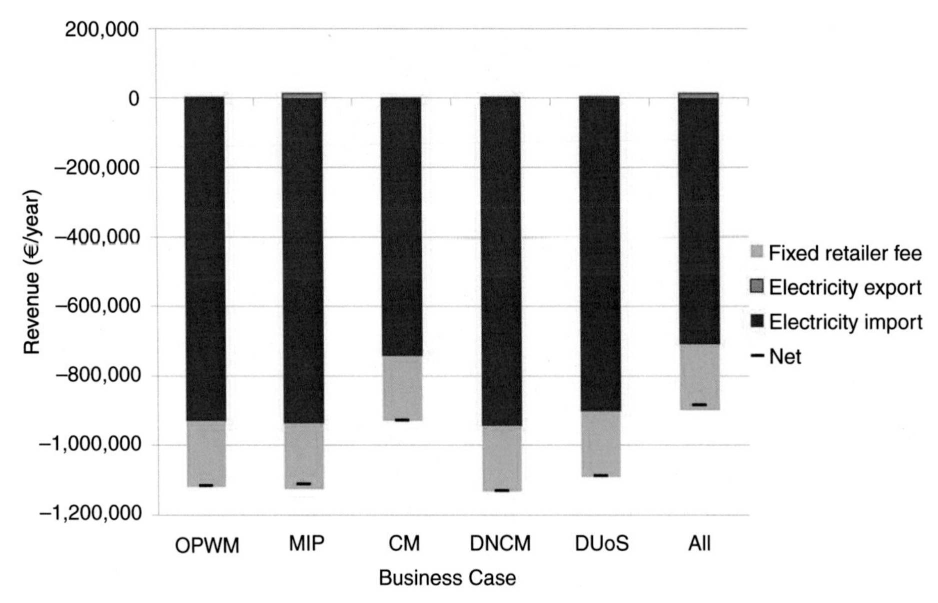

Figure 6.36 ***EPN revenue—various BCs—French case.***

4.3.3 Comparison of the RA-Led and the EPN-Led Methods

The difference between the two methods can be illustrated with the difference in the EPN profits for the different cases, as shown in Figs. 6.37 and 6.38. It can be seen that the EPN profits are highest for the EPN led method (as expected, as the EPN takes all of the benefits for this method). However, to evaluate if the BCs are beneficial to the EPN (and to the RA) the investments necessary to facilitate the different BCs must be accounted for, as shall be considered in Section 4.4.

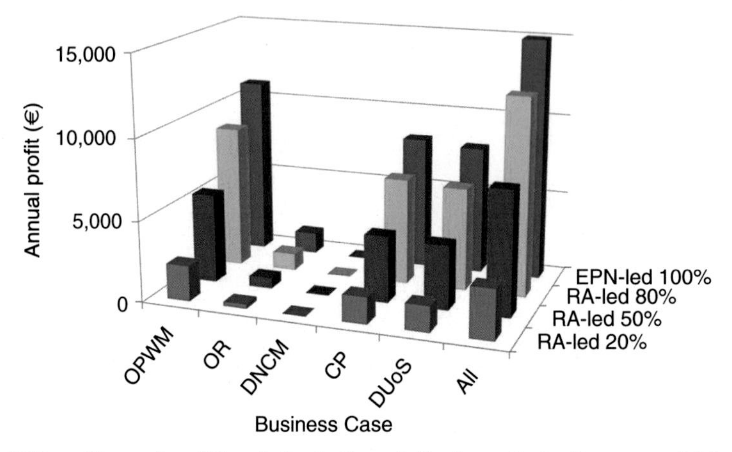

Figure 6.37 ***EPN profit—various BCs relative to the retail price optimization case—Irish case.***

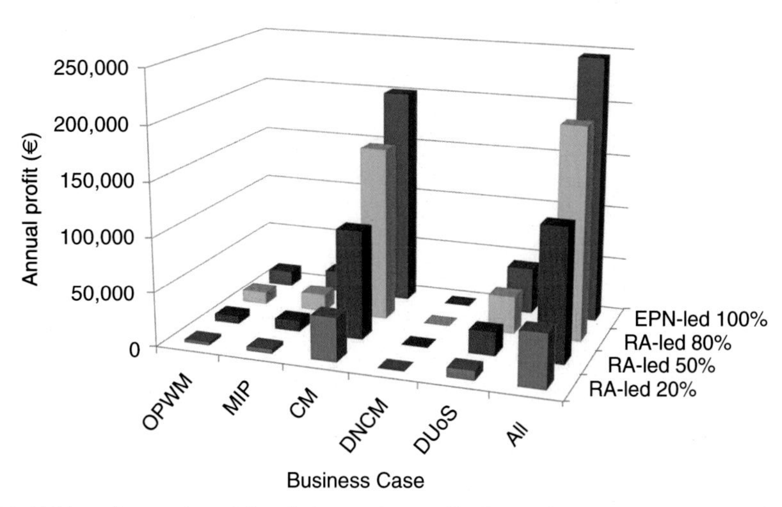

Figure 6.38 ***EPN profit—various BCs relative to the retail price optimization case—French case.***

4.4 Economic Assessment

In this section, an economic assessment for the various COOPERaTE BCs is conducted. This assessment is informed by the operational cash flow benefits obtained from Sections 4.2 and 4.3, and the associated gas and electricity flows (for calculation of emissions). To complete the assessment, the investments that are necessary to facilitate the different BCs are presented. These investments are assumed to be undertaken by the lead party (EPN or RA, see Section 4.3). As discussed in Section 2.5, a payment or income is valued differently depending on the time of which it occurs, as formulated by the time-value of money principle. This is accounted for by discounting cash flows, according to when they are expected to occur. The appropriate discount rate is dependent on the riskiness of the project and the cost of capital, for the actor in question. To reflect the smaller size of the EPN (compared to the RA), which results in less ability to handle risk and higher cost of capital, the EPN discount rate is taken as 10%, while the RA discount rate is taken as 7%. This means that, for the EPN, the benefits obtained from a BC 1 year ahead is worth 10% less than the benefits obtained from a BC today.

The investment costs of the equipment needed to facilitate some of the considered BCs are listed in Table 6.5. Most of the costs are given as a function of the system size, of which more information is given in Section 4.4.1 for the Irish case and Section 4.4.2 for the French case. Note that some of the costs are separated into an "industry average price" and a "large-scale price." This is to distinguish between the prices that the EPN might have access to (the industry average price), and the prices the RA might have access to (the large-scale price), due to the different economies of scale.

Given the uncertainty on the ICT costs needed to facilitate the EPN concept are very case specific and uncertain. Thus, the NPV is presented without consideration of

Table 6.5 Investment costs needed to facilitate some of the BCs

Equipment	Cost
2500 L hot water storage tank	Equipment: €15,500[a] Installation: €1,000[b]
PCM	€2.70 per liter[c]
Batteries	€370 per kWh (industry average price)[d] €270 per kWh (large-scale price)[e]
Power electronics to control batteries	€110 per kW[f]

[a]http://www.plumbcenter.co.uk/en/heatrae-sadia-hot-water-system-indirect-cylinder-2500ltr/
[b]http://www.nrel.gov/docs/fy03osti/32922.pdf
[c]http://www.nrel.gov/docs/fy13osti/55553.pdf
[d]http://www.nature.com/nclimate/journal/v5/n4/pdf/nclimate2564.pdf
[e]http://www.nature.com/nclimate/journal/v5/n4/pdf/nclimate2564.pdf
[f]http://www.paredox.com/foswiki/pub/Luichart/PaRedoxExamplePapers/Cost_Analysis_of_Electricity_Storage.pdf

the ICT costs. This gives an indication of the thresholds, under different BCs, for the NPV of ICT costs that would make each BC viable.

4.4.1 Ireland

For the Irish economic assessment part, the RO, All, IET, Increase Price Volatility (IPV), Increase Thermal Storage (ITS), and Increase Battery Storage (IBS) cases are evaluated and compared against the LF case. The investment costs for the EM, RO, All, IET, and IPV cases are the same, as these cases rely on the same equipment to optimize on the various signals. The investment costs for the ITS case includes the additional capital costs for a 2500 L hot water storage tank and a 1500 L PCM. The investment costs for the IBS case includes the additional capital costs for 35 kWh rated batteries and 85 kW rated power electronics. For the IET case the investment costs includes the additional capital costs of setting up a private network, and the costs of operation, maintenance, and possibly, expansion of the network. However, as this cost is very case-specific (it depends on the location in, and characteristics of, the distribution network, the regulations in the country and so on) it, like the ICT costs, is not included in the initial assessment. Therefore, it should be considered that the NPVs presented for the IET case are thresholds for ICT and private wire-related costs.

Accounting for the aforementioned investment costs and the change in the operational costs as a result of these investments (Section 4.2), the NPV has been calculated for the different Irish BCs. Fig. 6.39 shows the NPV for the different BCs for the EPN led method, for four different payback periods. Fig. 6.39 illustrates that the ICT costs (and, in case, private wire costs) can be higher for longer payback periods. Therefore, the level of ICT/private wire costs that can be considered acceptable vary according to the considered payback period. Fig. 6.39 also illustrates that the NPV does not vary considerably between the different BCs. This is because the substantial proportion of benefits

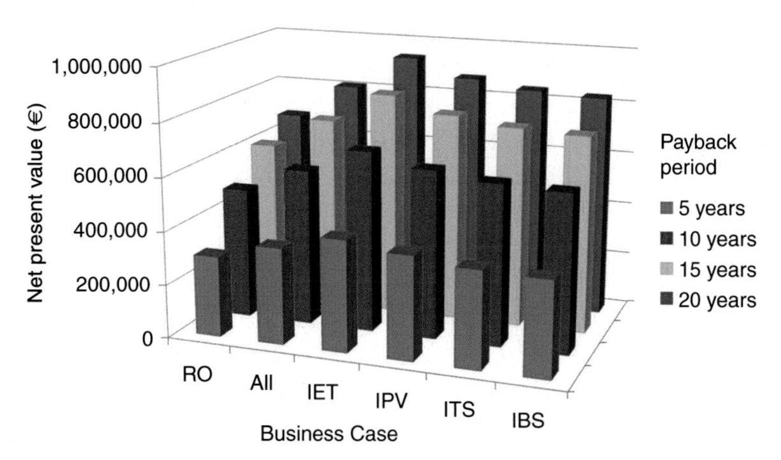

Figure 6.39 ***NPV for different payback periods—EPN led method—various cases relative to the load following case—Irish Case.***

are captured in the RO case, as import is avoided, particularly at high price times (due to the coincidence of base electricity and heat demand (and, hence, CHP generation), and high grid prices.

Fig. 6.40 shows the NPV for the different BCs for four of the previously considered sharing agreements. These are the RA led method (with 20, 50, and 80% of the benefits going to the RA) and the EPN led method (with all of the benefits going to the EPN). In each case, the assessment is for the lead actor (who is making the investment and controlling the EPN resources). It can be seen from Fig. 6.40 that the ICT costs would have to be very small for the RA led 20% case. It can also be seen that the ICT costs

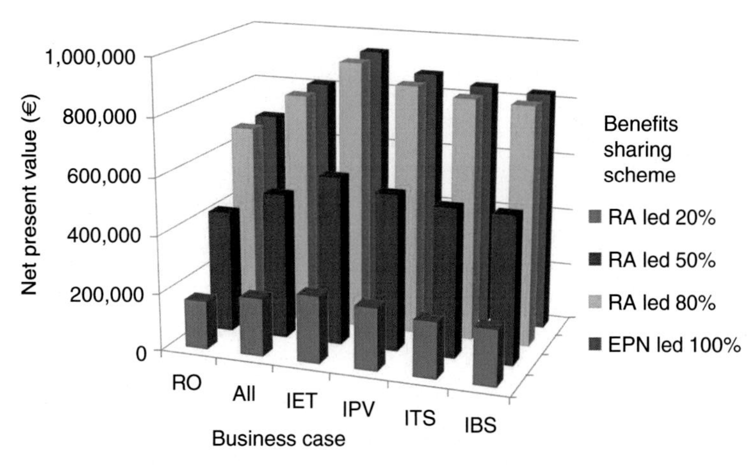

Figure 6.40 ***NPV for a 20-year payback period—various BCs relative to the load following case—Irish case.***

could be almost as high for the RA led 80% case as for the EPN led method. This is partly because the EPN has got a higher discount rate than the RA, and partly because the RA can use its economy of scale to reduce its investment costs. Indeed, it is possible that these factors could mean that it is more beneficial for the EPN for BCs to be led by the RA, dependent on the EPNs revenue share.

4.4.2 France

For the French economic assessment part, the RO, All, IPV, and IBS cases are evaluated and compared against the LF case. The investment costs for the EM, RO, All, and IPV are the same, as these cases rely on the same equipment to optimize on the various signals. The investment costs for the IBS case includes the additional capital costs for 660 kWh rated batteries and 720 kW rated power electronics. Fig. 6.41 shows the NPV for the different BCs for the EPN led method, for four different payback periods. As for the Irish case, the BC is only considered a good investment if the ICT costs are below the value of the NPV. Fig. 6.41 illustrates that the ICT costs can be more than twice as high for a 20-year payback period compared to a 5-year payback period. It can also be seen that the difference between the BC with the highest NPV (the IPV case) and the BC with the lowest NPV (the RO) case is higher than for the Irish case, with a difference of 470%.

Fig. 6.42 shows the NPV for the different BCs for the four sharing agreements considered for the Irish case. It can be seen that the difference between the different sharing agreements are larger than for the Irish case. This is due to the larger amounts of benefits resulting from the French BCs. For the French case it is therefore more likely that the EPN would be better off with the EPN led method than the RA led method, something that was not necessarily the case for the Irish case.

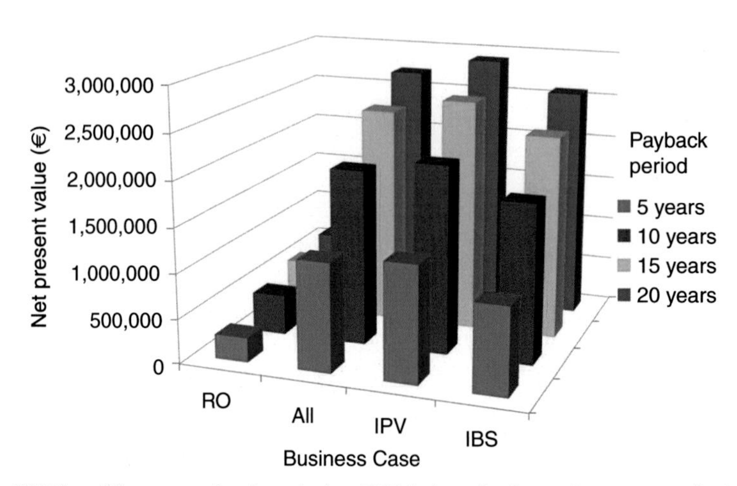

Figure 6.41 ***NPV for different payback periods—EPN led method—various cases relative to the load following case – French Case.***

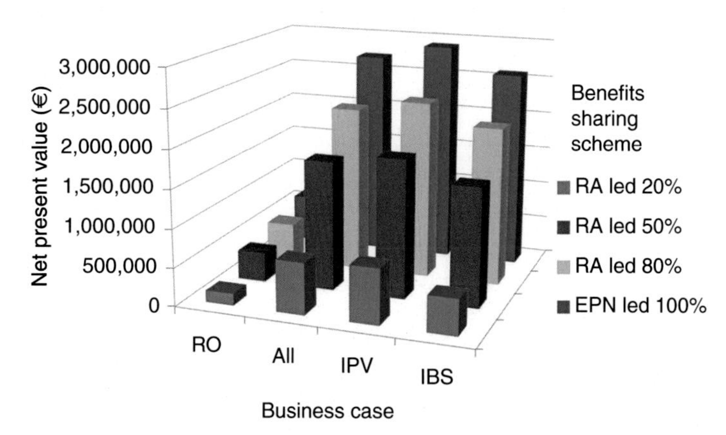

Figure 6.42 ***NPV for a 20-year payback period—various BCs relative to the load following case—French case.***

If one were to make a final decision about which BC to follow, the following points would have to be factored in:

- What payback period is appropriate for the project? Relevant factors to be considered include the expected lifetime of equipment and the applicable tax code, which will influence debt-related decisions, for example, by making interest payments tax deductible?
- What sharing agreement is most beneficial to both the RA and the EPN? (Accounting for the risks each of the actors is able to or willing to take with regard to investments and future payments.)
- Which BCs have NPVs that are higher than the ICT costs?
- Which BC yields the highest rate of return once the ICT costs are accounted for?

In particular, the IRR indicator is a salient metric for decision-makers. This metric is not shown here because its calculation requires all of the costs of a BC to be known (and, in these cases, ICT and private wire costs are not known). Once appropriate ICT (and, in case, private wire) costs are established, the IRR will also be a key metric for BC assessment.

5 CONCLUSIONS

This chapter has introduced the BC as the primary metric for assessing EPN interventions (and interventions by any actor) in a liberalized energy system and detailed the EPN BC assessment process. The first two steps of the process involved identification of EPN flexible resources, which can be exploited, and the markets in which those resources can partake. Fundamentally resources were identified as storage resources (*shifting* demand in time), alternative fuel resources (*substituting* fuel/devices), and curtailment resources (*trading* consumer *utility* for cash). Similarly, markets were identified as *energy*, *capacity*, or *flexibility* markets. For both resources and markets, specific examples, relevant to an EPN, were presented. The third and fourth steps involved understanding of the EPN regulatory context, and of the possible commercial arrangements (between the EPN and any commercial partner) that may be adopted to exploit the EPN flexible resources in relevant markets. For these steps the degree of liberalization of the energy system has been shown to be relevant. Subsequently, metrics for quantitative assessment were presented. Then the importance of recognizing the network interactions of EPN BCs, which impact other energy system actors, was discussed before a mapping methodology for identifying and quantifying those network effects was presented. Case studies in the French and Irish context were presented.

From the description of the BC assessment process, and from the results of the presented BCs, several key points relevant to EPN BCs are apparent.

- Understanding the nature of EPN flexible resources is clearly critical to accurately assessing EPN BCs. This includes district electricity/heat generation and storage, highlighting the importance of detailed EPN simulation (see Chapter 4).
- Further, understanding markets, and the network effects of BC adoption, on the counterparties in those markets and all network actors, is important. The case studies demonstrated how identifying and quantifying network effects revealed how, generally, the EPN concept results in the electricity producer suffering an erosion of operational revenue, and electricity network operators can be expected to lose operational revenue, as electricity consumption generally goes down. Clearly methods for mapping value in energy systems, as presented here, are crucial for energy system actors, in light of potentially severe disruption, from concepts, such as the EPN, to their BCs.
- New markets/methods of compensation may be required for EPNs who may provide new services, such as reliability. For example, in the case studies the EPN was shown to lose revenue from providing reliability services, which it should be compensated for.
- Finally, commercial arrangements, for exploitation of EPN flexible resources are relevant. In the case studies it was shown how the identity of the "lead actor" in EPN exploitation may affect the NPV of a scheme (as larger entities exploit economies of scale, to access bulk-buying discounts and cheaper capital), such that the value to the EPN of EPN and retailer–aggregator led schemes is nearly equivalent. This demonstrates the material effect of details, such as actor discount rates and size in determining BC viability.

Overall, this chapter has demonstrated that there is clear need for holistic BC assessment framework, within the wider EPN process (see Chapter 3), that:

- recognizes the technical capabilities of many and various district resources (as detailed in Chapter 4);

- can accommodate various markets (particularly new flexibility-related markets, to enable district flexibility to support energy systems, see Chapter 2); and
- can enable the effect of commercial arrangements between the EPN and commercial partners to be appreciated.

ABBREVIATIONS

BC Business Case
CBA Cost Benefit Analysis
CHP Combined Heat and Power
CM Capacity Market
CP Capacity Payments
DES Distributed Energy Sources
DG Distributed Generation
DNCM Distribution Network Constraint Management
DNO Distribution Network Operator
DR Demand Response
DSO Distribution System Operator
DUoS Distribution Use of System
EDUoS Electricity Distribution Use of System
EES Electrical Energy Storage
EHP Electric Heat Pump
EPN Energy Positive Neighborhood
ESCo Energy Service Company
ESO Environmental and Social Obligation
ETSO Electricity Transmission System Operator
ETUoS Electricity Transmission Use of System
EUETS European Union Emissions Trading Scheme
GCP Grid Connection Point
GDUoS Gas Distribution Use of System
GTUoS Gas Transmission Use of System
HP Heat Pump
IBS Increase Battery Storage
ICT Information and Communication Technology
IEC Intra-EPN Coordination
IET Intra-EPN Trading
IPV Increase Price Volatility
IRR Internal Rate of Return
ITS Increase Thermal Storage
LCI Low Carbon Incentive
LCM Local Carbon Market
LF Load Following
MIP Market Index Price
NEM Neighborhood Energy Manager
NPV Net Present Value
OPWM Optimized Purchase on the Wholesale Market

OR	Operating Reserve
PCM	Phase Change Material
POR	Primary Operating Reserve
PV	Photo-Voltaic
RA	Retailer–Aggregator
RAEPN	Retailer–Aggregator–EPN
RO	Retail price Optimization
RR	Regulating Reserve
SOR	Secondary Operating Reserve
SUDA	Sensing, Understanding, Deciding, Acting
TES	Thermal Energy Store
TOR	Tertiary Operating Reserve
TSO	Transmission System Operator
TUoS	Transmission Use of System
UoS	Use of System
VAT	Value-Added-Tax

REFERENCES

Aalami, H. A., Moghaddam, M. P., & Yousefi, G. R. (2010). Demand response modeling considering Interruptible/Curtailable loads and capacity market programs. *Applied Energy*, *87*(1), 243–250Available from http://linkinghub.elsevier.com/retrieve/pii/S030626190900244X.

Althaher, S., Mancarella, P., & Mutale, J. (2015). Automated demand response from home energy management system under dynamic pricing and power and comfort constraints. *IEEE Transactions on Smart Grid*, *6*(4), 1874–1883.

Capuder, T., & Mancarella, P. (2014). Techno-economic and environmental modelling and optimization of flexible distributed multi-generation options. *Energy*, *71*, 516–533, Available from http://linkinghub.elsevier.com/retrieve/pii/S0360544214005283.

Committee on Climate Change. (2014). *Meeting carbon budgets—2014 Progress report to parliament*. Committee on Climate Change.

DEFRA. (2010). *The carbon reduction commitment*. DEFRA.

DHC+ Technology Platform. (n.d.). *District Heating & Cooling: A vision towards 2020-2030-2050*. DHC+ Technology Platform.

EirGrid. (2013). *EirGrid grid code*. EirGrid.

EirGrid. (n.d.). *EirGrid, CO_2 emissions*. Available from http://www.eirgrid.com/operations/systemperformancedata/co2emissions/

Elexon. (2012). *The electricity trading arrangements: A beginner's guide*. Elexon.

EU. (2003). *Directive 2003/87/EC of the European Parliament and of the Council of 13 October 2003 establishing a scheme for greenhouse gas emission allowance trading within the Community and amending*. European Union.

Good, N., et al. (2015). Optimization under uncertainty of thermal storage based flexible demand response with quantification of residential users' discomfort. *IEEE Transactions on Smart Grid*, *6*(5), 2333–2342Available from http://www.scopus.com/inward/record.url?eid=2-s2.0-84923658441&partnerID=tZOtx3y1.

Good, N., et al. (2016). Techno-economic and business case assessment of low carbon technologies in distributed multi-energy systems. *Applied Energy*, *167*, 158–172, Available from http://www.sciencedirect.com/science/article/pii/S0306261915012155.

Haring, T. W., Kirschen, D. S., & Andersson, G. (2015). Incentive compatible imbalance settlement. *IEEE Transactions on Power Systems*, *30*(6), 3338–3346.

KEMA. (2009). *Study on methodologies for gas transmission network tariffs and gas balancing fees in Europe—Annex—submitted to: The European Commission, Directorate-General Energy and Transport*. Available from http://ec.europa.eu/energy/gas_electricity/studies/doc/gas/2009_12_gas_transmission_and_balancing_annex_fact_sheets.pdf

Keyaerts, N. (2012). Gas balancing and line-pack flexibility concepts and methodologies for organizing and regulating gas balancing in liberalized and integrated EU gas markets. PhD thesis. Katholieke Universiteit Leuven.

Kitapbayev, Y., Moriarty, J., & Mancarella, P. (2015). Stochastic control and real options valuation of thermal storage-enabled demand response from flexible district energy systems. *Applied Energy*, *137*, 823–831.

Losi, A., Mancarella, P., & Vicino, A. (2015). *Integration of demand response into the electricity chain: Challenges, opportunities, and Smart Grid solutions*. London, UK: Wiley-ISTE.

Ma, O., et al. (2013). Demand response for ancillary services. *IEEE Transactions on Smart Grid*, *4*(4), 1–8Available from http://ieeexplore.ieee.org/lpdocs/epic03/wrapper.htm?arnumber=6630115.

Mancarella, P. (2006). *From cogeneration to trigeneration: Energy planning and evaluation in a competitive market framework*. Turin: Politecnico di Torino.

Mancarella, P. (2014). MES (multi-energy systems): an overview of concepts and evaluation models. *Energy*, *65*, 1–17, Available from http://linkinghub.elsevier.com/retrieve/pii/S0360544213008931.

Mancarella, P., & Chicco, G. (2009). *Distributed multi-generation systems: Energy models and analyses*. Hauppauge, NY: Nova Science Publishers, Available from https://www.novapublishers.com/catalog/product_info.php?products_id=7335.

Mancarella, P., & Chicco, G. (2013a). Integrated energy and ancillary services provision in multi-energy systems. In *2013 IREP Symposium-Bulk Power System Dynamics and Control*. IEEE, New York, USA, pp. 1–19.

Mancarella, P., & Chicco, G. (2013b). Real-time demand response from energy shifting in distributed multi-generation. *IEEE Transactions on Smart Grid*, *4*(4), 1928–1938.

Martínez-Ceseña, E. A., Good, N., & Mancarella, P. (2015). Electrical network capacity support from demand side response: techno-economic assessment of potential business cases for small commercial and residential end-users. *Energy Policy*, *82*, 222–232, Available from http://www.sciencedirect.com/science/article/pii/S0301421515001184.

Martínez-Ceseña, E.A., & Mancarella, P. (2014a). Capacity to Customers (C2C) Development of Cost Benefit Analysis methodology for network expansion planning considering C2C interventions. Available from www.enwl.co.uk/docs/default-source/c2c-key-documents/economic-modelling-methodology.pdf?sfvrsn=4

Martínez-Ceseña, E.A., & Mancarella, P. (2014b). Distribution network reinforcement planning considering demand response support. In *Proceeding of the 18th Power Systems Computational Conference (PSCC)*.

Martínez-Ceseña, E. A., & Mancarella, P. (2014c). Economic assessment of distribution network reinforcement deferral through post-contingency demand response. In *IEEE PES innovative smart grid technologies* (pp. 1–6). Europe: IEEE.

National Grid. (2013). *The grid code*. National Grid.

OFGEM. (2014). *Feed-in Tariff payment rate table for photovoltaic eligible installations for FIT (1 April 2014 to 30 September 2014)*. Available from https://www.ofgem.gov.uk/ofgem-publications/58940/fit-tariff-table-1-april-2013-non-pv-only.pdf

Poudineh, R., & Jamasb, T. (2014). Distributed generation, storage, demand response and energy efficiency as alternatives to grid capacity enhancement. *Energy Policy*, *67*, 222–231, Available from http://linkinghub.elsevier.com/retrieve/pii/S0301421513012032.

Pudjianto, D., et al. (2010). Value of integrating Distributed Energy Resources in the UK electricity system. *Power and Energy Society General Meeting, 2010 IEEE*, pp.1–6.

Rte. (2015). *Mecanisme de capacite: Guide Pratique*, p 23. Available from https://clients.rte-france.com/htm/fr/mediatheque/telecharge/guide_mecapa.pdf

Saint-Pierre, A., & Mancarella, P. (2016). Active distribution system management: a dual-horizon scheduling framework for DSO/TSO interface under uncertainty. *IEEE Transactions on Smart Grid*, pp 1–12. Available from http://ieeexplore.ieee.org/xpls/abs_all.jsp?arnumber=7432045&tag=1

Schachter, J.A., & Mancarella, P. (2015). Demand response contracts as real options: a probabilistic evaluation framework under short-term and long-term uncertainties. *IEEE Transactions on Smart Grid*. Available from http://ieeexplore.ieee.org/lpdocs/epic03/wrapper.htm?arnumber=7055902

Siano, P. (2014). Demand response and smart grids—a survey. *Renewable and Sustainable Energy Reviews*, *30*, 461–478, Available from http://linkinghub.elsevier.com/retrieve/pii/S1364032113007211.

Six, D., et al. (2015). Techno-economic analysis of Demand Response. In A. Losi, P. Mancarella, & A. Vicino (Eds.), *Integration of demand response into the electricity chain: Challenges, opportunities, and Smart Grid solutions* (pp. 296). Hoboken, New Jersey, USA: Wiley-ISTE.

Sorrell, S., et al. (2009). White certificate schemes: economic analysis and interactions with the EU ETS. *Energy Policy*, *37*(1), 29–42Available from http://www.scopus.com/inward/record.url?eid=2-s2.0-56949083490&partnerID=tZOtx3y1.

Syrri, A.L.A., Martínez-Ceseña, E.A., & Mancarella, P. (2015). Contribution of microgrids to distribution network reliability. In *Proceedings of PowerTech*. Eindhoven.

Vlachos, A. G., & Biskas, P. N. (2013). Demand Response in a real-time balancing market clearing with pay-as-bid pricing. *IEEE Transactions on Smart Grid*, *4*, 1966–1975, Available from http://ieeexplore.ieee.org/lpdocs/epic03/wrapper.htm?arnumber=6520964.

CHAPTER SEVEN

Real Life Experience—Demonstration Sites

M. Boudon*, E. L'Helguen*, L. De Tommasi, J. Bynum**, K. Kouramas**, E.H. Ridouane****

*EMBIX, Issy-les-Moulineaux, Paris, France

**United Technologies Research Centre Ireland, Cork, Ireland

Contents

This chapter focuses on the implementation of the energy services in two demo sites. It describes the energy use-cases developed in COOPERaTE and outlines the evaluation methodology. A significant portion of the chapter is devoted to the presentation of the main results and details the energy-saving estimations.

Even if it was key to understand the results of COOPERaTE services on specific demo sites, the focus of these real-life experiences was to apply the COOPERaTE method having a system of systems approach. There was a clear focus on defining a baseline and a reporting period to make sure that the characteristics of each demo site would not have a too high impact on the COOPERaTE results. Appling the SUDA process (see Chapter 3) helped in understanding the obstacles and opportunities of each site and to make the most of them designing specific services, keeping in mind the overall purpose of the Energy Positive Neighborhood.

1 SITE DESCRIPTION

The CIT Campus Bishopstown is a multiownership neighborhood which includes three main buildings: NIMBUS, Leisure world, and Parchment square (Fig. 7.1).

They feature the energy consumption, production, and storage components listed in Table 7.1.

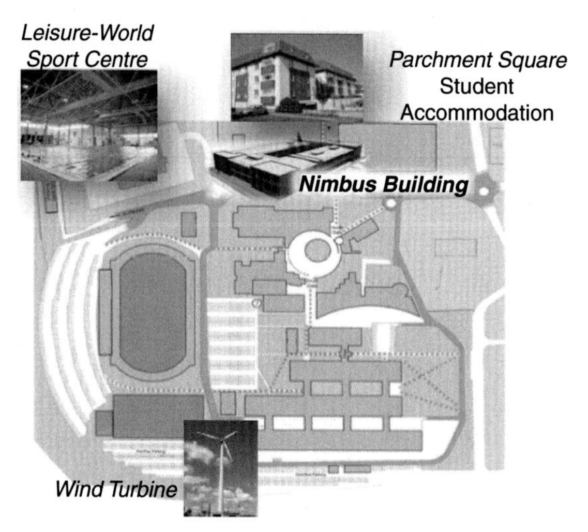

Figure 7.1 ***The Bishoptown Campus demo site.***

Being a decentralized optimization approach, each building of the CIT Campus Bishopstown performs optimal scheduling of flexible energy generation and storage components. From the three buildings composing the demo site, NIMBUS is a great example of a building with flexible generation (e.g., CHP), electrical storage, and renewable energy production (Wind) installed on-site.

Table 7.1 Energy components on the Bishoptown demo site

Energy Consumption Components

NIMBUS	Leisure world	Parchment square
Aggregated Electrical Load: • Heating Load • Computational Load • Lighting Load • Electrical Load	Aggregated Electrical Load (measurements only available since February 2015)	Aggregated Electrical Load (measurements not available for all the periods)

Energy Generation Components

NIMBUS			Leisure world		Parchment square
Wind power	Combined Heat and power	Gas Boilers	Combined Heat and power	Gas Boilers	(Not existing)

Energy Storage Components

NIMBUS		Leisure world	Parchment square
Battery	Thermal Storage (Water Tanks, PCM)	(Not existing)	(Not existing)

Figure 7.2 ***The Challenger Campus demo site.***

Table 7.2 Energy components on the Challenger demo site

Energy Consumption Components				
Heating/Cooling Load	Computational Load	Lighting Load	Electrical Load	Special Uses Load
Energy Generation Components				
Photovoltaics	(Reversible) Heat pumps	Electric Heating Rods		
Energy Storage Components				
Battery	Thermal Storage (Water Tanks)			

The Challenger single-ownership demo site (Fig. 7.2) features the energy consumption, production, and storage components listed in Table 7.2. The electrical storage installed on-site is used as energy optimization potential of the demo site.

The ability to store electrical energy on site is used by the neighborhood energy centralized optimization algorithm, developed by EMBIX, to provide flexibility, and balance the generation and consumption of the site. The size of the installed storage system capacity (66 kWh) is less compared to the size of the site's local generation [e.g., Photovoltaics approx. 3585 (kWp)] but was essential to understand complex mechanisms.

2 ENERGY USE CASES

The energy use-cases and the way they are assessed in the demonstration sites are summarized in this section.

The selected energy use-cases were intended to demonstrate the energy services and the functionality of the energy optimization algorithms described in Chapter 4. A subset of these use cases was validated via simulation. The four energy-related neighborhood use cases and application services are (Deliverable D1.1, 2013):

- *UC1 Real-Time Monitoring of the Consumption of a Neighborhood*: Measure, aggregate, and visualize the consumption of the neighborhood in real-time.
- *UC2 Energy Demand and Power Generation Forecasting*: Forecasting the consumption of the neighborhood and the local generation, hourly ahead, day ahead, or year ahead helps the neighborhood to manage efficiently the energy cost, the interaction with the grid and the market.
- *UC3 Optimization of Power Purchases Versus On-Site Generation* (implemented only in Bishoptown site): The Neighborhood is able to make the right decision at any time, between importing power from the grid and using the local generation.
- *UC4 Demand Response* (implemented only in Challenger site): The Neighborhood is providing a flexibility service to the energy market, participating to Demand Response programs, as a single entity, generating additional revenue.

Eight services have been implemented in the demonstration sites to test the different configurations of the use-cases.

Irish sites

- Real-Time Actuation of Optimized Set-Points enabled at Bishopstown *Nimbus* Building (*Supervisory Control Service*)
- *Decision Support Service* at Bishopstown *Leisure-world* Building
- *Analysis Service* for Building Upgrade at Bishopstown *Parchment Square*

Challenger site

- *Real-Time data acquisition* (more than 2 years of data)
- *Forecast Service (PV production, load)*
- *Optimization Service for batteries*
- *Energy flow representation*
- *Demand response Service*

In the Bishopstown demo site the prototype energy services based on the forecasting and optimization algorithms were evaluated. The optimization services are tailored to the neighborhood type and include the following for this demonstration site:

- Access to weather forecasts.
- Access to energy price forecasts.
- Anonymized aggregated residual demand.
- Partial ability to store and use historical data of the neighborhood usage (restricted to anonymized data).
- Ability to provide optimal schedules for flexible generation and storage components.

All of these functionalities are included as part of the CIT Bishopstown optimization services. Since real-time and historical data are required, use-case scenario UC1 was first implemented [see Deliverable D1.1 (2013) and Deliverable D1.2 (2013)].

Several additional features are also proposed for inclusion in the full-scale demonstration:

- Different energy profiles for working and nonworking days load forecasts.
- A method of including Parchment Square in demand response.

The first proposed functionality regarding different weekday and weekend energy profiles has been added to the optimization services. This was accomplished through the forecasting algorithm discussed in Chapter 4 (by implementing use-case UC2) to distinguish previous days of the same type (working/nonworking days). This functionality also implements use-case scenario UC3 (Deliverable D1.1, 2013).

The second functionality to enable demand response (use-case scenario UC4) can be realized both with the prototype optimization (Chapter 4) and with the Intel E2E platform (Deliverable D3.4, 2015). The prototype optimization addresses demand–response for the neighborhood level. The Intel E2E platform addresses the building-level demand response by exploiting the flexibility of the thermal load provided by the large building thermal mass and controlling the water heater installed in the building.

Currently at the CIT Bishopstown site energy injected into the grid is not rewarded. Additionally, the demo site does not participate in a demand–response scheme—such a case is not allowed by the current TSO/DSO and grid regulations. Therefore, services, such as selling energy to the grid or other actors in the neighborhood, and demand response that are relevant to the UC4 is limited and not possible to be evaluated. It tried to complement the UC4 demonstration by (1) demonstrating how the battery and local generation (CHP) in NIMBUS can help reduce or shift the electricity load based on energy prices, (2) revisiting the capability of the E2E platform to provide demand response based on the water heater controls in Deliverable D3.4 (2015) and (3) by showing here a day-ahead battery optimization in Parchment Square. Currently, there is no battery available in Parchment Square; however, an analysis of how a battery could be used is provided for day-ahead demand response services in Parchment Square, as this was also proved effective in the project by early demonstration in NIMBUS. It is expected that if both these current operational and regulatory limitations are surpassed there will be additional benefits to the demo site by exploiting the services described in UC4. A detailed analysis of the current regulatory and operational barriers, and cost/benefits analysis is given in Chapters 2 and 6.

The forecasting and optimization algorithms developed determine the optimal profiles, schedules, and equipment set-points for the Bishopstown neighborhood based on real-time and historical building data and prices gathered from the neighborhood. They are deployed in the NICORE platform as an optimization service at CIT (Fig. 7.3). Real-time and historical data necessary for the optimization are gathered by the CIT NICORE platform from NIMBUS and Leisure World, and the Intel E2E platform from Parchment Square, by first implementing Use-case UC1 (Deliverable D1.1, 2013). All real measurements are made available to the energy optimization algorithm via the

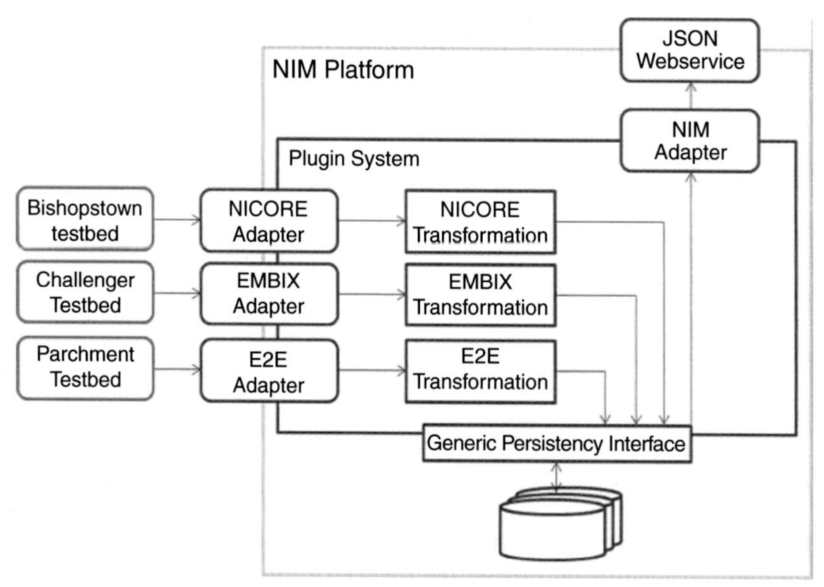

Figure 7.3 ***Integrating platform and services of COOPERaTE.***

NIM. The optimization results are sent either directly for the actuation of the building equipment or are stored in the relevant data-base and could potentially be leveraged as decision-support services to provide awareness and actionable information to the facility managers and the NEM. The data gathered from the demonstration were stored in NICORE and E2E platforms and used for the evaluation and testing.

In the Challenger demo site, the optimization services were tailored to the neighborhood type and include the following for this demonstration site:

- Access to weather forecasts.
- Access to energy price forecasts.
- Real-time neighborhood monitoring of local energy generation, energy consumption, and energy storage components (via the *EMBIX's meter data management platform*).
- Ability to provide optimal schedules for flexible local generation and storage components.

All of these functionalities are included as part of the Challenger optimization services. The integration of the *EMBIX's meter data management platform* with the COOPERaTE SoS is also shown in Fig. 7.3. The demonstration in the Challenger demo site includes all four use-case scenarios detailed earlier.

3 EVALUATION METHODOLOGY

The suggested KPIs for assessing the performance of an energy positive neighborhood are summarized in Table 7.3 (Deliverable D4.4, 2015). These KPIs were used in COOPERaTE to assess the performance of the energy services.

Table 7.3 COOPERaTE energy KPIs

Energy
Total energy consumption (kWh/day)
Local energy generation (kWh/day)
Max. power demand (kW)
Max. grid power demand (kW)
Environment
Energy savings (%, kWh)
Share of local and renewable energy generation (kWh, %/year)
Economics
Energy purchased from the grid (kWh)
Energy sold to the grid (kWh)

The evaluation effort focused on demonstrating how each of the KPIs is impacted in each neighborhood.

In the remainder of this section, the methodology for comparing the demonstration results against the baseline and the process for selecting the baseline and reporting periods based on the particular needs and limitations of the demo sites is explained.

In order to evaluate the savings in the two demonstration sites, the International Performance Measurement and Verification Protocol (IPMVP) was selected. It enables before and after intervention measurement and comparison. It involves the collection of data during a baseline period and then compares to a reporting period. Savings are defined as baseline period use—reporting period use ± adjustments.

The energy baseline for both sites was based on a 12-month historical data for each of the buildings in the neighborhood. For Bishopstown site, historical data of the NICORE platform was considered to determine day-ahead forecasts for the electrical and thermal consumption used to optimize the system. For Parchment Square (residential) data was pulled from the Intel system via the NIM into NICORE for evaluation.

Then a reporting period has been defined. In the Bishopstown demo site, for a neighborhood-level demonstration one needs to have access to all the buildings of the neighborhood during the same period of time and hence we look for a time window where all the buildings are accessible in terms of available data or actuation (where enabled).

The analysis was performed, based on real neighborhood data, for different periods covering high and low building demand. The following results section is focused on a representative example of performance assessment.

The scope of the energy study during the reporting period for Challenger is the whole set of buildings in the Challenger campus. A full year reporting period was used. The reporting period lasts from October 2014 to October 2015.

4 KPI EVALUATION AND ASSESSMENT

The KPI evaluation for CIT Bishopstown demonstration site will focus on a 1-week reporting period that corresponds to high heating demand (winter). The outcomes of the demonstration and the optimization implementation are explained later in the chapter. The overall objective as described in Section 1, is to minimize the energy cost for the overall neighborhood and for each of the individual buildings in it. Therefore, the overall neighborhood energy cost is optimized by optimizing the total energy cost of the three buildings collectively.

The KPIs are calculated and demonstrated both for the individual buildings and neighborhood as a whole. The main KPI in this case is of course the energy cost savings, nevertheless estimates of other KPIs listed in Table 7.1 is given to offer a more rigorous and comprehensive evaluation of the performance of the energy services in the neighborhood. Recommendations for the optimal operation of the equipment in the two main buildings of NIMBUS and Leisure World is discussed, where a significant capability for local heat and electricity generation and storage is installed. Potential demand response benefits in Parchment Square are also detailed here, as this building has a significant load shifting capability, due to the large thermal mass and number of apartments. These recommendations could potentially be used to optimize the future operation of the system.

In certain cases the on-line optimization of the NIMBUS battery and CHP is considered with the optimization service acting as an on-line optimization module, while the optimization of the Leisure World and Parchment equipment is performed off-line with the optimization services acting as a decision support tool rather than on-line optimization module.

The week from 13th to 17th February 2015 was chosen to demonstrate the energy benefit analysis. The baseline for this scenario is established considering the measured consumption for the chosen period. The reporting period results for all three buildings in this case are the results of the application of the energy optimization algorithm. Energy savings are evaluated as the optimized component set-points were implemented for the same period.

For the Nimbus building, the analysis is performed using real consumption data from the building, and the optimization algorithm working both for real-time actuation of the battery and the CHP, and as decision-support tool. All of the system components are active including the CHP since the heating demand is large enough to operate the CHP during the selected period.

Figs. 7.4–7.6 show the results of application of the energy optimization algorithm. In particular, it shows the electrical powers and the electrical storage state of charge. They are: the building load, the battery charging/discharging power, and the CHP power.

Fig. 7.5 shows the thermal powers that supply the building load. They are the boiler power and the output power of the CHP thermal storage. Fig. 7.4 shows the CHP thermal storage operation. The CHP thermal power is the thermal storage input power.

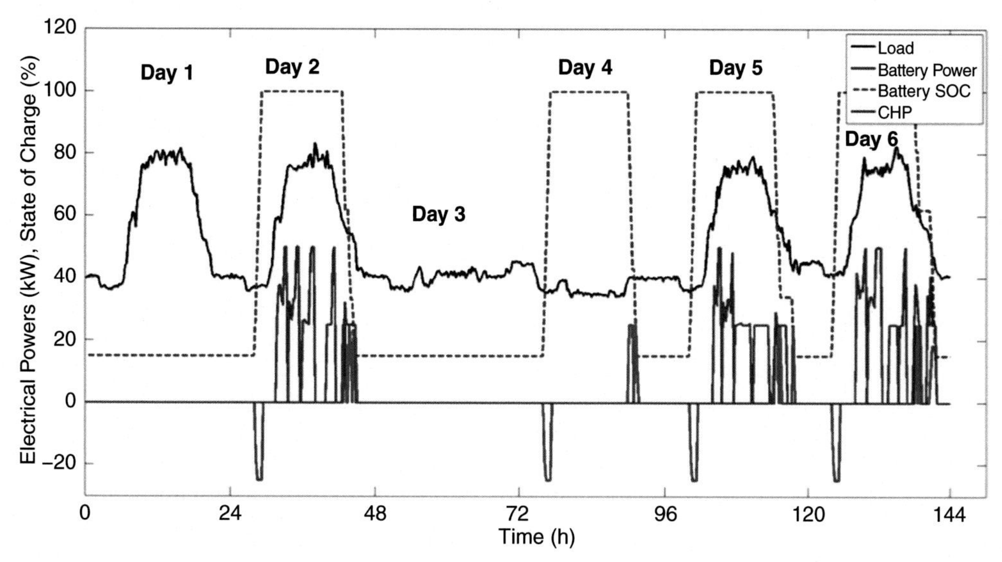

Figure 7.4 ***Electric Power consumption and generation for Nimbus.***

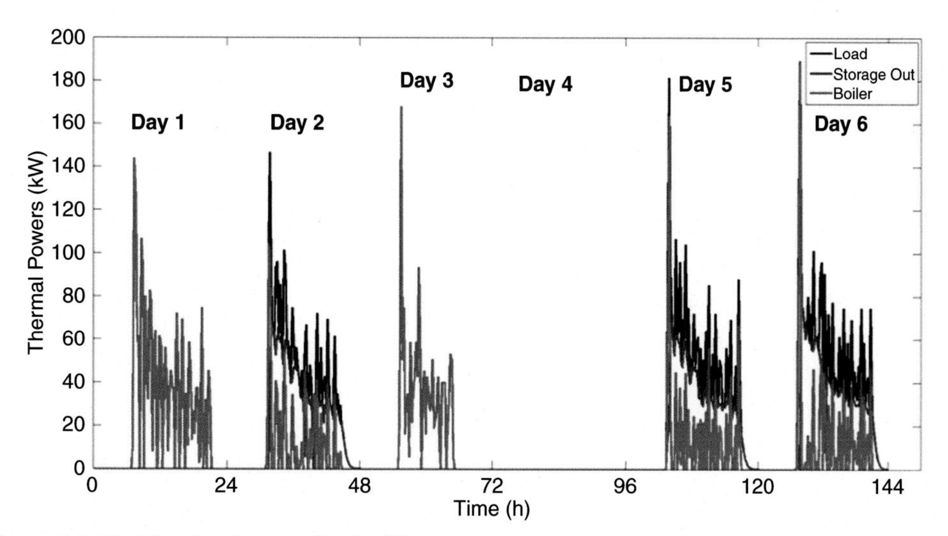

Figure 7.5 ***Heat load and generation for Nimbus.***

Power balance (difference between storage input and output powers) and energy losses determine the stored energy variations.

In order to compute the thermal load forecasts that have been used for optimizing the CHP set-point, the measured thermal demand of the building was considered. For most of the days of the selected period the CHP was not running. In particular, it was found from recorded data that only on February 13th the CHP was running for a few hours.

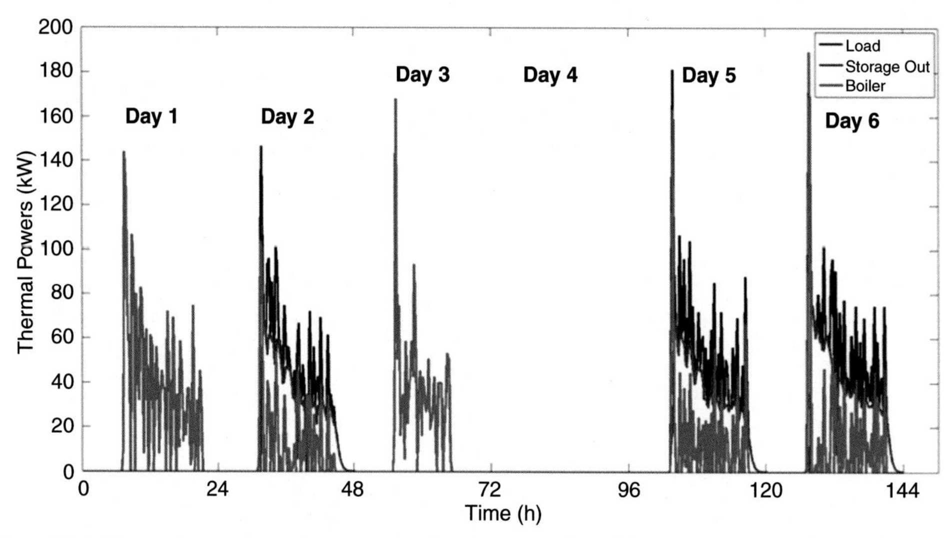

Figure 7.6 ***Thermal storage output power and energy stored, and CHP output power for Nimbus.***

This way of operating the system is likely to be an indicator of nonoptimal system operation. Furthermore, the supply of the thermal load using the boiler can give thermal load with several spikes and large variations in the supplied power—possibly showing that neither minimum output power nor minimum running time constraints which characterize the CHP operation are fulfilled. In order to optimize the CHP set-point, forecasts of the electrical and thermal consumption obtained were used by combining from 1 to 3 previous days of the same type (e.g., working or nonworking days) (Table 7.4).

For the LeisureWorld building, due to the presence of swimming pools, the heating load for this building is a constant base load and is not only dependent on the weather conditions; therefore the CHP can operate for much more of the year than in the

Table 7.4 Comparison of baseline and reporting period for NIMBUS

	Baseline	**Reporting period**
Energy		
Total energy consumption (kWh/day)	1641	1641
Local energy generation (kWh/day)	20 elec, 35th	182 elec, 326th
Max. power demand (kW)	83	83
Max. power grid power demand (kW)	83	83
Environment		
Energy savings (%, kWh)	—	0, 0
Share of local and renewable energy generation (%)	3.3%	31.0%
Economics		
Grid energy consumption and cost (kWh, €)	5867 kWh, €553.27	5053 kWh, €458.20
Energy cost savings (%,€)	—	4.9%, €33.00

Nimbus. For this building, however, it is not possible to send control signals to the equipment to modify their operation/actuation and the analysis was done offline.

The amount of thermal load allows a significant contribution of CHP to thermal and electrical load supply—with continuous operation for several hours of the day. Application of the proposed energy management tools enables to operate the CHP to achieve cost savings with respect to the building baseline. The building baseline considered here is the CHP operated in the standard thermal load following mode.

The reporting period, includes a weekend. The different energy profiles for working days and nonworking days load forecasts are considered separately in the optimization algorithm; it has been verified that this reduces the forecasting errors. On 12th February (1 day before the chosen reporting period), data gathering happened for forecasting purposes. The actual optimization started on 13th February. On that day the optimized operation of the CHP enabled to supply the thermal base-load and the electrical load for most of the hours of the day. The amount of electricity injected to the grid (not rewarded) is limited by the optimization—still present though—given the CHP output power constraints and load forecasting errors (although forecasts are pretty accurate since the load profiles are rather similar from one day to another).

On 14th February the daily consumption pattern changes because it is a weekend. The optimizer will gather data in order to forecast next weekend days. The next day (15th February) the actual optimization starts again and the CHP again supplies a significant fraction of the load. On 16th and 17th of February the data gathered in the previous days are used to forecast building consumption and optimize the CHP operation. It can be seen that the thermal load is higher than on weekend days and that is reflected in a longer utilization of the CHP. Application of the energy optimization resulted in energy cost reduction of approximately 15% over the period 13–17th with respect to the baseline (CHP operated in thermal load following mode). The reason for such savings is the fact that the optimizer takes into account energy prices and load forecasts to minimize the energy cost, thereby avoiding operating the CHP in a way such that a big amount of electrical energy is injected into the grid, since such energy export is currently not rewarded.

KPIs evaluations are summarized in Table 7.5. Fig. 7.7 shows the electrical and thermal powers for the period February 12–17th. Note that CHP is not operating on the 12th (data collection, first working day) and 14th (data collection, first weekend day).

For the Parchment Square buildings, there is no on-site electricity generation. Therefore, the local energy generation metric is not relevant as the maximal power used and maximal power asked to the grid are equal. As a result, for the energy metrics only the total site consumption and the maximal site power use are presented later in the chapter. Similarly, for the environment metrics the local resource utilization is not presented.

The energy optimization algorithm was used to analyze the demand response potential and related apartment-level energy benefits which can be achieved by installing

Table 7.5 Comparison of baseline and reporting period for LeisureWorld

	Baseline	Reporting period
Energy		
Total energy consumption (kWh/day)	4243	4243
Local energy generation (kWh/day)	1567 elec, 2267th	744 elec, 1077th
Max. power demand (kW)	91	91
Max. power grid power demand (kW)	75	91
Environment		
Energy savings (%, kWh)	—	—
Share of local and renewable energy generation (%)	90.4	42.9
Economics		
Grid energy consumption and cost (kWh, €)	1261 kWh, €95.46	2972 kWh, €263.95
Energy cost savings (%, €)	—	15.26%, €175.64

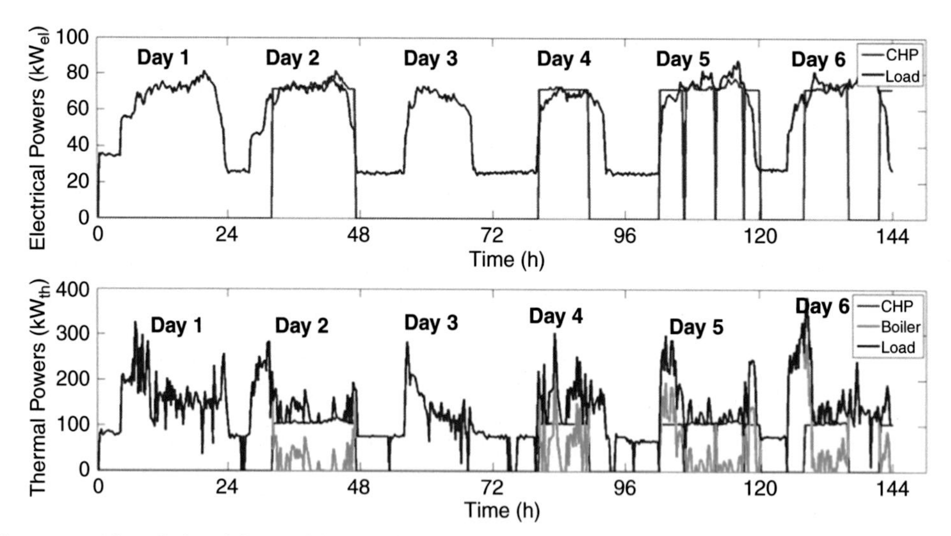

Figure 7.7 ***Electrical and thermal demand and generation for Leisure World.***

a battery unit in one of the apartments of Parchment Square. More specifically, the scenario of using a battery of adequate size for a single apartment was investigated to provide load-shifting capabilities that can support demand response services for Parchment Square. This supports the demonstration of the potential of UC4 scenario.

Economic considerations apply for the sizing, including purchasing costs and corresponding energy savings. This method has been proved effective in the project by the demonstration in the Nimbus. It does not require any modification of the consumption behavior of the building users. Furthermore, it has no impact on the thermal comfort (the building heating system is electric).

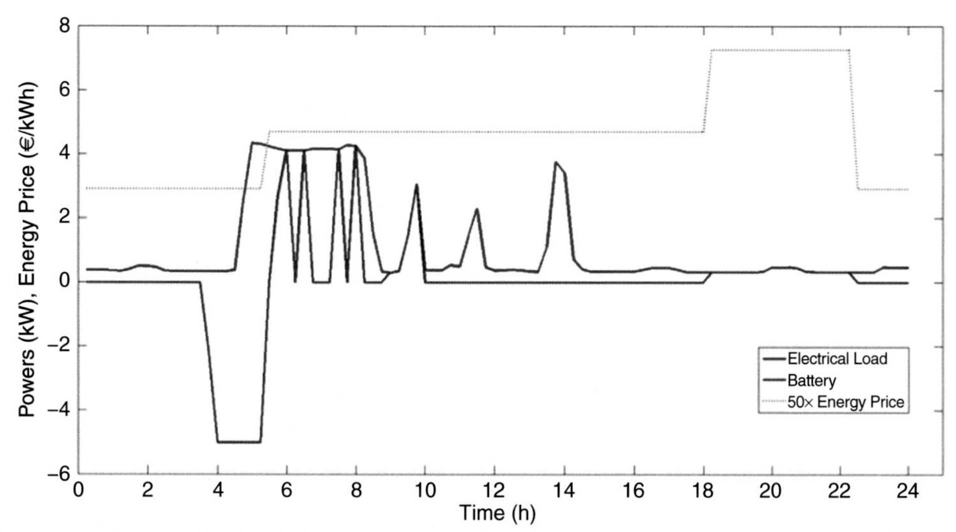

Figure 7.8 ***Electrical load and electrical storage power for Parchment Square.***

Addition of electrical storage further increases the flexibility needed to enable effective demand response services thereby enhancing load flexibility provided by water heaters and the slab heating. Furthermore, the use of electric storage (implying alteration of electric heating system temperature set-points to enable flexibility or hot water availability) does not decrease the thermal comfort of inhabitants.

Fig. 7.8 shows the electrical load and storage charging and discharging powers for the reporting period. A fixed three stages electrical tariff was used (Chapter 4). It shows that the battery is charged when the grid tariff is low (0.0583 €/kWh) and discharged when the tariff is high (0.0942 and 0.1453 €/kWh) to support the demand. The results show that the adoption of a battery of proper size can support the daily load of an apartment and lead to a sensible energy cost reduction. The optimization algorithm has been used to determine the optimal charging/discharging powers for a given load profile to minimize the energy cost (Fig. 7.8). Battery size considered is 9 kWh, which provides a reasonable load support. Corresponding energy cost reduction evaluated over the considered reporting period is about 13%. The previous analysis, is for a single apartment; however, it is expected that if extrapolated for all 171 apartments in the Parchment Square complex the benefits will be significantly higher in terms of total consumption.

KPIs evaluations for Parchment Square are illustrated in Table 7.6.

This section shows the evaluation results for the overall optimization of the neighborhood. As described before, the neighborhood objective is to minimize the overall energy cost (electricity and gas), thus combining all energy cost objectives of the three buildings in one single optimization objective (Table 7.7).

Table 7.6 Comparison of baseline and reporting period for Parchment Square

	Baseline	Reporting period
Energy		
Total energy consumption (kWh/day)	26.78	26.78
Max. power demand (kW)	4.34	9.34
Environment		
Energy savings (%, kWh)	—	
Economics		
Grid energy consumption and cost (kWh, €)	26.78 kWh, €2.41	26.78 kWh, €2.08
Energy cost savings (%, €)	—	13.7%, €0.33

Table 7.7 Comparison of baseline and reporting period for CIT Bishopstown Neighborhood

	Baseline	Reporting period
Energy		
Total energy consumption (kWh/day)	5884	5884
Local energy generation (kWh/day)	1587 elec 2302th	927 elec 1403th
Max. power demand (kW)	175	175
Max. power grid power demand (kW)	159	175
Environment		
Energy savings (%, kWh)	—	0, 0
Share of local and renewable energy generation (%)	66.1	39.6
Economics		
Grid energy consumption and cost (kWh, €)	7128 kWh, €648.73	8031 kWh, €722.15
Energy cost savings (%, €)	—	11.34%, €207.03

Concerning the Challenger demo site, the optimization process was a two parts process: first, EMBIX optimized the existing building systems (lighting, heating, cooling, ventilation, temperature, etc.) looking for the best operating point and schedule, then, the process was completed by the addition of an energy storage management system.

An example of improvement EMBIX provided is the regulation of the cooling system EMBIX optimized some of the existing systems in Challenger, such as the regulation of the cooling system. After the renovation, the need for cooling was lower. Thus, the regulation of the cooling system needed a finer adjustment in order to achieve significant savings. By shutting down the cooling system during the night and the weekends and having a better control during the day, a clear impact was stressed. In Fig. 7.9, the first weeks are late October 2014 (baseline) then the regulated weeks are the first weeks of the Reporting period. The estimated savings are calculated in the next section.

When the opportunity showed to work with batteries' storage (66 kWh capacity) installed at Challenger, EMBIX decided to explore the possibilities offered by this system. The complementarity between storage, renewable energy production and regulation

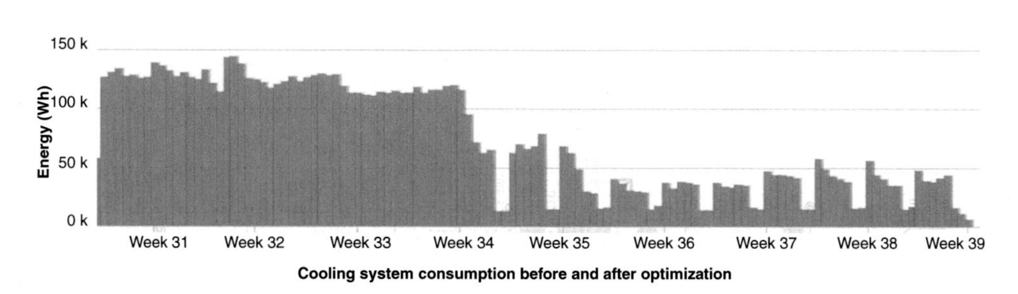

Figure 7.9 ***Cooling system consumption before and after fine regulation.***

actions was an interesting direction to study within the COOPERaTE program. The COOPERaTE tools developed until now were used in addition with new ones to tackle this issue. Mainly, actions of energy optimization and flexibility were conducted.

Firstly, batteries enable to reduce the power peak and consequently reduce the contract power. To do so, EMBIX developed an *optimization module*. The module generates a *smoothed curve of the power requested from the grid* everyday taking into account sunshine and occupancy.

Then, the EMBIX module is able to generate *battery power set values* [see Fig. 7.10 (adapted to the parameters of each batteries)]. On the same chart, it is possible to visualize the real power of the battery, following the set value.

Doing so, it is possible to reduce the amount of power ask to the grid. However, it should be recalled that the size of the batteries was too small to see a big impact on the consumption. Even if the batteries are small, the modules EMBIX developed are adjustable and helped having a better understanding of the process of reducing the power peaks.

In the Challenger demo site was also treated the demand response issue. Very linked to the regulatory framework, demand response is about pricing the flexibility of a site. The Challenger site can be flexible thanks to batteries and also thanks to peak shaving and peak shifting. EMBIX developed a module able to generate demand response

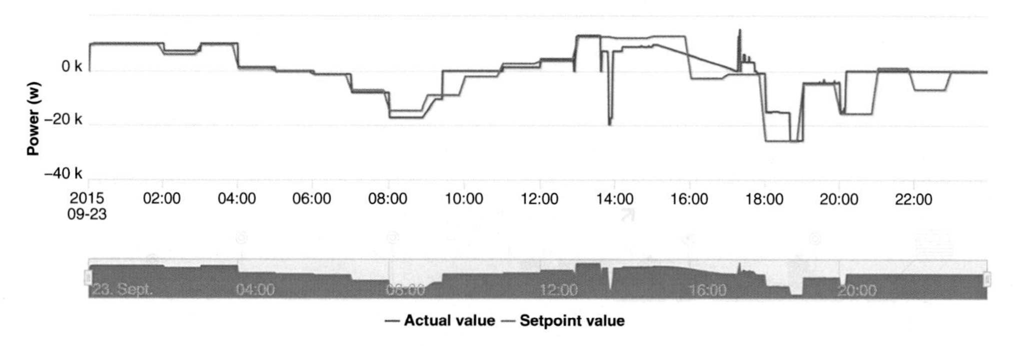

Figure 7.10 ***Batteries set value versus real power.***

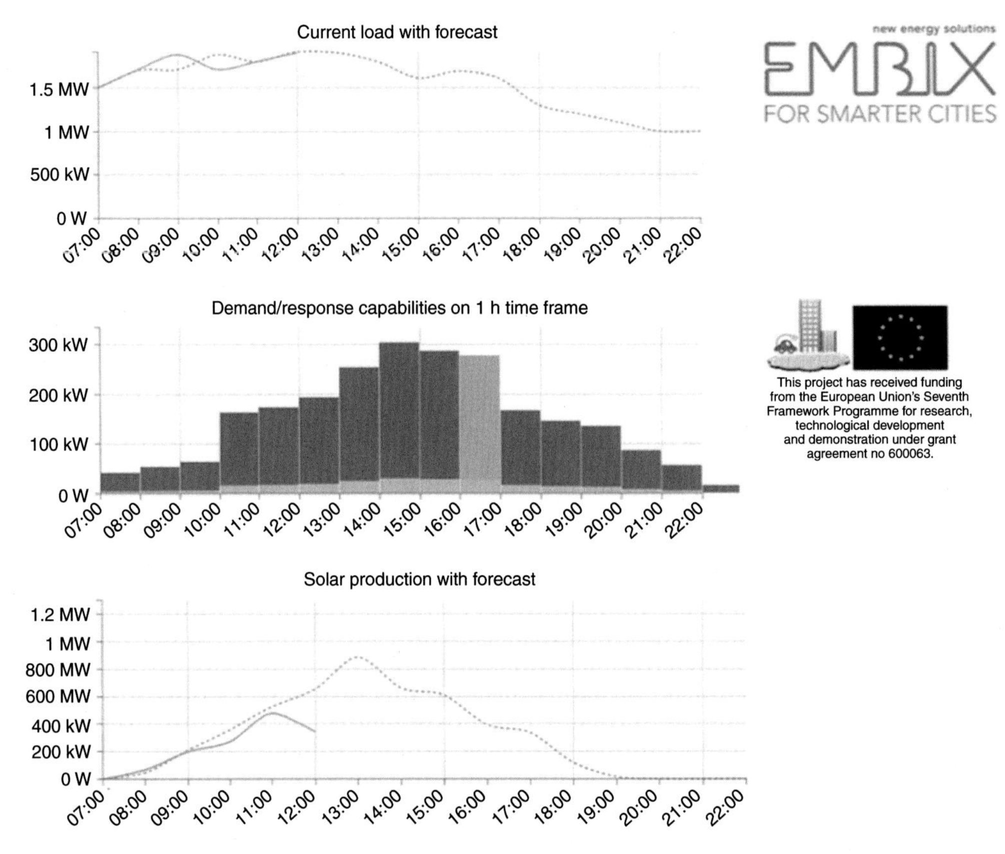

Figure 7.11 ***Platform of demand response by EMBIX.***

possibilities according to the consumption, PV production, and batteries' conditions. These possibilities are generated for the duration of 1 h. The first chart in Fig. 7.11 shows the current load (orange line) and the forecast load (pink line). The second chart displays the demand response possibilities in the 1-h time frame. The last chart displays the PV production (orange line) and forecast (pink line).

To go further, apart from reducing the peak power and helping pricing the flexibility of a site, batteries help in grasping new opportunities.

- The batteries generate arbitrage opportunities across the spot market. According to the price of electricity on the spot market and the selling price of PV electricity, charging and discharging cycles can be optimized. Concerning France, on 1st January 2016 every PV site >500 kW will have to sell its electricity on the spot market (Obersv'er, 2015). Thanks to the expertise developed through COOPERaTE, EMBIX is able to develop a module collecting the spot price and so to optimize in real time the electricity sales and purchases.
- Batteries, as PV inverters, enable to stabilize the voltage of the site by compensating the reactive energy required from the site. This can be valued on the market.

- Equipped with batteries and PV, the site can help to balance the grid at a national level. At the national level, it is crucial to balance electricity consumption and generation, particularly using primary, secondary, and tertiary reserves. It could be interesting to define the Challenger campus as a tertiary reserve valuable on the ancillary market. Nowadays, in France, a producer/consumer must have at least 10 MW to participate in the reserves system. It is not the case of Challenger but EMBIX estimates that projects like COOPERaTE are the occasion to discuss and challenge regulatory frameworks. It stimulates discussion and can lead to change.

Other actions were considered but as the campus was just renovated, these actions were already taken into account. For example, scaling of the boilers, all triggered in late afternoon every day (see the peaks from boilers in Fig. 7.12), was implemented and globally, energy consumptions were divided by 10 (from 310 to 31 kWhpe/m^2/year).

To sum up, in the Challenger site, two COOPERaTE actions were implemented:

- Cooling system optimization
- Batteries storage management

In this part, the KPIs defined earlier are evaluated. Table 7.8 summarizes the main numbers and shows the differences between baseline and reporting period (October 2013–October 2014 and October 2014–October 2015). Then, a description of the calculation during the reporting period is proposed.

Hereafter, the consumption of Challenger: close to 11.8 GWh during the whole reporting period. Fig. 7.13 shows the level of consumptions for every month for the period October 2014–October 2015. In green is the energy from the grid, in yellow the energy from PV and the black line represents the overall energy consumptions.

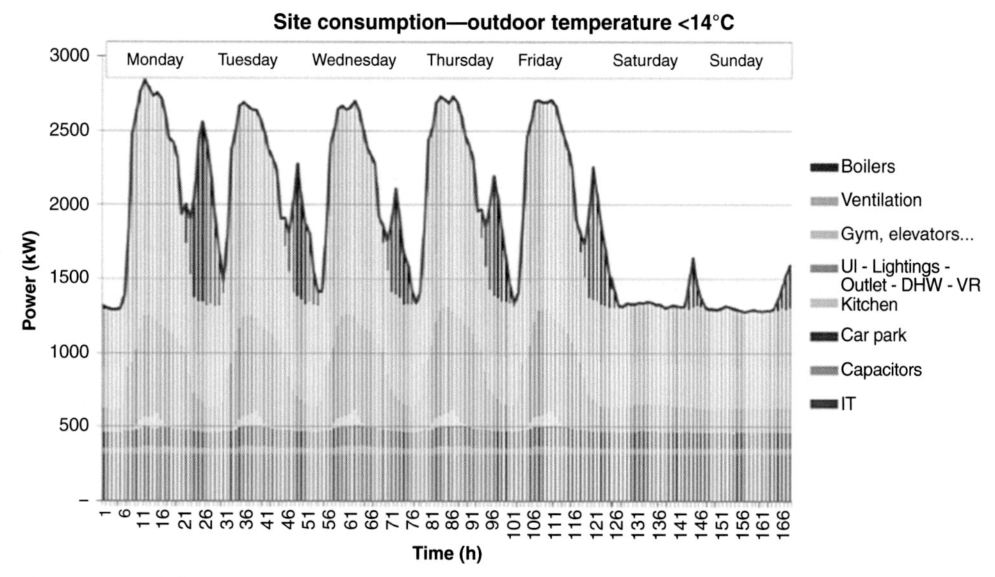

Figure 7.12 ***Challenger site electrical energy consumption.***

Table 7.8 Summary of the KPIs, comparison baseline/reporting period

	Baseline	Reporting period
Energy		
Total energy consumption (GWh/an)	12.6	11.8
Local energy generation (GWh/an)	2.48	2.53
Max. power demand (kW)	3800	2600
Max. power grid power demand (kW)	3600	2700
Environment		
Energy savings (%)	44.16 MWh, 0.4%	
Share of local and renewable energy generation (%)	19.7	21.5
Economics		
Grid energy consumption and cost (GWh)	10.1	8.7
Energy and cost of energy sold to grid (MWh)	133	123

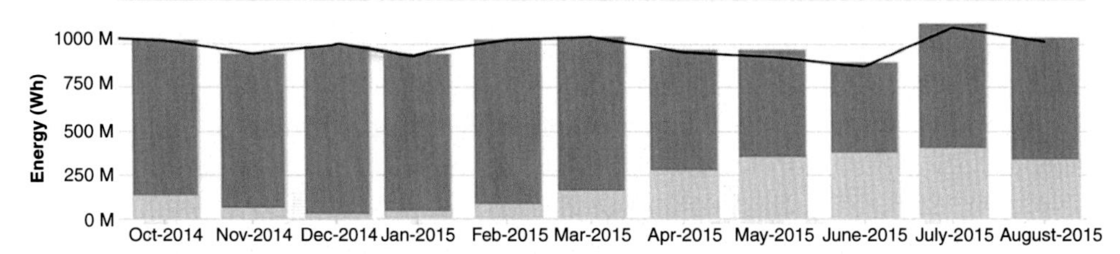

Figure 7.13 ***Levels of consumption per month.***

Fig. 7.14 shows the level of consumption per day for two weeks in September 2015. The green line shows the energy from the grid, the yellow line the energy from PV, and the black line the total consumptions.

Here is a focus on the local energy generation by solar panels. As a reminder, there are 21,500 m^2 of photovoltaic panels (2.5 MWpeak). Thanks to the *EMBIX's meter data management platform*, it is known that the KPI is 2.3 GWh for the reporting period (see Fig. 7.15). Fig. 7.16 shows the total PV energy generation per day in September, which amounts to approximately 10 MWh per day.

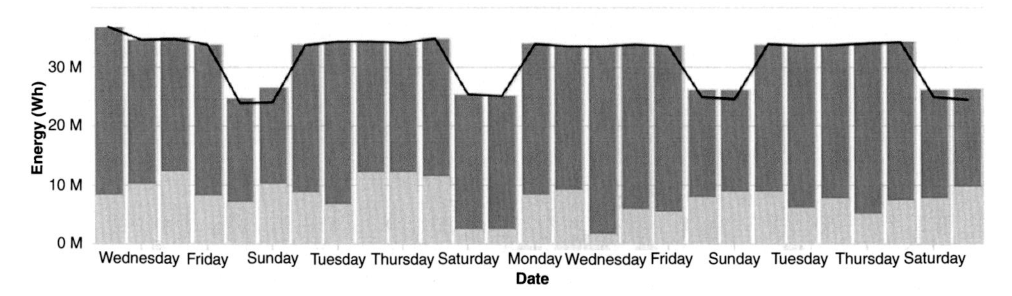

Figure 7.14 ***Levels of consumption per day.***

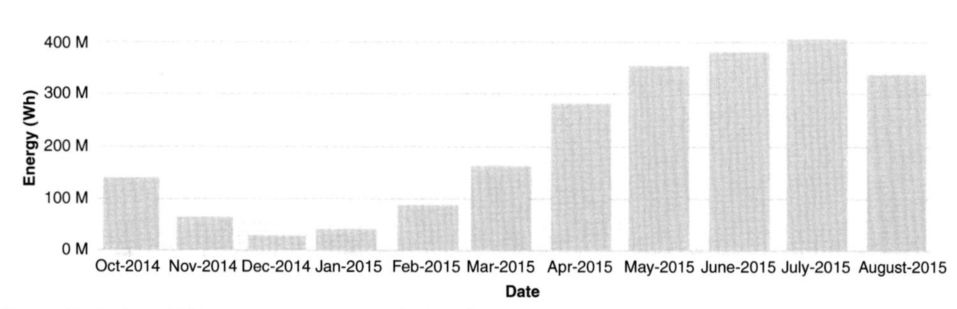

Figure 7.15 ***Local PV energy generation during the reporting period.***

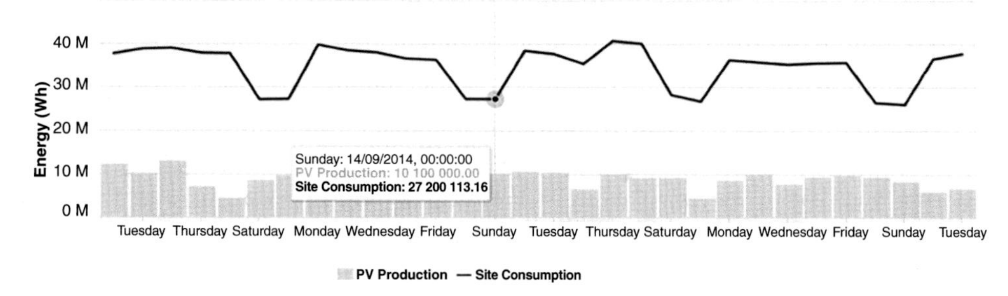

Figure 7.16 ***Local PV energy generation in September.***

Hereafter is the power consumption of Challenger site for the first week of September 2015 (maximal power use: 2.2 MW). For the whole reporting period, the maximum consumption of power was 2.6 MW on 3rd July 2015.

- *Results of the savings by controlling the cooling system*: a 70% decrease in the consumption (from 58 to 16 MWh).
- *Results of the savings by the batteries*: in 2015 (batteries were installed in May 2015), Challenger received 1.68 MWh from the batteries (Fig. 7.17).

The total energy savings achieved in Challenger both from the consumption of batteries and from managing the cooling system are *44.16 MWh.*

The local resources utilization for Challenger during the reporting period was up to 21.5%.

78.5% of the energy consumed on the Challenger Campus during the reporting period was purchased from the grid and amounts up to 8.7 GWh, as shown in Fig. 7.18.

On the Challenger campus, energy is sold to the grid when the PV production exceeds consumptions. During the reporting period, 123 MWh have been sold to the grid (Fig. 7.19).

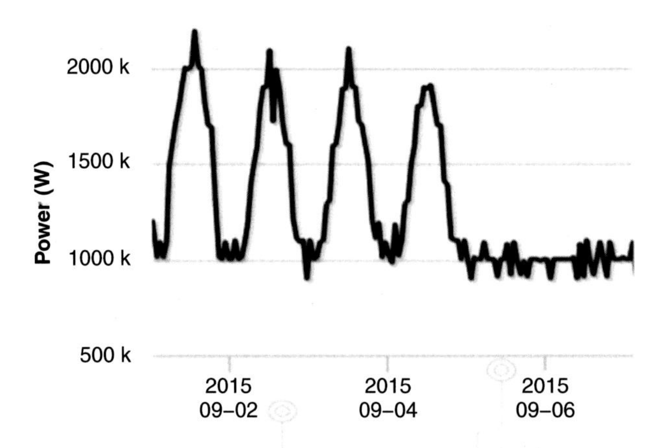

Figure 7.17 *Maximum power consumption for Challenger in July 2015.*

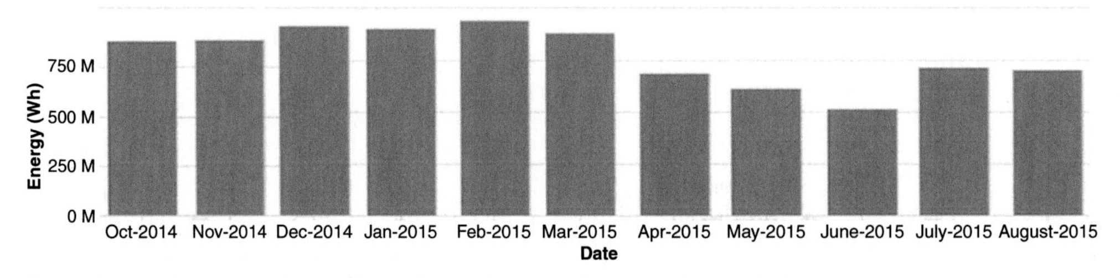

Figure 7.18 *Energy purchased from the grid during the reporting period.*

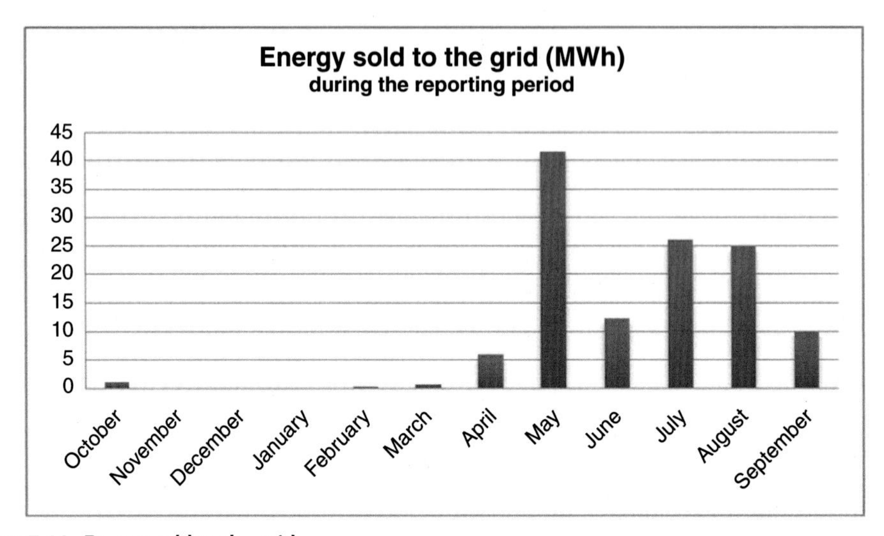

Figure 7.19 *Energy sold to the grid.*

5 USER AWARENESS

User awareness is also part of deploying services to demo sites. The user impact on energy consumptions is key to reach EPN goals, there is no further need to justify it. To address the user awareness issue, designing user interfaces was a part of the COOPERaTE project. In Challenger demo site, the users working in the campus can follow in real time the consumption of the site and the nature of energy flows (from the grid, from solar panels, from batteries, etc.).

The first user interface EMBIX designed (Deliverable D4.3, 2014) was showing the consumption of the buildings with details (multiple graphic elements), the local production and the main information about the building (meteo, events, etc.). The interface was displayed in the main hall of the building (general public) and in the facilities management offices (technical public).

This interface was appreciated by the users, mainly by the technical staff. EMBIX decided to go further by distinguishing the user interface for technical and general public. A new user interface was designed with a simplified message. This second interface showed in real time the energy consumption distribution. A screenshot of this platform is shown in Fig. 7.20. It is easier to understand it in a few seconds, the timeframe adapted to employees crossing the main hall many times during the day.

The interface is animated in order to catch the attention of the employees. It is easy to understand and piques curiosity. In a glance, the interface informs about the consumption of the site and where the energy is from (in the example, 75% from solar panels and 25% from the grid, the batteries are charging).

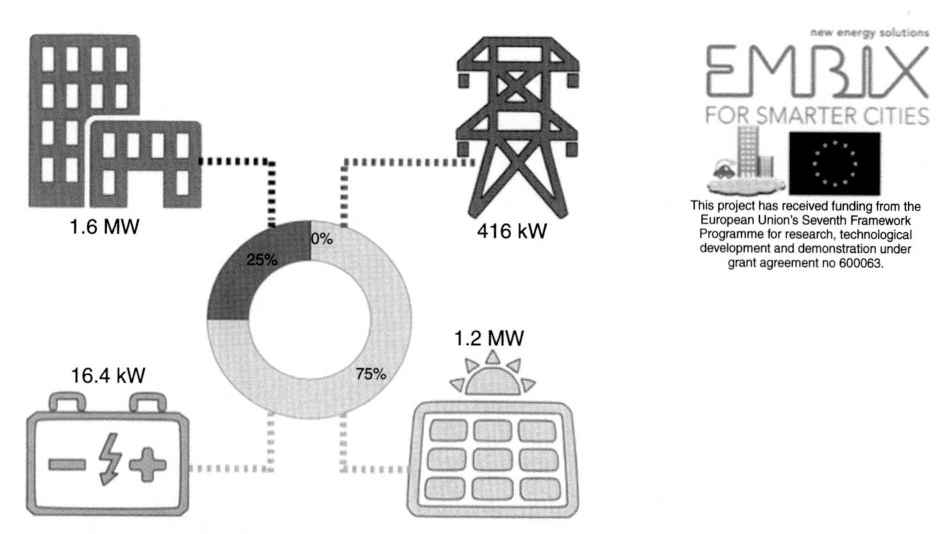

Figure 7.20 ***Improved neighborhood energy visualization platform for Challenger.***

Brainstorming with the future users is key to adapt the interface to their needs: it helps in increasing the awareness and the understanding of the project and approach.

6 GOING FURTHER

EMBIX applied the outputs of COOPERaTE in real life projects in order to test the technical and economical value. As an example, on a 240,000-m^2 neighborhood, EMBIX may expect a solid business model with a decent ROI in the incoming years (6 years). The energy savings may reach 15% on the energy bill, notably thanks to the energy from batteries becoming affordable (300€–400 €/MWh today) (Figs. 7.21 and 7.22).

7 CONCLUSIONS

This section presents the evaluation of the energy savings achieved from the demonstration of the COOPERaTE energy optimization services in the two project demo sites. This work not only demonstrated the potential to achieve energy savings in the two demo sites, but also the benefits of the COOPERaTE System-Of-Systems (SoS) ICT approach for the delivery of energy services to the neighborhood. The energy optimization algorithms offer improved performance for the demo sites in terms of the key energy KPIs for a neighborhood, and can be used for energy optimization both at the building and neighborhood level. The key to this is the optimized coordination of building-level and neighborhood-level supply, demand, and storage, based on energy prices and load and weather forecasts.

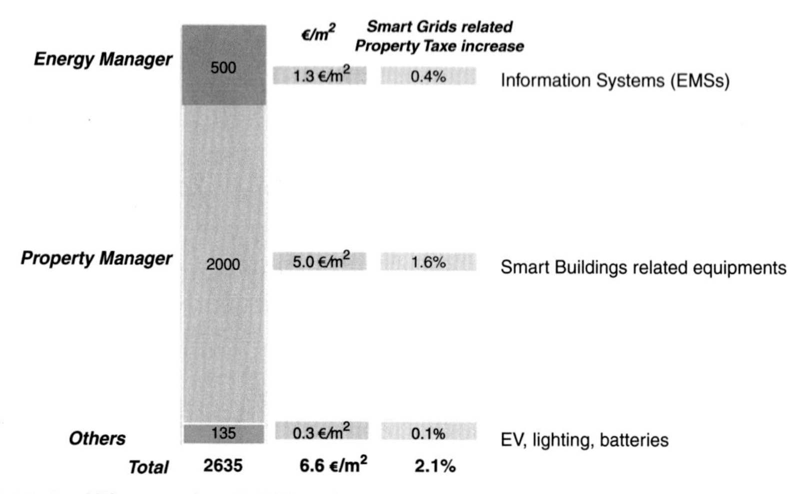

Figure 7.21 ***Real life example—EMBIX project.***

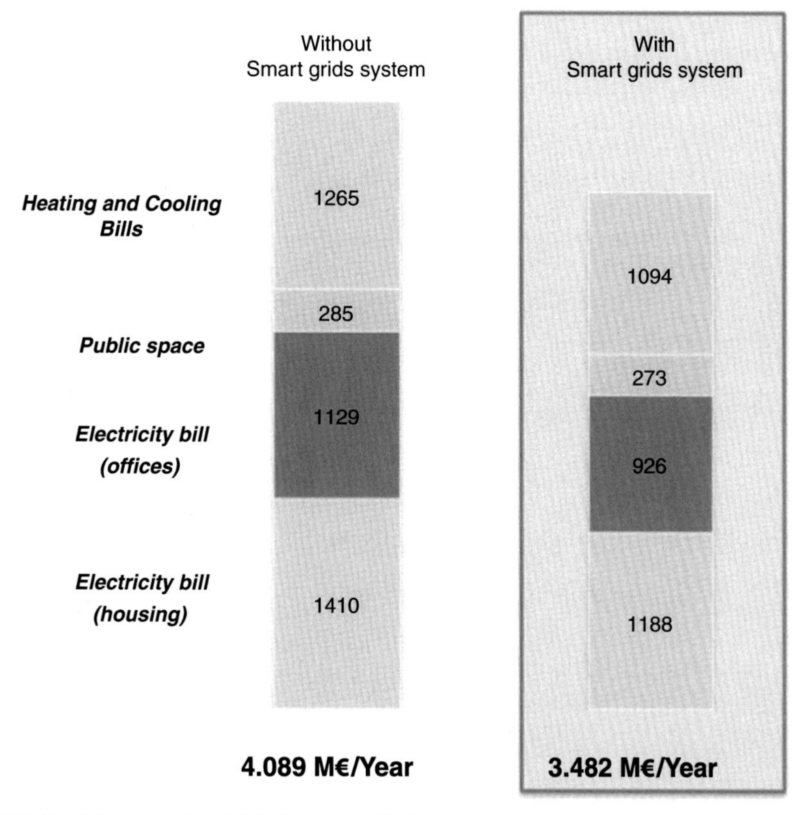

Figure 7.22 ***Real life example—EMBIX project : 15% energy savings.***

The integrated services platform, that was developed in this project by implementing the COOPERaTE SoS approach, has been key to the integration, implementation, deployment, and demonstration of the energy services. The platform allows the integration of different ICT platforms (such as NICORE, E2E, and the *EMBIX's meter data management platform*) and BMS systems currently available in the demo sites, and to deploy energy services in such a multitude of heterogeneous systems. It ensures that real measurements and historical data available in the neighborhood can be leveraged through the real-time monitoring, forecasting, and energy optimization services to improve neighborhood energy performance.

Overall, based on the demonstration results, it is fair to conclude that the proposed COOPERaTE SOS approach and energy services can impact positively the energy performance of a neighborhood toward establishing an EPN. This can be done first by maximizing the usage of local, renewable generation and storage resources to improve the energy benefit for the neighborhood and second by optimizing the performance of the neighborhood in order to participate in demand–response services. The key to this are the advanced energy optimization algorithms that were developed in this project and which were tested both in simulations and in real demonstrations.

REFERENCES

Deliverable D1.1 (2013). *Report on requirements and use-cases specification, COOPERaTE.*

Deliverable D1.2 (2013). *Report detailing neighbourhood information model semantics, COOPERaTE.*

Deliverable D3.4 (2015). *Report detailing integrated services for neighbourhood management, COOPERaTE.*

Deliverable D4.3 (2014). Mobile applications available, COOPERaTE.

Deliverable D4.4 (2015). *Report on the key performance indicators for performance evaluation of the service, COOPERaTE.*

Obersv'er (2015). Le baromètre 2015 des énergies renouvelables électriques en France http://www.energies-renouvelables.org/observ-er/html/energie_renouvelable_france/Observ-ER-Barometre-Electrique-2015-Integral.pdf

CHAPTER EIGHT

Barriers, Challenges, and Recommendations Related to Development of Energy Positive Neighborhoods and Smart Energy Districts

N. Good*, E.A. Martínez Ceseña*, P. Mancarella*, A. Monti, D. Pesch†, K.A. Ellis‡**
*The University of Manchester, School of Electrical and Electronic Engineering, Manchester, United Kingdom
**E.ON Energy Research Center, RWTH Aachen University, Aachen, Germany
†NIMBUS Centre for Embedded Systems Research, Cork Institute of Technology, Cork, Ireland
‡IoT Systems Research Lab, Intel Labs, Intel Corporation, Ireland

Contents

1 INTRODUCTION

Neighborhoods, or equivalent districts, are significant consumers of energy services and, thus, emitters of CO_2 emissions. Taking the most general view, which considers districts composed of residential, commercial, and public buildings (and any mix

Energy Positive Neighborhoods and Smart Energy Districts

thereof), energy services consumed by districts form a significant proportion of the society's consumption of energy services. These include services which can be intuitively allocated to a district, as they are consumed on-site, (e.g., heating, lighting, entertainment, computing, etc.), but also (personal vehicle) transportation services, which, if electrified, will become significant electricity loads at the district level. As a result, districts are significant consumers of energy and consequently emitters of CO_2 emissions. Based on this, and given legislative and political impetus for decarbonization of energy systems, districts clearly have a central role to play.

However, decarbonization must be considered within the real-world context, in which supply of energy services is expected to remain secure and affordable (World Energy Council, 2015). Indeed, this "trilemma" of decarbonization (sustainability), security, and affordability is central to the development of energy systems, including district level systems, which have, so far, not received adequate attention.

This book has addressed this oversight by systematically addressing the gaps, which prevent effective action on these issues. Addressing conceptual gaps, Chapter 2 examines the importance of selecting the appropriate objective, which recognizes the value of flexibility in energy system operation, when employing the EPN concept, while Chapter 3 presents clearly the process for transformation of neighborhoods to EPNs. From the technological perspective, Chapter 4 details models and optimization algorithms for EPNs, which address the crucial interaction between electrical and thermal domains at the physical level. Chapter 5 outlines the crucial role of the Neighborhood Information Model (NIM), as a facilitator of the EPN concept. From the economic perspective, Chapter 6 introduces a business case assessment process, which details how value can be extracted from the EPN concept. Experience of implementing the EPN concept is presented in Chapter 7.

This chapter first defines the general barriers to smart, demand-side interventions. Afterward, drawing on insights from the COOPERaTE project, specific challenges for the EPN concept in the key areas of modeling and simulation, performance assessment, information technology, and business model development, will be detailed. Finally, this chapter will describe the opportunities for EPNs, and recommendations for various energy system actors, to ensure full exploitation of the potential of EPNs, for various actor and system-level objectives.

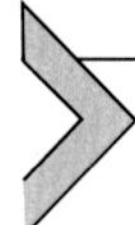

2 GENERAL BARRIERS TO SMART, DEMAND-SIDE INTERVENTIONS

The efficacy of demand-side interventions to reduce energy demand has long been recognized, as have the barriers to this aim. Under the classification of "Energy Efficiency" (EE) (and, to a lesser extent, "energy demand reduction") there is quite substantial literature on the subject. For a complete review of the literature on EE barriers, Thollander, Palm, and Rohdin (2010) and Chai and Yeo (2012) are useful references. An early work in the area (Jaffe & Stavins, 1994) studied the "paradox" of gradual diffusion of apparently cost-effective energy efficiency technologies. This work made the important observation, derived from the field of classical economics, that "barriers" could be fundamentally categorized as market failures or nonmarket failures. In the first case the barrier is due to a failure of a market to operate properly. Thus the barrier can be removed by improving the functioning of the market. In the second, the barrier is due to non-(classical) economic reasons. As observed in Jaffe and Stavins (1994), viewing EE through an economic lens highlights that policy should aim to encourage economic efficiency, rather than narrowly focusing on EE. This observation can be similarly applied to the EPN concept, under certain conditions, namely: liberalization of energy system markets or absence of significant market failures (such as nonincorporation of externalities related to CO_2 emission).

Sorrell (2004b) builds on this separation of barriers into market and nonmarket failures, by defining barriers as: (1) economic; (2) behavioral; and (3) organizational. Here (1) is equivalent to the market failure definition, with (2) and (3) mapping to the nonmarket failures of Jaffe and Stavins (1994). As highlighted by Sorrell (2004b), this typology is not exclusive, and barriers may have multiple aspects as well as be multiple and overlapping. Each of these three classifications has multiple examples, and can be informed by various fields, as shown in Table 8.1 [derived from Sorrell (2004b)].

Several further studies draw on the seminal work in Sorrell (2004b) and produce slightly different classifications due to differing perspectives and emphasis. Thollander et al. (2010) refers to work on sociotechnical change (Verbong & Geels, 2007), which argues that social and technological change is complex and interrelated. Here, the authors focus on the interaction of people and technology, dividing barriers into technical (relating directly to technologies), technological (related to human interaction with technologies), and sociotechnical (related to largely human factors). In UNEP (2006) a stakeholder consultation leads to a classification, which is related to that described in Table 8.1, but with a clearly more practical emphasis. The classifications proposed in UNEP (2006) are: management, knowledge/information, financing, and policy. This practical classification is useful as it highlights the financing (difficulty in obtaining financing for a project) and policy (weak legislation, limited or perverse incentives) nonmarket failures, which cannot be derived from the classification in Table 8.1. A final study on EE worth mentioning is Sorrell (2015), which is an evolution of Sorrell (2004b). Although not explicitly a study on barriers, it's formal recognition of the complexity of the issue (largely due to the significant "human" aspect) is a useful contribution. In particular, the identification of several apparently separate issues

Table 8.1 Perspectives on energy efficiency barriers

Perspective	Examples	Theory
Economic	Imperfect information, asymmetric information, hidden costs, risk	Neoclassical economics
Behavioral	Inability to process information, form of information, trust, inertia	Transaction cost economics, psychology, decision theory
Organizational	Energy manager lacks power and influence, organizational culture leads to neglect of energy/environmental issues	Organizational theory

(technology lock-in, emergence, user behavior, user preferences, and institutional design) may, in fact, be regarded as characteristics/factors of complex systems (Bale, Varga, & Foxon, 2015), which can draw attention to potentially important future directions for the study of demand-side interventions.

However, for the EPN concept, the literature on EE barriers is only of limited relevance. This is because smart, ICT-enabled solutions, which can monitor and control a district according to dynamic incentives, are central to the EPN concept (see Chapter 2). Recognizing the dynamic aspect of the EPN concept, there is additionally some literature of interest in the field of Demand Response (DR). In Nolan and O'Malley (2015), three broad types of barriers are detailed: market-related, behavior-related, and information-related. Here, the key barrier is defined as the "chicken and egg" situation where actors require evidence of value, which cannot be demonstrated due to the reticence of actors to pursue DR business cases. This barrier is also highlighted as significant in Strbac (2008). Cappers et al. (2013) regards barriers mostly as products of the required but unrealized changes to relevant institutions. For example, focus is put on the current definition of ancillary service products, which are suited to the capabilities of generators rather than demand. Similarly, conditions on minimum unit size or telemetry may be unnecessarily restrictive (particularly for small consumers with DR capabilities), given capabilities conferred by advancements in aggregator functions (Vafeas & Madina, 2008) and communication infrastructure (Güngör et al., 2011). Uncertainty on business models, required enabling infrastructure and restrictions due to retail-related regulation are also cited as barriers. Focusing on applications for industrial load shifting, Olsthoorn, Schleich, and Klobasa (2015) also explores barriers to DR, through surveys of consumers. This work defined seven types of barriers: technological, information, regulatory, economic, behavioral, organizational, and competences. Further, these barriers are classified as internal or external to the load-shifting organization, or both. The various identified barriers were then ranked to provide a rich (if subjective) description of the barriers.

Besides the academic literature, there is also interest in the barriers to DR from regulatory bodies. The UK's energy regulator (Ofgem) has demonstrated interest in the area (Daniell, 2013). As in Nolan and O'Malley (2015), Ofgem notes the necessity of demonstrating to relevant parties the potential value of DR. The other defined key requirements are the necessity of signaling value (from relevant markets) to consumers, and the necessity of consumers having access to the necessary information (on value and on their capabilities). The current failure to meet these preconditions constitutes a barrier to DR.

As demonstrated earlier, drawing on the literature of DR and EE, there are many and various classifications of barriers to the EPN concept that can be employed. As also shown, no matter which classification is adopted, barriers may span classes, and are frequently interrelated. In this context the "correct" classification system is open to debate. In this report barriers shall be split into four classes:

- political/regulatory,
- economic,
- social, and
- technological.

Given the importance of markets to the EPN concept, and the clear definition of classical economic barriers (market failures), it is clearly instructive to define a class of economic barriers. Although Sorrell (2004b) defines behavioral and organizational barriers separately, their common root in microeconomic theory motivates their common grouping. In this work they are covered by the "social barriers" class. Given the importance of technology in the EPN concept a technical class is also defined. Finally, although closely related to economic barriers, a separate political/regulatory class is defined to better detail the role of political decisions in forming and eroding barriers to the EPN concept. In the remainder of this section the barriers are further explored, within the framework of the previously defined classification. It should, however, be recognized that barriers may span classes, and should be viewed solely within the context of the assigned class.

2.1 Political/Regulatory

Political/regulatory barriers are those barriers, which exist as a result of government policies, usually enacted through regulation. For example, markets can be distorted by the applicable tax code, which may treat various expenditures differently. Discrepancy in the treatment of operational/capital costs, or between substitutable goods (such as electricity and gas, or types of heaters) can cause distortion. Another tax-related barrier can arise from the installation of electricity storage in an EPN. When such storage lies behind a meter, tax will be charged on electricity used for charging the battery (as this cannot be separated from actual consumption). This will create a barrier to the efficient use of the storage (European Commission, 2014). Regulation may also cause distortion in markets if goods that are practicably substitutable [i.e., generation and consumption based Operating Reserve (OR)] are precluded from competing with each other.

A more fundamental barrier for EPNs, or the demand-side in general, is the regulation which prevents market price signals from reaching ultimate consumers. As detailed in European Commission (2014), and demonstrated in Chapter 6, such regulation not only damages business cases for EPNs or similar demand-side parties, but also inhibits the efficiency of markets. In cases where transaction costs are thought to outweigh the benefits of full price-pass-through, or where net-metering, as an incentive for small-scale generation (European Commission, 2014) is appropriate, it may be justified to retain the regulation of consumer prices. Striking a balance between these motivations, so as to minimize overall barriers to the EPN concept is difficult.

Finally, given the heavily regulated nature of energy network operators, the barriers to the EPN concept posed by the regulation of network operators must be mentioned (Van Dievel, De Vos, & Belmans, 2014). These include: the focus on historical performance, rather than future requirements; short regulatory periods; focus on the network operator, rather than system-wide effects; and the lack of recognition of the value of research and development. In particular, the issue of different treatment of operational and capital cash flows is particularly relevant. Short regulatory periods, and the lack of uncertainty on the benefits of capital investment can encourage capital expenditure heavy grid expansion over operational expenditure heavy DR, leading to generally sub-optimal outcomes (Martínez-Ceseña, Good, & Mancarella, 2015). Such bias could result in a failure to consider the benefits of demand-side distribution network constraint management (see Chapter 6).

2.2 Economic

Partly due to the existence of a convenient framework for their analysis (i.e., classical economic theory) there is a large body of work on economic barriers to EPN-related concepts. This work generally involves study of market failures (i.e., flaws in the way a market operates) and market barriers (i.e., other obstacles to the given objective) (Jaffe & Stavins, 1994; Hirst & Brown, 1990; Brown, 2001; Thollander et al., 2010; Sorrell, 2004a; Chai & Yeo, 2012).

2.2.1 Market Failures

Significant market failures can occur if market participants do not have access to *perfect information*. This situation occurs as there are (cash and time) costs associated with collecting and processing information (Brown, 2001). Imperfect information may also arise if markets are so immature that the demand for certain types of information is not sufficient to motivate its collection and distribution by market participants (Thollander et al., 2010). Another information-related failure is defined as the "principal-agent problem" (Sorrell, 2004a; Thollander et al., 2010), which can result in an agent, such as an aggregator, behaving in a way which is not in the interest of the principal (i.e., the EPN).

Another market failure is that of *incomplete markets* (Sorrell, 2004a), which may arise when property rights are not well defined (i.e., comprehensively assigned, exclusive, transferable, and secure). For example, the costs of unregulated CO_2 emission are not exclusive; they accrue to many (through increased atmospheric warming and associated implications). Another example of an incomplete market is where benefits of an asset are not excludable, as can be the case with DR (Nguyen, Negnevitsky, & Groot, 2011). This can result in some parties free-riding, which is a clear market failure.

Lastly, a clear market failure occurs when a party/parties have such a large market share that they are able to exert market power, creating *imperfect competition*. In this scenario, parties can charge prices in excess of their marginal costs, resulting in an inefficient market.

2.2.2 Market Barriers

Access to capital may be a barrier to the EPN concept, depending on the degree of investment required (which may be small, if the flexible infrastructure is largely already present) and the party making the investment.

Hidden costs related to the costs associated with participation in markets, may also be a barrier. These include negotiation and enforcement transaction costs. If these hidden costs are excessive, they could represent a barrier to the EPN concept.

Value is not considered a barrier in the literature on EE, as the value of reducing energy consumption is assumed to be logical. However, for the EPN concept this is not so, as the value of flexibility is not given. Although all systems are likely to have some value for flexibility (as demand is always uncertain, and generator tripping is always possible) some systems will have more value than others. For example, flexibility may be more valuable in systems with highly variable and unreliable generation than in systems with a predictable and reliable generation.

2.3 Social

Social barriers may, in the first instance, be usefully classified following the example in Sorrell (2004a), as organizational and behavioral. Organizational barriers may be relevant to commercial parties within an EPN, as such barriers relate to the social systems of such structured organizations. Arguably of greater importance are behavioral barriers. Given the large number of decisions involved in utilizing energy (particularly in an EPN where a higher level of engagement is required of users), behavioral barriers may be very significant.

2.3.1 Organizational Barriers

Power (or lack of it) may be a barrier where it relates to the power of the person within an organization who has a responsibility for implementing the EPN concept (Sorrell, 2004a). If that person does not wield enough power within their organization [e.g., to install necessary enabling technology, to instruct (to the degree possible) behavior change, or to invest in increased flexibility], then the EPN concept may not be viable. Power, as a barrier, is closely linked to the less precise barrier of *organizational culture*. Specifically, if energy, environmental and even economic concerns (outside of the core business) are not generally regarded as important within the organization, then this will form a general "soft" barrier to the EPN concept.

2.3.2 Behavioral Barriers

Behavioral barriers may be described as those factors which explain why the behavior of any individual deviates from that of the ideal, fully rational (in the classical economic sense) agent (Bowles, 2004). For a firm, rational means profit-maximizing, while for an individual it means utility-maximizing. This latter definition is more complex, as an individual must consider factors, such as convenience and comfort, as well as cash. This can lead to barriers that are often considered to be behavioral [e.g., the requirement of veto over third party control of devices and general aversion due to perceived inconvenience (Losi, Mancarella, & Vicino, 2015)], to instead be classified as (micro)economic. Given that the definition of rational behavior is open [e.g., is inertia, in fact, rational, as it economizes on cognitive exertion (Zhu, 2013)], such barriers shall be considered as behavioral here.

Form of information may be a barrier if information is not regarded as intended by the sender. An example can be a poor user interface design that result in unexpected behavior (Losi et al., 2015). This issue may be exacerbated if a commercial partner is not *credible* or *trustworthy* in the eyes of an individual decision maker, as the partner will struggle to encourage desirable behavior. More fundamentally, there may be a barrier to the EPN concept if EPN consumers hold *values* that do not align with the EPN objective. For example, those who value privacy, autonomy, ownership, power, or control highly may not welcome control of devices in their home (Parkhill et al., 2013). Even if all other barriers are not relevant, there may still be issues in the adoption of the EPN concept due to *inertia*, as entrenched behavior may result in rejection of an intervention (Losi et al., 2015; Balta-Ozkan et al., 2013). Alternatively, there may just be a delay in adoption, as such barriers decline in relevance over time, as institutions [e.g., "hard" institutions, such as markets and laws, and "soft" institutions, such as social norms] change over time (Bowles, 2004). Such change may be aided if the amount of benefit the user can expect from sharing the data is shown to be significant. A related issue here is *bounded rationality*; that is, customers may not be willing to change if issues are simply too complex for EPN consumers to understand (with the cognitive resources they are willing to commit to the task). Concrete examples that they can easily relate to can help here.

2.4 Technological

The EPN concept relies on state-of-the-art Information and Communication Technology (ICT)/Internet of Things (IoT) technology, which leverages both cloud and embedded computation. There may be barriers associated with misalignment between the design/development of such technology and its required application and deployment, as a result of a poor or hampered requirement elicitation process. There may be genuine and/or perceived technology readiness issues, or there may be misconceptions and/or uncertainty as to what the technology can deliver. For example, without content experts to interpret data, the potential of "Big Data" to identify new business opportunities cannot be realized. Even with the necessary experts, results are not deterministic, which must be appreciated by users.

Within the district, barriers to the EPN concept may exist where data acquisition and actuation infrastructure are inadequate. This can be the case for metering infrastructure, which may not be sufficiently disaggregated (temporally and/or spatially). This may be relevant as typically, energy markets trade in 15 min to 1-h periods. Further, depending on particular commercial arrangements, there may be a requirement for extensive submetering within the EPN. This may be required if it is deemed necessary to separate flexible and nonflexible resources, for example. It is also possible that, for some markets (e.g., for "fast" OR products, such as primary and secondary OR, in the Irish context, see Chapter 6) even finer metering may be required, down to the scale of minutes or less. More broadly, barriers may exist due to the traditional classification of operation technology as "built environment" rather than "energy" infrastructure. In essence the EPN is a cross-sectorial multiownership concept, which adds to complexity.

Many of the business cases, which require action against a suitable "baseline" profile, will require a suitable baseline methodology. To provide a baseline for measuring the delivery of certain products (such as operational reserve or distribution network constraint management, see Chapter 6), adequate metering is a necessity, but not a sufficient condition; it cannot quantify what the profile would have been *without* the call. But, to determine compliance with respect to the relevant contract, it is necessary that some baseline can be agreed between the buyer and seller. Indeed, as discussed in Nolan and O'Malley (2015), establishing a baseline may be a barrier to deployment of an EPN as it can impede proper valuation of a product and raising of necessary project capital.

Linking to barriers related to perceived lack of privacy, insufficient data security may be a barrier to widespread adoption of the EPN concept, particularly given the tendency of EPN/IoT systems to be:

- physically distributed;
- a mixture of very small to very large devices;
- dependent on closed and open or untrusted networks; and
- large-scale deployments, which may extend to tens of thousands of components.

Further, the diversity of vendors and changes in the use and functionality of components can also impact system security.

Diversity of vendors is also likely to result in a lack of standardization in enabling technologies, which can form a barrier to EPN-type concepts. Without the necessary standardization here, the components of the EPN cannot talk to each other, and flexibility cannot be exploited (Losi et al., 2015), or they may interfere with each other (Balta-Ozkan et al., 2013). Further, lack of standardization of EPN components may prove to be a barrier if there is concern on the part of those responsible for investing in equipment that they may become "locked-in" to a particular supplier. This may result in constraints on future decisions, which lead to suboptimal outcomes.

As opposed to the barrier of too little standardization described earlier, too much standardization may also be a barrier. Although energy and capacity (the basic commodities which are being traded in all markets in which the EPN may partake) are continuous in nature, they are typically traded as defined products. If the definitions of the standardized products are too restrictive, they may preclude provision by an EPN, or may mean that the full value of the EPN concept cannot be realized and hence result in suboptimal system efficiency. On the other hand, however, standardization of products generally reduces transaction costs, as the definition of a restricted number of products can reduce transaction costs (Cordella, 2006). Thus there may be tension between the motivation to reduce standardization (to increase realized EPN value and system efficiency) and to increase standardization (to reduce transaction costs).

With respect to the impact of the physical electricity network, network limitations can be a barrier if increasing levels of resources and flexible loads responding to dynamic price signals cause technical issues, as controllers shift large portions of power consumption toward the lowest price periods, overloading distribution network assets and lead to voltage rise/drop issues. The traditional solution to these issues is (expensive) network reinforcement. This is clearly a technological barrier, but there is also a regulatory aspect as financing of Distribution Network Operators (DNOs), and the incentives the financing rules produce, is regulated. Such issues may encourage some kind of locational marginal pricing (Elexon, Smarter Grid Solutions & Baringa, 2014), which, as it affects retail pricing, will also have regulatory aspects.

Another barrier which crosses classes is that of complexity related to ownership and control of district devices. If there are multiple owners in a district, who do not wish to contract out provision of energy services then decentralized optimization is required. This can increase complexity due to the increased number of agents and increased amount of communication required in decentralized optimization approaches, for example, Lagrangian relaxation based methods (Papadaskalopoulos, Pudjianto, & Strbac, 2014), or game theory based intradistrict markets (Mohsenian-Rad et al., 2010).

More generally, the complexity of the district can become a barrier. This is because, as the complexity of the system increases, so must the complexity of the infrastructure to control it. In cybernetics (the science of communications and automatic control of systems in both machines and living things), Ashby's law of requisite variety (Ashby, 1956, 1962) essentially states that "only variety can destroy variety." Thus any proposed ICT system for controlling an EPN must be equal to the variety or complexity of the EPN in order to control it. This may prove a barrier if the complexity of the system increases the computational burden of any optimization to the extent that it cannot be practically executed in required timescales. However, in practical terms any regulator of a system only requires a level of sophistication that can respond to the most likely stimuli. Nevertheless constructing an ICT system complex enough to reasonably control an EPN is an arduous task, primarily due to the variety introduced by a myriad of existing in-situ systems. This is why a system-of-systems approach is posited within COOPERaTE.

3 SPECIFIC CHALLENGES FOR THE EPN CONCEPT

Drawing on the general barriers to smart, demand-side interventions detailed earlier, and from experience in the COOPERaTE project, this section presents an overview of the specific challenges to the EPN concept, in the areas of modeling and simulation, performance assessment, information technology, and business model development.

3.1 Energy Modeling and Simulation

Modeling and simulation tools serve as decision support for planning and operation of energy systems in buildings as well as proof of concept of energy optimization algorithms before carrying out field tests. Drawing on the work on modeling and simulation conducted in Chapter 4, it is apparent that the availability of equipment and consumption data must be specified in order to understand the neighborhood infrastructure and capabilities. In this context, several challenges are encountered. These challenges are related to the users' *uncertainty of their technology capability*, *lack of appropriate metering*, and the *complexity of the energy systems*.

The users' *uncertainty of technology capability* may lead to a neighborhood model that does not capture the deployed thermal and electrical generation and storage technologies. Furthermore, *lack of appropriate metering* complicates the process of model validation to real data; therefore, several assumptions must be taken and later evaluated and corrected in an iterative process until the model captures the actual dynamic behavior of the neighborhood energy system. In the real-life experience, we observed that weather data, such as direct solar irradiation and wind speed are rarely metered. Thus, based on the location of the test sites we used test reference year weather data which enabled us to test the models of photovoltaic modules and wind turbine, respectively; when comparing the results to recorded data, the output power for this equipment was between tolerable range (±5%).

The simulation environment for the thermal and electrical domains should be integrated in order to study joint effects. This represents a technological challenge given the physical *complexity of the energy systems*, as very different dynamic characteristics must be considered. While thermal processes, such as heat exchange are rather slow and changes take place in the range of minutes, electrical effects are much faster and range from

10^{-5} s for electromagnetic transients to a few seconds for electromechanical or longer term transients. The proposed simulation environment avoids neglecting fast effects or to carry out a disproportional amount of calculations for slower phenomena.

In the context of energy optimization algorithms, different data-privacy barriers are encountered with different neighborhood types. These barriers are related to users' *unwillingness to share data* and depend on the degree to which they trust the party they are sharing with (see Section 2.3.2). In particular, users may be concerned about the ability to derive nonenergy-related information (e.g., building occupancy and activity patterns) from such energy data. A solution is to optimize only with the mutually agreed level of transparency for each party; either high level of transparency with all generation, storage, and consumption data available to the energy management service, or low level of transparency only with anonymous or encrypted data, such as comfort measures and corresponding CO_2 emissions associated to a set of equipment schedules. The actual sets of building schedules do not have to be necessarily shared with the neighborhood central optimization engine.

An adequate *metering infrastructure* is also relevant for the optimization algorithms as typically, energy markets trade in 15 min to 1-h periods. Therefore, to partake in these markets, metering at this level is required (see Section 2.4). The possibility of storing metered data should be available (e.g., in a database), allowing the energy optimization algorithms to exploit historical energy consumption and generation data and forecast or estimate 24-h ahead of the electrical/thermal neighborhood consumption and generation. Furthermore, depending on the contractual agreement with the energy supplier, both energy price forecasts based on real-time recorded data should be available. In cases when there is lack of appropriate metering infrastructure, input data for the energy optimization algorithms providing forecasts of the neighborhood energy consumption and generation are taken from modeling and simulation tools.

As expanded upon further in Section 3.4, *economic and regulatory* barriers can be significant for energy optimization algorithms, if these prevent price signals reaching the neighborhood.

3.2 Performance Assessment

Based on data collection, performance assessment gives an accurate picture of a site's capacity to fulfill its commitments, which may relate to the EPN concept or other special considerations. In order to assess the performance of a neighborhood, a proper evaluation methodology and priorities have to be established in advance.

As detailed in Chapter 7, the methodology boils down to four key elements:

- Definition of the baseline and reporting period.
- Assessment protocol and adjustments.
- Key performance indicator (KPI) definition.
- Parameters listing.

Defining a baseline enables focusing only on the savings due to EPN services. The complexity of a neighborhood makes it almost impossible to have an overview of performance without a baseline. As encountered within the COOPERaTE project, defining a baseline and an associated reporting period has two main issues: defining a time frame and a perimeter. On the Irish demo site, definition of specific time frames was complex. It was decided to catch the specificities of the year (winter/summer time) because we couldn't monitor a full year. On the Challenger demo site, the perimeter of the baseline was complex as buildings were being renovated during the assessment.

For both demo sites, UTRC and EMBIX used the IPMVP protocol in order to have a process to analyze performance and perform adjustments (to the baseline), if needed. For the Challenger demo site, adjustments helped to account for the renovation issue.

Defining KPI enabled a big picture of the neighborhood and production of clear objectives to focus on. For the demo sites within the COOPERaTE project, it was decided to have energy, environment, and business case related KPI in order to have an overview of the sites. However, there was no performance commitment based on these KPIs as the project was a demonstrator.

Collecting data from consumption and production units is essential to assess performance (as well as calibrate neighborhood models, see Section 3.1). However, it is also important to collect data from parameters and influencing factors like weather forecast (temperature, sunshine, etc.). Most of the time, as in the COOPERaTE project, simulation tools and energy management platforms are limited to single parameters (like weather) and don't include complex and systemic modeling (economic, governance parameters, behaviors parameters, etc.). Including these complex modeling parameters could be an interesting next step to reach EPN goals [see the Aspern Smart City project in Vienna (http://www.ascr.at/en/research-infrastructure/smart-ict/)].

3.2.1 High-Quality Data

High-quality data collection is crucial for a good performance assessment, which relies on the deployment of platforms and equipments that must be properly maintained. Indeed, high-quality data collection requires efficient maintenance of a large number of devices that may be installed in buildings, housings, offices, public spaces, etc. Thus strong coordination of maintenance activities, within a contractual framework is crucial to maintain these devices in good conditions.

The data collection infrastructure should be regularly checked (commissioning) to make sure that the data collected is correct. This commissioning is more difficult to implement when there are many stakeholders on site. For instance, we observed that commissioning was easier on the Challenger demo site where there is simplified governance (i.e., single ownership, single operator, single property developer). Commissioning is also facilitated if there is a simplified data collection (less data sources) and easier social acceptance (less social diversity). However, the tendency of new neighborhoods is to promote social diversity and the innovative aspect of data collection leads to an interlocked and complex infrastructure involving many suppliers using different technologies.

Taking a step further, data collection may facilitate associating facility managers to the EPN project. More specifically, managers are, most of the time, not aware of the objectives and processes implemented. Communication and awareness could help them have a better understanding of the issues at stake in the neighborhood and how to take them into account. Indeed, these processes and commitments involve more sensitive adjustments that are more time consuming.

3.2.2 Make a Common Use of Performance Assessment

In the COOPERaTE project assessment of performance with KPI calculation was considered. There were no commitments (contractual or moral), and it was not planned to gather stakeholders and help them understand their impact and how to take efficient actions to improve the performance. An opportunity after the COOPERaTE project is the production of a governance framework to help reach the EPN goal. Indeed, the performance assessment should overcome current limits through a new governance framework, based on both technological (real-time data collection and systemic analysis) and contractual (performance commitments) innovations.

Due to the nature of the demonstration and the focus on ICT, the COOPERaTE project took limited account of the shared responsibility by stakeholders on energy targets. However, involving buildings' constructors, owners, managers, and occupants is key to achieving efficient and long-term performances. Ideally, thanks to performance assessment, stakeholders could move toward a shared responsibility.

This concept of coresponsibility could lead to the creation of a new role of EPN technical coordinator, who will give technical recommendations to energy operators in order to coordinate and improve their performance. This technical coordinator could possibly be in charge of data collection and performance assessment in order to achieve more transparency, high expertise, and cohesiveness. This role could be supported by the NEM (see Chapter 6).

3.2.3 Empower Inhabitants and Workers

Empowerment of inhabitants and workers is a major impediment to smart grid projects. The use of personal data and the appropriation of digital tools in a socially diverse context present a major challenge, as social acceptance must be guaranteed (see also Sections 2.3.2 and 3.1). To tackle this, an opportunity is to work on a new deal on data, which gives to the consumer the right to possess its data and to have full control over its use. The aim is to let them realize that they are contributing to create a better and more sustainable neighborhood.

With the COOPERaTE project we understood that giving the inhabitants and workers a holistic and simplified vision of the neighborhood through use of dashboards or similar user-friendly interfaces (to ensure information presented clearly, see Section 2.3.2) is key: it helps them understand the performance assessment, pique their curiosity, and even promote action. To go further, an opportunity could be organizing challenges, serious games, or events that deliver tangible benefits to the often abstract concept of performance.

3.3 Information Technology

As discussed in Chapter 5, urban infrastructure, for example, homes, commercial and industrial buildings, outdoor spaces, streets, transport infrastructure, and energy grids are instrumented with a variety of monitoring and control systems. These systems have been embedded into our environment to manage everyday tasks. These systems have traditionally been standalone siloes, for example, a building monitoring and control system manages just the building it is installed in. However, our increasingly digitalized world now offers opportunities to connect these individual systems via the Internet into a much larger system. The opportunities for energy security, flexibility, and efficiency are immense as is the complexity involved.

The emergence of the Internet of Things (IoT) provides opportunities to augment existing systems connecting hitherto unconnected assets and to create a system-of-systems that goes beyond individual isolated monitoring and control systems. The focus within COOPERaTE was to develop concepts and services for EPN as outlined in Chapter 2. COOPERaTE, like other initiatives focused on making urban neighborhoods more energy efficient and energy flexible, envisioning new energy services at a district level (see Chapter 5, Section 5) and business models (see Chapter 6).

However, there are many road blocks to achieving an integrated IoT-based EPN. The COOPERaTE project identified key challenges, namely—interoperability, security, privacy, and complexity, proposing an interoperable data-driven approach that overcomes some of the problems without forcing major changes to existing systems or demanding implementations of a single technology based solution. What follows briefly reviews these key challenges.

3.3.1 Interoperability and Complexity

As outlined in Chapter 5 there are many different standards for machine-to-machine communications, radio frequencies, networking protocols, data formats, management platforms, service formats, and security protocols which lead to fragmentation and inaction. This plethora of options has hindered integration across existing systems at both the system and the data layer. Techniques to overcome this are the focus of much current research. Chapter 5, Section 2 "state-of-the-art analysis" identifies technologies and standards to consider with respect to building an end-to-end system that can support services within the built environment, smart city, and smart grid sector at scale. But, conversely, it is also highlighted that navigating in a bourgeoning landscape of multiple standards, technologies, and alliances would be extremely challenging.

Figure 8.1 ***Internet of Things (IoT) alliance round-up.*** *(March 2015 http://postscapes.com/internet-of-things-alliances-roundup).*

Specifically, the COOPERaTE experience has been that interoperability, as it has been historically, is a major challenge within EPNs. That challenge is somewhat complicated in the case of EPNs given the need to offer services and make decisions at the district level and given the multiownership of the systems and assets involved.

That is not to say there is no complexity in terms of interconnectivity at all levels. Technical interoperability, syntactic and semantic at all levels is an issue with individual layer and full end-to-end solutions offered, aptly illustrated by Fig. 8.1 from Chapter 5 (Postscapes, 2015).

Any end-to-end solution essentially resolves interoperability issues and system-level complexity by defining system-specific standards and there are three system examples within COOPERaTE. But the issue is then one of adoption at scale. Interoperability within that individual system is no longer an issue, but the system does not represent the district and is unlikely to offer marked advantages that would warrant multiple system owners, with existing systems, to migrate. There may also be reluctance in moving toward a position of perceived lock-in or the criticality of the service maybe such that any change is strongly resisted. That said, older systems at lower levels of sophistication might well, either fully or partly, migrate.

But while the issue of heterogeneity, scale, and multiownership is a complication it can also be seen as an opportunity. If dealing with a brownfield scenario of existing systems, technologies, and protocols one can take the position that what the current systems control, they continue to control and that you offer augmentations that they can adopt, but you also offer a means by which they can interconnect with other existing systems. This was the approach taken within COOPERaTE in tackling interoperability. By enforcing the NIM as a means of supporting data exchange between systems the project enabled federation of the information required to deliver an EPN, but allowed the in-situ systems to continue controlling the assets of the EPN.

3.3.2 Security and Privacy

Cloud technology is arguably fit for purpose today, in terms of delivering the scale, security, and flexibility required at the neighborhood level. However, there are still challenges of ownership and interoperability with legacy in-situ systems and new increasingly distributed resources. Again Chapter 5 establishes the importance of data security and privacy for the delivery of an EPN. Chapter 5 outlines the nervousness in terms of cloud computing which the authors suggest is a proxy for the importance of security and privacy in the context of EPNs. Whereby EPN solutions are a combination of traditional embedded monitoring and control systems and cloud computing based data services, that is, solutions increasingly being described as IoT systems. Again, the experience within COOPERaTE has been that the closer one comes to the physical control of on the ground assets the higher the reluctance to adopt new systems, despite apparent advantages. Again, the COOPERaTE approach has been a pragmatic one in allowing existing systems to take what aspects of IoT technology they wish in modularly augmenting their existing systems, but also enforced a standard means of system-to-system communication in the form of the NIM. The NIM itself addresses issues of data privacy by including fields at the meta-data level that can be used for enacting polices on how that data is managed, namely: an "expiry date," a "physical location" as in point of origin, and an "agreed usage" field. Security is still maintained by the existing systems in terms of the assets they hold, what data they make available is their choice based on business service level agreements. Then one simply applies best practice in terms of data transmission layer security and authentication between system-to-system services.

In short, the COOPERaTE strategy has been to allow systems to evolve in terms of IoT sophistication, at the appropriate system-specific pace, while allowing those same systems to be part of a cooperating system-of-systems delivering advantages at scale within an EPN.

3.4 Business Model Development

Development and quantification of a business case is crucial to implementation of the EPN concept. Without this, districts have no motivation to exploit their inherent flexibility, as either that flexibility is not valued highly enough, or the barriers to its exploitation are too high. Drawing on experience in the COOPERaTE project, several specific challenges related to the development of business models can be identified.

First, there is the fundamental issue of price-signal visibility. Districts, being largely composed of consumers are often, especially in the case of domestic consumers, subject to significant regulatory control with respect to energy pricing. Further, going beyond regulatory requirements, retailers typically remove all risk and variability from the retail tariffs, to suit perceived user requirements. This is a fundamental challenge for the EPN concept, as exploitation of flexibility requires favorable behavior to be signaled to the district, which is done, in liberalized systems, through price signals. Although consumers may be willing to contract-out energy service provision and, hence, responsibility for operation of flexible devices, regulation which prevents market price signals from reaching ultimate consumers may be a challenge for the EPN concept. Further, they inhibit the efficiency of markets. In cases where transaction costs are thought to outweigh the benefits of full price-pass-through, or where net-metering, as an incentive for small-scale generation (European Commission, 2014) is appropriate, it may be justified to retain regulation of consumer prices. Striking a balance between these motivations, so as to minimize overall barriers to the EPN concept is difficult.

A related challenge is the regulation of network operators in the energy system. Traditionally, regulation of such actors has focused on historical performance (rather than future requirements), been based on short regulatory periods (disadvantaging long term thinking and research and development), and lacked appreciation of system-wide effects. In particular, short regulatory periods, and the lack of uncertainty on the benefits of capital investment can encourage significant capital expenditure (i.e., heavy grid expansion) and/or over operational expenditure (i.e., heavy DR), leading to generally suboptimal outcomes (Martínez-Ceseña et al., 2015). With reference to Chapter 6, such bias could result in a failure to consider the benefits of the distribution network constraint management business cases, or of the improvement to reliability if the EPN is able to act as a microgrid.

In Section 2.2.1 incomplete markets were highlighted as a general economic barrier to smart, demand-side interventions. This issue may be a specific challenge for the EPN concept, as it means that the benefit of the EPN concept is not fully recognized and remunerated. For example, the costs of unregulated CO_2 emission are not exclusive, as they accrue to all the population through increased atmospheric warming, and associated implications. Hence the potential for an EPN to reduce system emissions is not recognized and any reduction in emissions is not remunerated. The existence of this "externality" constitutes a market failure. This failure can lead to increased levels of CO_2 emission (with the associated increased cost to society), under certain business cases (e.g., those Irish business cases not involving wholesale electricity price signals). Incomplete markets also pose a challenge to the EPN concept when the benefits of an asset are not excludable, as can be the case with DR (Strbac, 2008; Nguyen et al., 2011). An example is demonstrated in Chapter 6, where retail prices reduce peak grid import compared to the base, load-following case. This benefits the DNO, by reducing peak load, without the DNO having to remunerate the EPN for this advantage. A further example may be where cost socialization results in a lack of signals to market participants to adjust their behavior in an appropriate way. A specific example is the lack of imbalance penalties (in Ireland) and the not-fully cost-reflective imbalance penalties (France), which prevent districts from perceiving signals indicating system-beneficial behavior (Haring, Kirschen, & Andersson, 2015).

Chapter 6 also demonstrated how access to capital may be a challenge for the EPN concept. Depending on the identity of the investing party, this issue may be a significant challenge. Given that larger parties tend to be able to access capital on more favorable terms, this may be a particular issue for smaller parties, such as a district seeking to make investments on their own.

In addition to the specific challenges detailed earlier, a more general, but perhaps more significant challenge may be the lack of tools for the assessment of business cases. As detailed in Chapter 6, such business cases should be assessed through rigorous cost-benefit analysis. Although such analysis is familiar to any commercial party, the difficulty lies in quantifying the benefits which due to several factors (e.g., constraints associated with operating on the distribution network, interactions with heterogeneous, unpredictable people, complexity of regulation, and commercial interactions) are uncertain and variable, by case.

4 RECOMMENDATIONS

Considering the previously described general barriers to smart, demand-side interventions, and the specific challenges highlighted in the course of the COOPERaTE project, several opportunities for improving the conditions for implementation of the EPN concept, and smart district and demand-side interventions more generally, can be formulated. To conclude this chapter, this section presents these opportunities, and details several key recommendations for policy-makers to promote smart district interventions, to address the energy trilemma of affordability, sustainability, and security. These recommendations include:

1. Enabling passing price signals relevant to various markets and charging regimes to the district.
2. Relaxing existing overly restrictive limits on minimum resource size for power system services subject to the availability of adequate metering infrastructure to quantify the services.
3. Facilitating energy service based business models.
4. Developing new models to assess traditional asset-based solutions and novel nonasset-based solutions (e.g., storage and other demand-side solutions) on comparable basis.

5. Providing financial support (e.g., low interest rates for loans) for investment in EPN enabling infrastructure.
6. Developing effective EPN interfaces to operate controllable devices using standardized protocols and clear rules on data availability and privacy.

4.1 Enabling Passing Price Signals Relevant to Various Markets and Charging Regimes to the District

One of the most fundamental recommendations to facilitate the EPN concept is enablement of the *pass-through of price signals, from various markets and charging regimes, to the district.* This would provide the owners and operators of flexible devices the information and motivation to exploit their flexibility to provide valuable services, such as (1) increasing the efficiency of various markets by reacting to high prices, which indicate scarcity and (2) supporting efficient operation of energy systems by providing explicit services, such as reserve of distribution network constraint management services. To enable this, consumers should be given the option to access cost-reflective tariffs. This should benefit consumers by removing the risk premium that retailers add to their flat-rate tariffs. However, a corollary of this is that consumers, particularly if they do not manage their flexible resources effectively, may face highly variable bills. For some risk-averse consumers this risk may be unacceptable. Hence, adoption of cost-reflective tariffs should remain optional.

4.2 Relaxing Existing Limits on Minimum Resource Sizes for Power System Services

To offer explicit services, districts will need to partner with a suitable third-party, such as an aggregator. To facilitate such arrangements, *overly prescriptive and restrictive rules on minimum resource sizes and communication protocols should be relaxed.* Within the district, to enable participation in markets which require granular metering, and also to enhance modeling of districts, *adequate metering infrastructure is a key requirement.*

4.3 Facilitating Energy Service Based Business Models

Aggregator business models typically assume a "transactional" relationship with the consumer, whereby price or volume signals, reflecting market signals, are sent to the consumer by the aggregator, with the consumer assuming responsibility for reacting to these signals (and may choose not to do so, if too inconvenient for them). This may lead to unreliable service provision, or may prompt the aggregator to procure much more flexibility than it offers to the market, to ensure reliable delivery. One potential remedy for this situation is an "aggrESCo" business model, whereby an Energy Services Company (ESCo) takes responsibility for delivering some useful energy vector (e.g., heat), or some energy service (e.g., thermal comfort). In these situations, the ESCo adopts control, and possibly ownership, of flexible devices (e.g., a combined heat and power unit, or electric heat pump, with storage). This allows the ESCo to exploit this flexibility in various markets, constrained only by the service requirements of the clients. This eliminates the issue of unreliable response from potentially disinterested consumers to an aggregator signal. This business model would be aided *by large-scale investment in enabling infrastructure, such as heat networks,* in addition to a *general cultural shift to purchasing energy services, rather than energy*.

4.4 Developing Comparable Means to Assess Asset Based and Demand-Side Based Solutions

More generally, efficient pass-through of price signals to districts requires a paradigm shift to *put demand-side resources on an equal footing with generation resources*. This will take time, as policy-makers and flexibility buyers are less familiar with demand-side resources, and may view them as less reliable. An important measure to ameliorate this lack of confidence is *development of advanced modeling and simulation approaches*, such as described in Chapter 4, which enable understanding of complex demand-side resources. Moving beyond the generation/demand dichotomy, advances in the feasibility of storage present an opportunity at the district level. However, to fully realize the potential of storage, *regulators and grid operators must develop tax and UoS policies which do not inappropriately disadvantage storage*, for example, by charging tax and UoS fees on electricity import to charge an electricity store (energy which will ultimately be used to provide a service, or respond to price signals). To further aid efficient network operation, *UoS fees should also become more cost reflective, by varying temporally and spatially*, in response to varying levels of losses and capacity. Such measures would recognize the specific benefits (such as reduced losses and potential for network constraint management) that demand-side resources, such as an EPN, can bring. However, as with all measures to increase cost reflectivity, the costs of developing such measures should be considered. The administration costs of, for example, fully cost-reflective UoS fees may well outweigh the benefits, though this is less and less likely to be true as monitoring and metering equipment become ubiquitous, and networks increasingly operate near limits, as electrification of heating and transport increases.

Considering the contracts between third-party intermediaries, such as aggregators or ESCos, and district consumers, *development of suitable metrics (and baselines) to measure performances are required*, as highlighted in Chapter 7. Particularly in relation to DR services, definition of well-defined metrics is crucial for specification of service levels (e.g., relating to heat or thermal comfort provided or to reliability of electricity supply) in contracts. This is crucial to avoid the "principal-agent" problem, which can result in inadequate performance (see Section 2.2.1). This is linked to the wider issue of governance. As discussed in Section 3.1, a simplified governance framework in a district will make introduction of the EPN concept easier, which will be aided by clear definition of preferences and requirements in contracts between district consumers and TPIs.

Development of suitable metrics is also necessary to be able to quantify how the cost of some DR action should be shared between beneficiaries, if multiple parties benefit from the exercise of DR, or how parties should be compensated, if activation of DR disadvantages some party. Another externality that deserves specific attention is CO_2 emission. The EPN concept is likely to reduce system emissions, by encouraging local generation (avoiding losses) and by pushing carbon intensive generation out of various electricity system markets (e.g., wholesale electricity markets and reserve markets). This represents a benefit to society and *districts who embrace the EPN concept should be rewarded accordingly*, for example, *through incentives, tax breaks, or through carbon market operation.*

4.5 Providing Support for Investments in EPN Enabling Infrastructure

Although the development of an EPN is likely to require relatively small investment (if only ICT investment is required), for cases where larger investments are required problems accessing capital (compounded by uncertain returns due to regulatory complexity and energy price uncertainty) may be a significant barrier to implementing the EPN concept, particularly for smaller actors. To counter this *standardized assessment and valuation protocols should be developed*. Given the potential to benefit society (e.g., through reduced emissions and through generally reducing system/network operation costs), *governments should consider ameliorating problems with accessing capital*, for example, through loan guarantee schemes, or simply through contributions to project costs. Similarly, smaller actors may struggle to deal with the hidden costs (e.g., search, negotiation, and enforcement transaction costs), which result from becoming active providers of energy/services.

4.6 Developing Effective Interfaces

Focusing on the interaction with the consumer, experience within the COOPERaTE project has shown the importance of well-designed interfaces/dashboards (see Section 3.2.3). Therefore *much attention should be focused on designing interfaces which communicate favorable behavior to district energy users*, in a way which maximizes the likelihood that these prompts are acted upon. This can be aided by maximizing trust between the sender of the information and the recipient. This is a difficult task, for which there are no quick fixes, but it may be easier for independent TPIs whose business cases are focused on partnering with consumers to exploit their flexibility for mutual gain, than for parties, such as retailers, for whom exploitation of DR is not their main business.

Beyond the interface with the consumer, *automation will be a significant enabler of DR and the EPN concept.* Consumer demand will motivate much development in this realm, primarily for increased comfort and convenience. However, the increased deployment of sensing, computation, and actuation devices will introduce functionality, which can also enable delivery of DR services. To facilitate this deployment of ICT, *progress should be made on standardization to enable interoperability and substitutability*, to promote an open market in ICT devices, to facilitate innovation and competition.

A major barrier to automation, and also to effective modeling of EPNs (see Section 3.1), could be concerns on privacy and security. *Privacy concerns should be assuaged by increasing trust between consumers and TPIs; specifically TPIs should be very clear on who owns data (such as meter data), and how any party may use such data.* Security may be a more difficult concern to address. *Ensuring data are hosted on secure servers, protected by encryption*, are the minimum measures that should adopted. Further measures, to enhance data anonymization may need to be developed, so that stored data cannot be linked to individual meters.

REFERENCES

Ashby, W. R. (1956). *An introduction to cybernetics*. New York: J Wiley.

Ashby, W. R. (1962). Principles of the self-organizing dynamic system. *Transactions of the University of Illinois Symposium*, pp. 255–278. http://www.tandfonline.com/doi/abs/10.1080/00221309.1947.9918144

Bale, C. S. E., Varga, L., & Foxon, T. J. (2015). Energy and complexity: new ways forward. *Applied Energy*, *138*, 150–159.

Balta-Ozkan, N., Davidson, R., Bicket, M., & Whitmarsh, L. (2013). Social barriers to the adoption of smart homes. *Energy Policy*, *63*, 363–374.

Bowles, S. (2004). *Microeconomics: Behaviour, institutions and evolution* (3rd ed.). Princeton, New Jersey: Princeton University Press.

Brown, M. A. (2001). Market failures and barriers as a basis for clean energy policies. *Energy Policy*, *29*, 1197–1207.

Cappers, P., MacDonald, J., Goldman, C., & Ma, O. (2013). An assessment of market and policy barriers for demand response providing ancillary services in U.S. electricity markets. *Energy Policy*, *62*, 1031–1039.

Chai, K. H., & Yeo, C. (2012). Overcoming energy efficiency barriers through systems approach—a conceptual framework. *Energy Policy*, *46*, 460–472.

Cordella, A. (2006). Transaction costs and information systems: does IT add up? *Journal of Information Technology*, *21*(3), 195–202.

Daniell, R. (2013). Creating the right environment for demand-side response: next steps, Ofgem. Available from: https://www.ofgem.gov.uk/ofgem-publications/75245/creatingtherightenvironmentfordemand-sideresponse.pdf

Van Dievel, P., De Vos, K., & Belmans, R. (2014). Demand Response in Electricity Distribution Grids: Regulatory Framework and Barriers (pp. 1–5). Krakow, Poland: 11th International Conference on the European Energy Markets.

Elexon, Smarter Grid Solutions & Baringa (2014). Actively managed distributed generation and the BSC. Ofgem. Available from: https://www.ofgem.gov.uk/ofgem-publications/88954/sgf-ws6-17juneactivelymanageddistributedgenerationandthebscv1.0.pdf

SWECO (2014). *Study on the effective integration of demand energy recourses for providing flexibility to the electricity system*, Final report to The European Commission. Available from: http://ec.europa.eu/energy/en/studies/study-effective-integration-distributed-energy-resources-providing-flexibility-electricity

Güngör, V. C., Sahin, D., Kocak, T., Ergüt, S., Buccella, C., Cecati, C., & Hancke, G. P. (2011). Smart Grid Technologies: communication technologies and standards. *IEEE Transactions on Industrial Informatics*, *7*(4), 529–539.

Haring, T. W., Kirschen, D. S., & Andersson, G. (2015). Incentive compatible imbalance settlement. *IEEE Transactions on Power Systems*, *30*(6), 3338–3346.

Hirst, E., & Brown, M. (1990). Closing the efficiency gap: barriers to the efficient use of energy. *Resources, Conservation and Recycling*, *3*(4), 267–281.

Jaffe, A. B., & Stavins, R. N. (1994). The energy paradox and the diffusion of conservation technology. *Resource and Energy Economics*, *16*(2), 91–122.

Losi, A., Mancarella, P., & Vicino, A. (2015). In A. Losi, P. Mancarella, & A. Vicino (Eds.), *Integration of demand response into the electricity chain: Challenges, opportunities, and Smart Grid solutions*. Wiley-ISTE.

Martínez-Ceseña, E. A., Good, N., & Mancarella, P. (2015). Electrical network capacity support from demand side response: techno-economic assessment of potential business cases for small commercial and residential end-users. *Energy Policy*, *82*, 222–232.

Mohsenian-Rad, A. -H., Wong, V. W. S., Jatskevich, J., Schober, R., & Leon-Garcia, A. (2010). Autonomous demand-side management based on game-theoretic energy consumption scheduling for the future Smart Grid. *IEEE Transactions on Smart Grid*, *1*(3), 320 331.

Nguyen, D. T., Negnevitsky, M., & Groot, M. D. (2011). Pool-based demand response exchange—concept and modeling. *IEEE Transactions on Power Systems*, *26*(3), 1677–1685.

Nolan, S., & O'Malley, M. (2015). Challenges and barriers to demand response deployment and evaluation. *Applied Energy*, *152*, 1–10.

Olsthoorn, M., Schleich, J., & Klobasa, M. (2015). Barriers to electricity load shift in companies: a survey-based exploration of the end-user perspective. *Energy Policy*, *76*, 32–42.

Papadaskalopoulos, D., Pudjianto, D., & Strbac, G. (2014). Decentralized coordination of microgrids with flexible demand and energy storage. *IEEE Transactions on Sustainable Energy*, *5*(4), 1406–1414.

Parkhill, K., et al. (2013). Transforming the UK Energy System: Public values, attitudes and acceptability Synthesis Report, UKERC. Available from: http://www.ukerc.ac.uk/asset/835A77F5-62DA-4062-925917DF4D288372/

Postscapes (2015). *IoT Alliances round-up*. http://postscapes.com/internet-of-things-alliances-roundup

Sorrell, S. (2004a). *The economics of energy efficiency: Barriers to cost-effective investment*. Cheltenham: Edward Elgar.

Sorrell, S. (2004b). Understanding barriers to energy efficiency. *The Economics of Energy Efficiency*. Cheltenham: Edward Elgar, pp. 11–62.

Sorrell, S. (2015). Reducing energy demand: a review of issues, challenges and approaches. *Renewable and Sustainable Energy Reviews*, *47*, 74–82.

Strbac, G. (2008). Demand side management: benefits and challenges. *Energy Policy*, *36*(12), 4419–4426.

Thollander, P., Palm, J., & Rohdin, P. (2010). Categorizing barriers to energy efficiency: an interdisciplinary perspective. *Energy Efficiency*, 49–63, http://www.intechopen.com/books/energy-efficiency/categorizing-barriers-to-energy-efficiency-aninterdisciplinary-\nperspective.

UNEP (2006). Barriers to energy efficiency in industry in Asia. *Review and policy guidance*, UNEP. Available from: http://www.energyefficiencyasia.org/brochure_pub.html

Vafeas, A., & Madina, C. (2008). Developing aggregation business models: first application in the real cases, Deliverable D9.

Verbong, G., & Geels, F. (2007). The ongoing energy transition: lessons from a socio-technical, multi-level analysis of the Dutch electricity system (1960–2004). *Energy Policy*, *35*(2), 1025–1037.

World Energy Council (2015). *World Energy Trilemma Priority actions on climate change and how to balance the trilemma*. http://www.worldenergy.org/wp-content/uploads/2015/05/2015-World-Energy-Trilemma-Priority-actions-on-climate-change-and-how-to-balance-the-trilemma.pdf

Zhu, M. (2013). Searching and switching across markets: is consumer "inertia" the result of a mistake or a preference? In J. Mehta (Ed.), *Behavioural economics in competition and consumer policy*. Norwich: ESRC Centre for Competition Policy, p. 118.

INDEX

N

S

Printed in the United States
By Bookmasters